RENEWALS: 691-4574
DATE DUE
MAY 07
FEB 17
Demco, Inc. 38-293

D0772856

Clipped Wings

Clipped Wings
The American SST Conflict

Mel Horwitch

The MIT Press
Cambridge, Massachusetts
London, England

This book was set in Times Roman by Achorn Graphic Services, Inc., and printed and bound in the United States of America.

Library of Congress Cataloging in Publication Data

Horwitch, Mel.
Clipped wings.

Bibliography: p.
Includes index.
1. Supersonic transport planes—Government policy—United States.
2. Technology and civilization. I. Title.
HE9770.S9H67 387.7′3349 82-83
ISBN 0-262-08115-6 AACR2

To
Adeline Schwartzman Horwitch
and
the memory of Louis S. Horwitch

Contents

Acknowledgments

Many people and institutions contributed directly or indirectly to this work. For the sake of brevity (and also to keep a promise of confidentiality to many of the individuals who went out of their way to help me, especially by participating in interviews), I am specifically acknowledging only those who were extensively and directly involved in assisting me.

Several institutions provided me with access to their archives and files. I must especially thank the officers and staffs of the National Academy of Sciences (NAS) and particularly the NAS Archives; the Systems Research and Development Service in the FAA; the Office of the FAA Historian and especially FAA historian Nick Komons; the Federal Records Center in Suitland, Maryland; the National Archives in Washington, D.C.; the Office of Presidential Libraries in the National Archives; the Eisenhower, JFK, and LBJ Presidential Libraries; and the Citizens League Against the Sonic Boom in Cambridge, Massachusetts, and particularly its founder William Shurcliff.

This study could never have been written without generous support from several sources. The William Barclay Harding Memorial Fellowship for Aerospace Industry Studies and the Division of Research of the Harvard Business School provided funding for most of this book's primary research. The Alfred P. Sloan School of Management at MIT provided an excellent environment and support for writing the final drafts of this work.

I must express my gratitude to the innumerable colleagues who reviewed and discussed this work in various stages and who generously shared their ideas. In particular, I must thank William J. Abernathy, James P. Baughman, Alfred D. Chandler, Jr., Ram Charan, Balaji Chakravarthi, Frank Davidson, Dan H. Fenn, Jr., R. Edward Freeman, the late John D. Glover, Modesto A. Maidique, Thomas K. McCraw, Lenore and Rafe Pomerance, C. K. Prahalad, Richard S. Rosenbloom, Richard S. Tedlow, and Eric Von Hippel. I must also express my enduring appreciation of the late Raymond A. Bauer, who always had great faith in this endeavor. Although a participant himself in the SST conflict, he encouraged total candor and honesty. The editorial suggestions of Max Hall and Maria Kawecki increased the clarity and logic of this work.

Many people were involved in the production aspects of this work. I am particularly indebted to Julie Brownell, Regina Collinsgru, Margaret Godwin, Meriel Greenhalgh, Lenner Laval, and Billie Lawrence.

My most severe (yet enthusiastic) critic, my wife Sally Schwager, forced me to reach for new concepts, and she was invaluable in bringing this work to fruition. My son, Gregory, at the age of six, searched newspapers for articles on the "SST." Finally the imminent birth of our second son, Adam, presented me with the most rigid of all deadlines for completing the manuscript.

Illustrations

1. Najeeb E. Halaby, FAA administrator from 1961 to 1965. (Courtesy: Records of the FAA historian)

2. Secretary of Defense Robert S. McNamara who was chairman of the President's Advisory Committee on Supersonic Transport, which met from 1964 to 1966. (Courtesy: The Lyndon B. Johnson Presidential Library)

3. President Lyndon B. Johnson swears in Air Force General William F. "Bozo" McKee as FAA administrator on July 1, 1965. (Courtesy: The Lyndon B. Johnson Presidential Library)

4. Air Force General Jewell C. Maxwell, who was director of the SST program from 1965 to 1969. (Courtesy: Boeing Aerospace Company)

5. William A. Shurcliff, director of the Citizens League Against the Sonic Boom, at his home, the headquarters of the Citizens League, in Cambridge, Massachusetts. (Courtesy: Richard Howard)

6. William M. Magruder, the last SST program director, meets in the Oval Office in the White House with President Richard M. Nixon and presidential aide John Ehrlichman. (Courtesy: The photograph archives of the Richard M. Nixon papers, National Archives, Washington, D.C.)

7. Mock-up of the Lockheed fixed-wing SST design that lost the SST design competition in 1966. (Courtesy: Records of the FAA historian)

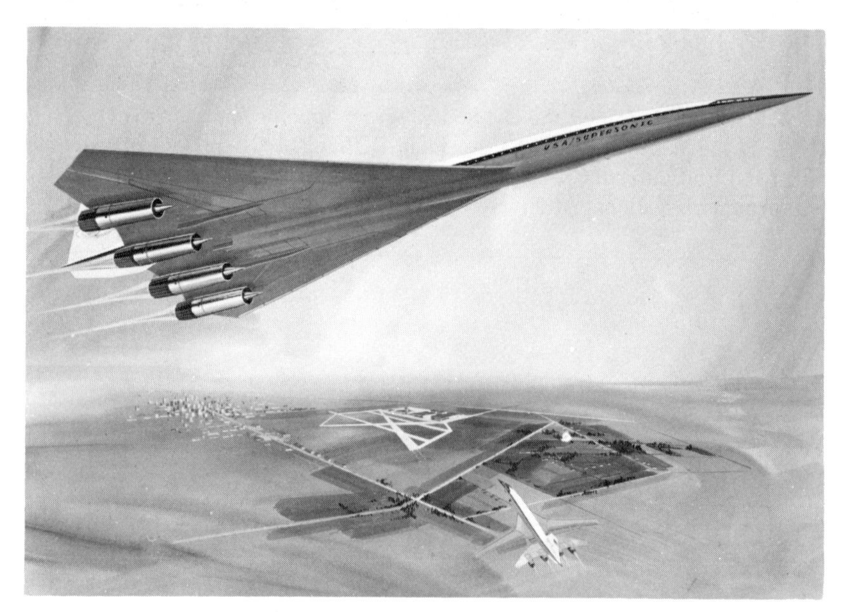

8. Boeing swing-wing SST design that won the SST design competition in 1966. (Courtesy: Boeing Commercial Airplane Company)

9. Mock-up of the Boeing fixed-wing design that was selected in 1969. (Courtesy: Boeing Commercial Airplane Company)

10. Artist rendition of a 1971 version of the basic Boeing fixed-wing SST design that was selected by the U.S. Government in 1969 after the swing-wing model was judged unacceptable. (Courtesy: The Boeing Commercial Airplane Company)

11. Artist rendition of a second-generation SST design by Lockheed called the "Arrow Wing." (Courtesy: Lockheed-California Company)

12. Artist rendition of a second-generation SST design by Boeing called the "Blended Wing." (Courtesy: The Boeing Commercial Airplane Company)

Clipped Wings

1 Introduction

The debate over whether the United States would build a commercial aircraft that would fly at supersonic speeds encompassed more than the weighing of the strengths and weaknesses of competing designs. Of course, a large part of the supersonic transport (SST) effort involved assessing the substantive issues of economics and technology. The SST conflict, however, also grew from a limited debate into a general battle about how to develop technology; the traditional decision-making rules that prescribed contained, technocratic, cost-benefit methods were challenged; new kinds of participants pushed for the incorporation of socioenvironmental considerations in arriving at technological decisions. The SST affair also seemed to be a war over economic growth—representing, to some, an attempt to block a natural and beneficial development in one of our most successful industries. To others, the controversy was over the need to protect and enhance American prestige, which appeared to be under severe economic and ideological attack from abroad. The project took on an intense and highly personal meaning for a small number of champions within the government who saw the program as immensely important for the country, their career, or their ultimate place in history. Conversely key resisters also emerged who sought to delay or terminate the program for a similar blend of patriotic, professional, and personal reasons. The SST conflict involved ambitious officials, huge corporations, and large public bureaucracies that were vying for power, prestige, and the control of massive funds. SST development also posed important management challenges—challenges that were never successfully met. In the end the SST conflict became a symbolic battle for many Americans—a war in which people outside the formal power structure fought to gain control over decisions that had hitherto been out of their reach.

The SST conflict spanned more than a decade. It involved a changing cast of characters, diverse locales, and radical shifts of plot. It evolved according to a pattern that has become particularly important since the 1960s. At first the development of a commercial supersonic transport was a matter of limited and conventional concern, mainly to those who traditionally had a stake in such endeavors: the aerospace industry, related government agencies, and allied congressional interests. They viewed the SST basically as a logical development in the aerospace field. But in less than a decade the SST was totally transformed as an issue. By 1970 the controversy over the aircraft's development had come to involve large numbers and diverse kinds of people and organizations across the country; it had changed from an internal and rather technical debate into an all-out societal war. Ultimately the SST became to environmentalists a symbol of all that was wrong with technology and to SST proponents a symbol of technological progress and technology's benefits to humanity.

But this dramatic change did not happen abruptly. The SST conflict is not a simple rise-and-fall controversy. The story is as much, if not more, about bureaucratic intrigue, managerial weakness, and economic and technological failures as it is about the dramatic emergence of widespread protest. The program's protection and support gradually were chipped away. When the conflict came into public view at the end of the decade, it already had been seriously weakened from within.

Although the evolution of the SST conflict, like that of many other historical and social processes, was at times uneven, certain discrete stages are apparent. The first phase, which took place during the late 1950s and very early 1960s, was a kind of "prelude" in which various political and technological forces coalesced and in which the possibility of a commercial supersonic transport was first discussed and studied. The next phase of the American SST program was one of "containment," beginning with the election of John F. Kennedy as president in 1960 and lasting until the autumn of 1963. During this phase an SST program was announced by Kennedy and a foundation for the program was laid by a vigorous, determined SST champion, FAA administrator Najeeb E. Halaby.

But Halaby's achievement soon began to crumble. The SST program entered a lengthy period of "fragmentation" that lasted from the fall of 1963 to about 1968. During this period it was battered by bureaucratic jealousies, economic doubt, lack of managerial confidence, changing loci of policy-making power, technical uncertainty, ambiguous presidential support, a growing number of participants, the beginnings of criticism in the media, and the emergence of organized public opposition.

The SST conflict then entered its final and most dramatic stage: a period of "explosion" lasting from about 1968 to 1971 when the SST program was finally terminated by Congress. During this phase the protest against the SST, which previously had involved isolated individuals and a small organized protest group, exploded into a national campaign.

In many ways the SST conflict was destined to be visible, significant, and controversial. The role of government was almost immediately the subject of intense debate. Unlike space or defense programs, the SST was one of the first post–World War II examples of large-scale governmental funding of civilian projects with explicit commercialization of a new technology as their ultimate goal. Indeed most thoughtful retrospective observations about the SST usually occur within the context of discussing the larger issue of the appropriate strategy for government in advancing commercialization technologies. Such discussions usually conclude by urging a limited, cautious, and flexible governmental role in backing massive civilian technological projects.[1] Defense Secretary Robert S. McNamara, who was the most powerful decision maker concerning the SST between 1964 and 1966, and Halaby took diametrically opposite positions over this question of effective public strategies for managing large-scale commercialization technologies.

Other factors also worked to make the SST in the end a matter of widespread societal concern. The aircraft, if built, would have directly affected a large number of people—many more than, say, a space program or a weapons system. Because the SST program was sponsored largely by the federal government, public hearings were a requirement. It also involved the aircraft manufacturing industry, a sector that the United States had totally dominated since the end of World War II and an important and highly visible balance-of-payments contributor. But the airlines, especially in private or off the record, gave the SST only halfhearted and ambiguous support partly because of their financial burden in expanding their fleets with subsonic jumbo jets and partly because of their own doubts about the SST's economic performance. The issues of aircraft engine noise and sonic boom effects naturally involved many communities near airports or under the potential SST flight paths. The SST program also gained attention as a potential stimulator of large-scale employment and of a favorable balance of payments for the nation. Toward the end of the program's life the SST's potential as a polluter of the upper atmosphere and as a destroyer of the earth's protective ozone layer—hazards on a global scale—was publicized. The SST also had the limelight because from the beginning it was promoted and defended in such grandiose terms: as a means to maintain America's prestige, to advance the technological state of the art, and to protect America's hegemony in the aerospace field. Finally the proliferation of uncertainties in practically every relevant area—economic, environmental, and technological—worked to delay the program substantially and to increase the intensity of the controversy.

The SST was significant as a barometer and as a catalyst for larger societal changes that occurred during the 1960s. When the conflict began technological decisions as a rule were not considered appropriate matters for debate in the public arena. These decisions tended to be made by experts in government agencies, research organizations, and industry. It was implicitly accepted that the general public had neither the competence, the right, nor the interest to question seriously the actions resulting in major technological advance (or economic advance, which was seen as a derivative). Frequently, in fact, the technological imperative was sufficient: if something could be done, it should be done. The early SST program was the result of such thinking.

But during the 1960s, as seen in the evolution of the SST conflict, a new mindset emerged. Gradually disciplines and professions that had hitherto been absent from technological decision making entered the various technological debates. In time, traditional technical experts, behavioral scientists, community leaders, politicians, and representatives of public interest groups were demanding to be heard. Decision making for many large-scale endeavors also became politicized; experts and managers from agencies, research organizations, and corporations gave up more and more decision-making responsibility to politicians who operated more or less within the public view. The public at large (as distinguished from special interest groups, corporations, or government agencies) grew more eager to enter discussions on technological issues. At the same time, organizations like the Sierra Club, Friends of the Earth, and ad hoc groups (such as the Citizens League Against the Sonic Boom), which claimed to represent the general public and which mounted attacks on the traditional technological evaluation process, grew increasingly effective in promoting their ideas in the public arena.

Underlying this shift in attitude toward technology was a fundamental cultural change in American society: the old attitude based on the notion that technological advance equaled progress was greeted with increasing suspicion. The environmental and social consequences of a program—a program that once had been considered primarily in technological terms—became more important in reaching a decision.

Public opinion surveys documented this general change. Although confidence in scientists and engineers remained high and although the public still tended to look with favor on technological developments like computers and automation (though not on more remote endeavors like the space program or the antiballistic missile), technological decisions were greeted with a growing amount of skepticism. A nationwide poll in 1965 showed that air and water pollution was felt to be a "very" or "somewhat" serious problem by 28 percent and 35 percent of the sample, respectively. By 1970 the comparable figures had more than doubled. Also during that year, for the first time, a statistically significant number of respondents to a Gallup Poll (6 percent) named pollution and deteriorating environmental quality as "the most important problem facing America."[2]

Mass protest and mass political action also became part of the 1960s, beginning with the civil rights movement in the early part of the decade and continuing with the protest over the Vietnam war and with the 1968 presidential campaigns of Senators Eugene McCarthy and Robert F. Kennedy. Grassroots political activism became common, and this style of political opposition soon characterized the environmental movement as well. A whole cadre of effective leaders and workers emerged, and a network of organizations surfaced to support their activities. Manifestations of this new activism included widespread protests over the construction of a dam on the Colorado River near the Grand Canyon and the near defeat in the Senate in 1969 of the antiballistic missile program. Both cases indicated a willingness on the part of masses of people to become involved in matters that previously had been considered distant and intimidating. In both situations, established and politically skillful national groups coordinated the opposition's attack.[3]

It is not, then, the specific defeat of the SST that gives the SST conflict its key historical significance. Indeed a different style of management, more effective advocacy at key junctures, a different kind of planning, and a greater inclination to compromise might have saved the program in some form. Rather it is the entire evolutionary pattern of the SST conflict—from a limited, conventional matter to a massive, societal concern—that provides it with its more general relevance. Its various stages—from prelude to containment to fragmentation to explosion—define a pattern of change that has become increasingly common.[4]

The SST conflict signaled the rise of new values, of new kinds of participants, and of new rules for decision making. By the time the program was defeated, a new age had emerged, and the conflict both mirrored and promoted its birth.

The aftershock of the SST defeat is still being felt. Although the environmental movement has lost some of its vigor and public appeal and has become more institutionalized, the public is no longer reluctant to speak out on technological issues, and many of the individual opponents of the SST have moved on to new protests. The demise of the SST program was a warning to technological developers that significant, novel dangers threatened their plans and enterprises and that broad segments of society, which hitherto had not actively participated in key technological decisions because they had no expertise or direct interest, were now intimately involved. Such opponents now often possess the power and the will to make or break a large-scale technological enterprise and to exert an enormous amount of influence over the evolution of such efforts.

2 The Beginnings

The origins of the American SST conflict go back at least to the late 1950s. Even at that time—the beginning of the commercial jet transport age—there was considerable interest in developing an SST. And many of the major trends and characteristics of the SST conflict in the United States that emerged in the 1960s were already evident. From the start the Americans exhibited an almost absolute faith in their ability to develop advanced SST technology; the United States was determined to avoid any kind of international SST collaboration in order to maintain its hegemony in commercial aviation; and such persistent features as individual advocacy, bureaucratic conflict, contradictory technological evaluations, equating the SST with national prestige, debates over SST financing, and skepticism about the program management capability of the Federal Aviation Agency (FAA)* appeared during this early period. These policies and issues remained remarkably intact until the end of the SST conflict in 1971.

Some of the basic American views on the SST were in evidence even before the Boeing 707, the first American commercial subsonic jet, initially flew scheduled routes in October 1958. In May 1958 Boeing's competitors, Douglas and Lockheed, were reportedly already working on SST designs, and according to one public forecast Douglas would fly a Mach 2** SST prototype within two years.[1] Rand Corporation experts confidently predicted that SSTs with operating costs comparable to commercial subsonic jets could be developed by about 1970.[2]

American firms were clearly and perhaps fatally attracted to the notion of bypassing a Mach 2 aircraft and developing a more advanced Mach 3 model that would use a steel-titanium airframe rather than an aluminum one. In early 1959 officials from General Dynamics, Boeing, and Douglas publicly stressed the economic and long-term technical advantages of a Mach 3 strategy.[3] Lockheed later echoed similarly optimistic views. One Lockheed official, writing that the SST was "completely feasible," exuded confidence: "We know we can build such an airplane and that, if we were to begin now, we could meet a

*The Federal Aviation Agency became the Federal Aviation Administration in 1967 when it became part of the new Department of Transportation, which was signed into law in October 1966.

***Mach,* named for the Austrian physicist Ernst Mach, refers to the speed of a body, measured in relation to the speed of sound that itself is equal to Mach 1.0—about 760 mph at sea level.

certification date of 1965." The only hint of trouble at this time came from a Pan American official who warned of extremely difficult problems from the sonic boom.[4]

This generally bullish mood continued into late 1959. An air-force-sponsored study concluded that the most economical and profitable SST would cruise at speeds of about Mach 2.5, would be more productive than subsonic vehicles for long distances, and perhaps would even be competitive for routes in the 1,000- to- 2,000-mile range.[5] The International Air Transport Association in October 1959 proposed a "supersonic transport symposium" for the following year; the association's president believed that the SST was inevitable and warned against actions that would "discourage or retard" the aircraft.[6]

This aggressive American attitude was being spurred by foreign rivalry in SST development. Although the Soviet Union and France had exhibited little interest in the SST, Great Britain had seriously considered building an SST. The British postwar experience accounted in part for this attitude. During World War II Britain had produced a number of aviation successes, including the Hurricane and Spitfire fighters, radar, and the jet engine. But after the war had come costly failures: the eight-engine Bristol Brabazon, the ten-engine flying boat, the Saunders-Roe Princess, the Avro Tudor, and, above all, the Comet.

The Comet was the world's first long-range commercial jet aircraft and represented a spectacular attempt by the British firm de Havillands to challenge American dominance of commercial aviation. But this aircraft was relatively small and possessed limited range; it carried only about eighty-five passengers, compared with 140 passengers for the forthcoming Boeing 707, and could not fly from New York to London nonstop. Also the Comet suffered a fatal flaw in its construction (fatigue failure in its fuselage) that led to three tragic crashes during 1953–1954. The entire fleet of Comets was grounded. Subsequent testing and redesign took about three years, during which the American-made Boeing 707 and Douglas DC-8 captured the long-range commercial jet market. Early British interest to develop an SST stemmed partially from the same frustration that led Britain to try to leapfrog the Americans with the Comet.

British SST research began in the early 1950s at the Royal Aircraft Establishment at Farnborough, England. Within a few years Farnborough had established an active SST "working party" that soon urged a comprehensive inquiry into SST feasibility. The group, which feared that the Americans might soon begin serious SST work, was successful. In October 1956 the British government created a diverse and high-level committee to conduct a full-scale SST inquiry for determining the feasibility of building an SST, and several technical subcommittees also were established.

The committee's investigation lasted about two and a half years, and in an important report released in March 1959 the committee came out in favor of SST development (actually of two SSTs: a long-range and a short-range aircraft), rejected the concept of an advanced Mach 2.6 steel-titanium aircraft in favor of a long-range aluminum-based Mach 1.8 vehicle, and supported a more conventional delta-wing design rather than the more sophisticated swing-wing design that was being advocated by some in the United States. The recommended payload for the long-range aircraft was rather small—about 150 passengers. Finally, prefiguring what was to become a consistent European attitude, the report paid little attention to the potential problems of the sonic boom.[7]

In some ways paralleling the evolution of American SST views, British thinking on SST development had gone beyond technical feasibility and cost-effectiveness considerations. Officials defended the SST as crucial to the survival of the British aviation industry and to the maintenance of Britain's status as a global power. In 1959 the managing director of Bristol Air Craft, Ltd. publicly warned that if the British aviation industry did not give the SST immediate attention, Britain would "fall behind forever" and disappear from the ranks of first-class world powers: "The basic issue—whether we like it or not—is whether the aircraft industry in the United Kingdom can remain much longer in the 'big league' of military and civil aeronautics." "Big league" meant developing ballistic missiles, antiballistic missiles, SSTs, vertical takeoff and landing aircraft, and perhaps nuclear power plants. He envisioned the 1970 long-haul transatlantic passenger market as belonging to SSTs, charging premium fares, and to jumbo subsonic jets, charging low fares. In September 1959 the Ministry of Aviation authorized two British firms to make separate feasibility studies on a Mach 2 SST.[8]

Throughout the international aviation community interest in SSTs also began to grow. In July 1959 a major item on the agenda at the annual session of the International Civil Aviation Organization, which was affiliated with the United Nations, was the SST's impact on international air transportation. The British representatives, intensely concerned about American SST efforts, successfully introduced a resolution calling for a preliminary international study by September 1960 on the prospects for developing and then flying a commercial SST by the mid-1970s.[9]

Ironically the British resolution stimulated the first serious governmental inquiry on the SST in the United States. Responding to the British action, the air coordinating committee, an American government interagency body that deliberated national aviation policy, established in November 1959 an interagency ad hoc group to prepare the official American submission. A few weeks later, in mid-December, FAA administrator Elwood Quesada established a small SST study group within the FAA. Before the decade ended therefore an ongoing American governmental unit was functioning to deal solely with SST matters. The FAA was also taking the lead in coordinating and designing American SST policy, and the agency would wear this mantle of SST bureaucratic champion for years to come.

The interagency ad hoc group's rather optimistic report was sent to the International Civil Aviation Organization in February 1960. Reflecting prevalent American views, the group called for an advanced Mach 3 design. The group expressed little sense of urgency and advocated introducing the SST in the 1970s rather than the mid-1960s.[10]

But with a permanent bureaucratic base for SST matters now established, the FAA's enthusiasm for developing the aircraft grew. Its SST study group discussed military plans and research on large supersonic aircraft with officials at the National Aeronautics and Space Administration (NASA), and it conferred with the air force on the B-70 bomber, a Mach 3 aircraft being developed by North American Aviation. Even after the Eisenhower administration substantially reduced its B-70 appropriations request for fiscal 1961 (which slowed down the B-70 effort and eliminated subsystem development work), the study group still maintained a close relationship with the air force and monitored work on the B-70 and the supersonic B-58. The study group also met frequently with representatives from such firms as General Dynamics (the manufacturer of the B-58), McDonald Aviation, Lockheed, and Boeing. In late January 1960 the group optimistically reported that a Mach 2–2.5 SST was indeed feasible, though an advanced Mach 3 aircraft would require governmental financial support. Quesada then enhanced the FAA's position as SST champion by establishing in early March 1960 a supersonic planning team within the FAA to conduct further SST studies and to act as an SST information clearinghouse for the federal government. This team operated with thoroughness and vigor. At its first formal meeting it reported that "a large amount of investigation, preliminary planning, and organization has taken place."

Pro-SST sentiment surfaced elsewhere in the government. In February 1960 Civil Aeronautics Board (CAB) chairman James R. Durfee warned that a one- to three-year delay in developing an American vehicle could force American-flag air carriers to order foreign SSTs to meet the competition. He based this view on CAB intelligence that both Britain and the Soviet Union planned to begin SST operations between 1965 and 1968, which, if true, Durfee declared, would mean that American hegemony in aviation could be lost for decades. (According to one wild rumor at the time, one of these countries was already "cutting metal.")

Industry's public statements were also extremely favorable at this time. Both North American and Lockheed called for a firm SST go-ahead on the grounds that the United States needed the aircraft to maintain its aviation supremacy.

Airline views, however, were more reticent and less uniform, and the air carriers' position would continue to be ambivalent throughout the SST program. At least part of the reason for this ambiguity was the increasing financial burden on the airlines as they replaced their piston aircraft with jet transports. Also the airlines were quite varied in terms of competition markets and routes. For example, while TWA confidently predicted an operational SST by 1967, American Airlines, which used domestic and overland routes, projected a much later date and expressed concern about such issues as safety, engine noise, and the sonic boom.[11]

Still by mid-1960 a considerable amount of agreement already existed in both government and industry about the design of an American SST and the nature of program financing, and this agreement was visible in a crucial and illuminating set of hearings in May 1960 before the Special Investigating Subcommittee of the House Committee on Science and Astronautics, under the chairmanship of Congressman Overton

Brooks (D-Louisiana). The extremely positive and detailed opinions expressed at these hearings are all the more remarkable and important considering the relatively limited, introductory, and haphazard nature of the SST investigations then taking place. In many ways, the comments reflect more a general belief in advanced technology and an American confidence in aviation than specific findings on the SST itself.

At the hearings, which were devoted entirely to the SST, it was generally agreed that an SST program would significantly increase national prestige, a view that Brooks expressed at the outset. He cited global tension and instability, the apparent progress of communism, the far-flung nature of American military and commercial operations, and the threat of foreign SST competition as key reasons for launching an SST program.

To a greater or lesser degree, other witnesses echoed these views. Ira Abbott, NASA's director of advanced research programs, cautioned not to "turn our backs on supersonic air development." General Victor Haugan, the air force director of development and planning, called the SST technically feasible and a desirable national objective. Elwood Quesada warned that foreign competition already was beginning to make inroads into the country's leading position as a producer of civil transport aircraft. Industry and airline representatives agreed.

Regarding SST technology the prevailing mood was a curious mixture of optimism and caution. Ira Abbott declared that more than 50 percent of the required basic research already had been completed and, expressing America's consistent commitment to an advanced SST, called for continued research and a "vigorous" program for developing a Mach 3 SST; due to the temperature limit of aluminum alloys, Abbott claimed, a Mach 2 vehicle would have "little or no growth potential as far as speed is concerned" and might be prone to metal fatigue. He minimized the SST's engine noise problem, stating that it would be "not much different than for existing jet transports." On the other hand, Abbott warned that program costs would be high, that SST technology would present far more difficulties than subsonic development, and that engine noise and the sonic boom would be crucial problems.

General Haugan too was ambivalent; the SST faced serious prob-

lems, mainly due to its excessive weight (300,000 to 600,000 pounds); airport surfaces would have to be strengthened; and the need for large fuel reserves would reduce efficiency. But on the whole, according to Haugan, the SST was "technically and economically feasible"; the short flight time of a Mach 3 craft would make possible extremely accurate weather forecasting at destination terminals; and the plane's high altitude would reduce sonic boom disturbances on the ground to a level that would be "no more objectionable than a momentary distant thunder."

Elwood Quesada reinforced the positive views. In referring to initial FAA investigations in 1959 and early 1960, Quesada also indicated that Mach 3 SSTs were technically feasible and possibly possessed superior efficiency compared to some current jet aircraft. He minimized the problem of engine noise and rejected the notion that an SST would cause any difficulty for modern airports. Although public acceptance of the sonic boom might be a problem, Quesada admitted, restricting supersonic flight to high altitudes would minimize such disruptions.

As would be true during most of the SST program, SST financing and economic feasibility appeared more uncertain and controversial. Indeed the greatest debate during this early period arose over the question of government support for what was ostensibly a private commercial endeavor. Abbott correctly termed financing "one of the [SST's] most serious problems." Stressing the great technological leap from subsonic to supersonic commercial flight, the high development costs, and the relatively small production run, he urged the government to finance the SST as it had supported shipbuilding and nuclear reactor construction.

Quesada too called for government support, especially since the military now emphasized ballistic missiles and NASA's major focus was on space; consequently the traditional military-supported research and development foundation in civil aviation no longer existed. Wanting to limit the government's managerial involvement, however, he called for financial contributions from industry and proposed that the government's investment be repaid in the form of "royalties," which he defined as some reasonable percentage of the development cost. Thus began a protracted debate on SST financing that would continue for at least seven more years. (Quesada, of course, was confident that the

SST was a good investment; one FAA consultant, Paul Cherington of the Harvard Business School, had already projected a free-world SST market of about two hundred aircraft.)

Another area of disagreement, which also would continue for many years to come, was over which agency should run an SST program. The FAA under Quesada had taken the lead, but during the next several years that agency's program management's qualifications would be seriously questioned. At the Brooks hearings this issue of appropriate management responsibility first emerged, at least in public. Stressing the air force's enormous management capability, General Haugan suggested that it would be cheaper to develop an SST as a spinoff of the current B-70 bomber program rather than designing the plane from scratch; the air force already had spent three years and $350 million to develop the B-70. Moreover, in contrast to NASA, which offered technical support while eschewing lead administrative responsibility, the air force claimed to be ready to serve as executive manager for an SST program. The air force's public willingness to coordinate an American SST program challenged the FAA. Quesada reacted by reminding the subcommittee that under the Federal Aviation Act of 1958 the FAA administrator was directed to foster and to supervise the development of civil aeronautics, as well as to certify civil aircraft for safety. Again reminding committee members of the Defense Department's preoccupation with unrelated technologies and NASA's lack of interest, Quesada declared that the FAA was uniquely qualified and virtually required by law to run the SST development effort. "I don't like it," Quesada added, "but by the law this is our job."

The manufacturers generally continued to express strong overall backing for an SST program, though also calling for substantial government financial support and holding somewhat differing views on certain items. North American Aviation stressed that its B-70 bomber could serve as a basis for an SST prototype, while General Dynamics promoted its B-58 as a cheaper alternative. (The B-58, it was claimed, could be converted into an "interim supersonic transport" with a gross weight of about 200,000 pounds, a capacity of fifty-two passengers, a cruising speed of Mach 2 to Mach 2.4, and a transatlantic flight range with one stop.) The airline representatives were similarly favorable in public, though SST economics clearly worried them. Government

financing was necessary, they argued, because private industry would be unable to raise sufficient capital.[12]

The Brooks committee hearings exhibited an extremely positive consensus: the SST was technically feasible and commercial flights would be possible by the late 1960s; SST development was necessary for continued American leadership in aviation, and that leadership in turn was necessary for maintaining national prestige; a Mach 3 aircraft of steel and titanium was preferred; significant government financing was urged; the SST was an important next logical step in aviation; and although certain technical problems such as engine noise and the sonic boom were mentioned and although the need for the government to recoup at least part of its investment was discussed, these matters could be resolved. The committee's final report, released in June 1960, reflected this favorable climate of opinion by calling for an American SST program backed by strong government financial support. But surprisingly the committee also recommended that NASA run such a program, which indicated continuing unease with the FAA's new role as program manager.[13]

Fortunately for the FAA, however, NASA showed little interest in a lead role. In late July 1960 NASA administrator Keith Glennan held an important meeting on the SST with representatives of NASA, the FAA, and the Defense Department where he made it clear, in view of the increasing significance of space programs and the correspondingly lower priority of aeronautics at NASA, that NASA would not run an SST program. Instead the three agencies agreed that the FAA would be responsible for overall program leadership, budgeting, and cost control; NASA would provide basic research and technical support; and the Defense Department would oversee "development management." All favored establishing a high-level SST advisory group that would include nongovernmental representatives.

Armed with this informal agreement, Quesada acted quickly in an effort to establish an official SST program under FAA control before the Eisenhower administration ended. He submitted a draft letter for President Eisenhower's signature, lobbied with White House aides in late August 1960 for quick presidential approval, and anticipating presidential approval organized an FAA-NASA-Defense Department SST working group to prepare an overall report.[14] The document was ready by early autumn 1960 and was everything that Quesada could have wished. For a program not yet approved by the president the report presented a remarkably detailed schedule and budget, looking forward to a commercial aircraft by 1969 after spending $450 million in direct costs over nine years. The FAA was given overall responsibility for directing the enterprise, and a number of strongly worded reasons were presented to justify the project.[15]

But Quesada then suffered a series of severe setbacks. Although both Quesada and Glennan quickly signed the draft report, Secretary of Defense Thomas Gates refused to sign. The Pentagon had recently expanded the B-70 program and had ordered the construction of two B-70 prototypes. Gates felt that another major undertaking in the supersonic area would be untimely and too costly, and he urged a more modest initial SST budget request.

Quesada was equally unsuccessful at the White House. He had pushed for the inclusion of SST funding as part of Eisenhower's last budget in the president's state of the union message. But White House aides wanted to avoid anything that appeared to commit the next administration. Also Eisenhower, who warned his fellow citizens of a growing military-industrial complex, was not likely at this late date to approve an expensive and grandiose effort like the SST. The aircraft was not even mentioned in the message, nor was it listed as one of the budget items.

Quesada was forced to retrench. He allocated a small amount of FAA funds to the SST and asked the air force for $100,000 to investigate SST engine designs.[16] He also modified the recent draft report. The final document, which was issued in December 1960, still backed establishing an SST program, but the original budget recommendations were missing. The report called for airline financial participation, recovery of the government's direct SST investment, and governmental financial assistance for only the program's development phase. The report also appeared to favor a Mach 3 vehicle since it used mostly Mach 3 assumptions in its technical discussions.[17]

In spite of frustrating setbacks in the administration, the continuing threat of foreign SST competition was acting as an incentive for greater American SST activity. In addition, although the Americans and the British backed radically different SST designs, the British pushed for large-scale collaboration. They made it clear that they wanted a full

SST partnership and even declared their willingness to reconsider their design approach if they found American reasoning convincing. During the autumn of 1960 Peter Thorneycroft, British minister of aviation, announced that a limited SST design-study contract with the British Aircraft Corporation would include exploring possibilities for international collaboration. Thorneycroft invited Quesada to take part in this analysis. At about the same time respected British aviation expert Sir George Edwards, executive director of the British Aircraft Corporation, publicly declared that the Americans "were not unsympathetic" to international collaboration.

But such British hopes were unfounded. The Americans had no desire for meaningful SST collaboration, a position that was clear before Thorneycroft's invitation. At a meeting with British officials in London in early September 1960, Quesada urged that Britain and the United States share SST-related information only in such areas as the sonic boom and engine noise. The Americans also clung to their faith in advanced SST technology. At this meeting one of the leading American technical experts, John Stack of NASA, stressed the doubtful value of an aluminum body. He also bluntly and rather inhospitably reminded the British of their disastrous experience with the Comet. A little later Quesada in effect rejected Thorneycroft's invitation by vaguely replying that no action could be taken until a formal American SST program was authorized; such an authorization never took place during the Eisenhower administration. The British aviation journal *Flight,* in November, had already discerned the futility of pursuing SST collaboration with the Americans, correctly perceiving "no shred of evidence" to support collaboration hopes. Another long-term pattern for the SST program had surfaced: the Americans were committed to developing an advanced SST alone and would use the existence of foreign SST efforts to further this end.

By the end of 1960 the basic shape of an American SST program and of the international SST rivalry was already apparent. In all likelihood, the United States and Britain would go their separate ways on SST matters. (By early December Thorneycroft was discussing SST collaboration with the French.)[18] The program's enduring multifaceted nature was already clear, and by the end of 1960 many of the policies, problems, and diverse, conflicting views that were to plague the program for a decade had been exhibited. The United States would aim for an advanced Mach 3 aircraft, making it virtually impossible to work with the British. In terms of funding, all parties acknowledged that substantial governmental financial assistance would be required. Finally, with the professed lack of interest of NASA and the preoccupation of the air force with military programs, the FAA was given a clear field to direct any future SST program.

At the same time, issues that would continue to hound the American SST effort—financing, engine noise, and the sonic boom—had also surfaced. Although *Aviation Week and Space Technology*, reporting on the December 1960 SST review, enthusiastically, though prematurely, proclaimed the "supersonic transport program launched,"[19] a program did not yet exist. Quesada was only a part-time champion, and support for such an effort was limited and unfocused. The ingredients were all in place, but it was left to the incoming Kennedy administration to commit the United States to a specific and permanent structure for SST development.

I CONTAINMENT

3 A Champion Emerges

The SST's prospects rose with the coming of the Kennedy administration. Such a venture seemed consistent with John F. Kennedy's 1960 campaign theme of "getting the country moving again." A postcampaign task force even reported to the president elect that NASA had ignored aeronautical research, with the Russians and the British possibly ahead of the United States in developing an SST; it called for substantially increased aeronautical research, perhaps outside NASA altogether, a view that would obviously help the FAA.[1]

In addition, in the new FAA administrator Najeeb Halaby the SST now had an ardent champion, perhaps the critical element for the success of any radical new endeavor. A Stanford graduate and a Yale-educated lawyer, Halaby had experience in practically all aspects of the aviation industry. He was, using his own term, a "natural born" pilot, and he tested aircraft for the navy and for Lockheed during World War II. He also had served as an air corps flight instructor and as an aviation intelligence officer for the Department of State. In 1948 he became a foreign affairs adviser to the Department of Defense and later assumed the post of deputy assistant secretary of defense for international security affairs, which he held until 1953. He then entered private industry, working for a few years with Laurence Rockefeller in the venture capital business but always remaining close to the aerospace field. In the mid-1950s he had served as vice-chairman of a White House study group on aviation facilities.

Halaby was a true believer in the benefits of aviation. Indeed he viewed aviation in almost spiritual terms. As he wrote in his autobiography, "Flight *can* be a poetical, mystical, almost religious experience; for me, flight has always evoked the biblical Genesis, in which 'God gave man dominion over the earth and over every creeping thing upon the face of the earth.'" To Halaby at the time of his appointment as FAA administrator, the SST seemed to be "the inevitable airplane of the immediate future." He called the aircraft "a logical expression of faith in aviation's progress—the faith of the airman himself." Halaby explained, "To the airman, progress and speed had been synonymous; growth and compressed time, identical. The airman regarded swift transportation as a logical goal; economics and politics perceived it as merely complex, costly and difficult. To me in 1961, a man of aviation, the supersonic transport was just as sensible as the

DC-6 and Constellation replacing the DC-3, and then being supplanted by the 707 and DC-8."[2]

Appointed in mid-January 1961, Halaby lost little time in pressing energetically for large-scale research on SST feasibility. Even more vigorously than his predecessor, Elwood Quesada, he championed an SST program under FAA control, promoting the idea to such influential figures as Roswell L. Gilpatrick, the deputy secretary of defense; James E. Webb, the new NASA administrator; and David E. Bell, the director of the Bureau of the Budget. Halaby also quickly moved to establish closer contacts with airframe and engine manufacturers. In March he informed these firms of his "earnest conviction" that an American SST was "essential to our continued leadership in commercial aviation," and he requested briefings on corporate SST activities and plans,[3] which were subsequently held in May 1961. Still during 1961 and 1962 Halaby faced a number of obstacles to SST development within and outside the government.

Bureaucratic and Political Ordeals

President Kennedy's approval of a $12 million FAA budget request in late March, for the first year of a two-year $50 million SST research effort, which both the Defense Department and NASA strongly supported, demonstrated Halaby's persistence and effectiveness as a bureaucratic champion.[4] Halaby, however, then had to defend this request in House committee hearings. Although some on Capitol Hill considered the request odd after having recently seen the B-70 bomber budget cut, Halaby (like Quesada before) argued that "we cannot count on military development" to produce an "adequate" SST; therefore he planned an FAA-directed SST research effort comprised of feasibility studies by various engine and airframe manufacturers; with this work available, along with military data, the FAA would determine whether SST development "makes sense and how much it will cost." Halaby called for substantial government funding. Like Quesada, however, he also expected the manufacturers to finance some portion of development, and he too believed in partial recovery of the government's investment through royalties.[5]

Not entirely persuaded, the House approved only an initial $10 million and a $20 million ceiling for the proposed two-year effort. But claiming in June that "we cannot live with" the $20 million limit, Halaby succeeded in obtaining from the Senate Subcommittee on Independent Offices Appropriations a recommendation for his original budget request of $12 million with no ceiling.[6] A key force in promoting the FAA's position was Senator Warren G. Magnuson (D-Washington), a consistent and important SST backer in the Senate.

Despite this support, the Senate vote was close. In late July 1961 Stuart Symington (D-Missouri), a respected Senate aviation expert, offered an amendment to eliminate the $12 million to protest funding the SST while the administration was dropping work on the B-70 bomber. This amendment lost on a tie vote, which turned out to be the closest Congress ever came to defeating an SST funding bill during the 1960s. Reflecting Magnuson's enormous influence, the Senate went on to vote for the $12 million appropriation without a ceiling. A House-Senate conference then reached a compromise, and, finally, in mid-August 1961 Congress granted $11 million to the FAA for SST feasibility studies, with NASA funding an additional $8.5 million for SST research.[7] Halaby passed a key test as a champion.

Halaby's effectiveness lay in his flexibility as well as commitment. In April and May 1961 on Capitol Hill he stressed the need for the United States to beat its SST competition in order to "capture the world [SST] market." But after meeting with airline presidents and the CAB in June, Halaby's urgency to be first diminished considerably. He explained, "There is no immediacy in their [the airlines'] demand for a commercial supersonic transport due to the heavy debt load and their low profit margin," though most airline presidents realized that "this new airplane is coming." Similarly Halaby testified two months later that he had rejected an offer by General Dynamics to convert its B-58 bomber into an SST because "we are going to get the best supersonic transport . . . [although] it may not be the first."[8]

Halaby skillfully mobilized bureaucratic support. He secured letters from NASA and the Defense Department that backed the FAA's SST budgetary request. A joint SST report in June 1961 by the three agencies, signed by Secretary of Defense Robert S. McNamara, NASA administrator Webb, and Halaby, formally gave "leadership and fiscal support" to the FAA, "administrative and technical support" to Defense, and "basic research and technical support" to NASA. Where Quesada had failed to obtain the Defense Department's formal approval, Halaby succeeded.

The document was extremely positive, finding the SST technically feasible, projecting a free-world SST market of one hundred to two hundred aircraft, and favoring an advanced long-range steel-titanium Mach 3 vehicle, probably based on a swing-wing design. The report called for extensive government financial assistance, some return on government funds from royalties, significant contributions from private industry, an immediate government-financed $40 million to $50 million technical and research effort of one to two years, minimal government control, full airline involvement, possible multifirm cooperation, and a "vigorous effort" to build an SST, with certification in 1970 or 1971.[9]

Kennedy also appointed Halaby head of a presidential task force on national aviation goals for the 1960s, which provided Halaby with another opportunity to promote the SST. In September 1961 this group released its report, *Project Horizon,* that not surprisingly presented a forceful case for developing an American SST. Reflecting Halaby's general views and goals, *Project Horizon* categorically declared the SST technically feasible and, like Halaby earlier in the spring, dwelled on the tangible and intangible advantages of being the first to develop such an aircraft. The report favored a Mach 3 aircraft, arguing that an aluminum-based Mach 2 aircraft would soon be obsolete. The document called for a largely government-financed $1 billion program and generally placed emphasis (as the recent report of the three agencies had) on technical rather than economic feasibility. Finally the task force provided perhaps the strongest statement to date linking, in cold war–era terms, SST development with national prestige: "The loss of this Nation's preeminent position in the production and sale of transport aircraft would be a stunning setback. In the light of Russian accomplishments in space technology, it is imperative that the United States retain its leadership in aviation."[10]

The pro-SST thrust of *Project Horizon* was not totally accepted, even by certain members of Halaby's own task force. William Littlewood of American Airlines, chairman of *Project Horizon*'s technical review committee, warned against a crash program "like the Manhattan Project which produced the atom bomb without regard to cost." He was also skeptical of developing immediately a Mach 3 SST, favored further debate and testing, and stressed the need to prove that economical SSTs could be built or "the bankers will never give us the money."[11]

But Littlewood's opinions simply exhibited the cautious views found in certain airline quarters and had little impact at that time. Halaby was already at work building a support network with friendlier nongovernmental groups. In late October 1961 he established the Airline Advisory Group and a month later the more prestigious and influential Supersonic Transport Advisory Group (STAG), which was to undertake independent studies at Halaby's request and to suggest areas for further review. STAG's chairman was General Orville R. Cook, president of the Aerospace Industries Association, and its membership represented various areas of the aviation field. It met periodically throughout 1962, and Halaby would consistently use it as a key vehicle for legitimizing pro-SST positions.[12]

Halaby also strengthened the interagency relationships, which would become increasingly important in performing SST feasibility studies. After the joint FAA-NASA-Defense June 1961 report was presented, a high-level triagency coordinating body, the Supersonic Transport Steering Group, was formed. After funding approval in August 1961 the joint Supersonic Transport Working Group (a staff-level body established in 1960) was designated the SST Task Group, and an air force SST Support Office was established at Wright Patterson Air Force Base in Ohio. By early November 1961 the FAA had transferred about $10.8 million to the air force for technical research, a move indicating the continuing emphasis on technical issues during these early years.[13]

Halaby's dilemma as a champion was that in building interagency SST cooperation, the FAA's control over the SST effort might diminish. In July 1961 his deputy administrator for administration, Alan L. Dean, in an extremely prophetic review of a proposed management plan, informed Halaby of such "serious misgivings." Dean cautioned that "the tools of management are largely outside his [the FAA administrator's] control," with the "practical effect" of leaving the "FAA Administrator responsible for the program in the eyes of Congress, the President and the Public without his having the capability of assuring that the job gets done."

Dean also focused on perhaps the Achilles' heel of Halaby's ambitions: the FAA's lack of experience in running massive technological projects such as the SST. Throughout the summer he urged that the FAA rather than the air force administer all research contracts in order to develop the necessary management skills. As he advised Halaby, "This [the SST] is a big project with which the Agency [the FAA] will be concerned for some years, and there are strong arguments for equipping ourselves to do the job of managing required."[14]

Halaby did take some action to upgrade the SST effort in the FAA hierarchy in late September 1961 when he created the new Supersonic Transport Program Management Office to replace the old internal arrangement that Quesada had established. Colonel Lucien S. Rochte, Jr., was made its director, and four other FAA officers were assigned to his staff.[15] A focal point for SST development finally existed. No longer was the effort a loose collection of ad hoc interagency groups and FAA teams without a congressionally approved budget.

During the late 1961–early 1962 period technical issues were the major concerns for those involved in the SST effort. In particular the NASA director of aeronautical research, John Stack, was something of a technological champion at this time. In early 1962, as he had during the Quesada era, he again energetically promoted a Mach 3 variable sweep- or swing-wing SST design with wings that can be moved in flight as an "all purpose aircraft." Such an SST could allegedly fly efficiently at supersonic speeds with its wings fully swept back and at subsonic speeds for takeoff and landing and possibly over land with its wings only at a moderate degree of sweepback, probably less sweepback than that of the subsonic commercial jets of the 1960s. This "subsonic-supersonic" airliner, Stack claimed, in addition to being efficient in both speed ranges, would possess a lower weight than a Mach 3 SST with a conventional fixed-wing delta design. Stack confidently estimated the market at six hundred to seven hundred aircraft and predicted that the European Mach 2 fixed-wing concept would operate poorly in the subsonic range. The chairman of the recently formed Airline Advisory Committee also echoed Stack's opinions.

Although he was an SST enthusiast, Stack was also a NASA researcher, and he cautioned against setting deadlines for state-of-the-art technological development projects like the SST. Moreover his optimistic technical views were criticized by a STAG member, Theodore P. Wright of the Cornell Aeronautical Laboratories, who in January

1962 outlined several serious technical problems that still needed intensive study, among them the sonic boom, solar radiation effects, and engine design. Another STAG member, James Mitchell, vice-president of the Chase Manhattan National Bank, also voiced doubts on the economics and financing of such an advanced aircraft since it would be unable to compete in the near term with an earlier, less-advanced European aircraft.

After listening to such cautious sentiments, STAG counseled the FAA in February 1962 to portray the current American SST effort as primarily a research program and to stress that the effort was not part of a race with any other country.[16] The technical orientation of the program continued.

A major concern for Halaby and the FAA in early 1962 was organizing SST feasibility studies, which had been allocated $11 million for fiscal 1962. Proposals from industry were evaluated by a team of experts from the FAA, NASA, and the Defense Department, and contracts were to be awarded in the spring. Looking ahead, Halaby was also intent on securing as much as possible of the $25 million SST allocation that was mentioned in the president's fiscal 1963 budget message. To Halaby the SST work in the next fiscal year would be crucial. "In fiscal year 1963," he remarked to Senator Magnuson, "the scope of the work will shift from exploration and concept to applied research. . . . The results of small-scale tests and experimentation, using fiscal year 1962 funds, will be translated into large-scale research."

Halaby went to great lengths to protect his enterprise. He attempted to demonstrate that the FAA's SST work was efficient and did not duplicate NASA or Defense Department activities, reiterating particularly that military supersonic aircraft could not be transformed into commercial SSTs because of the former's excessive noise, short range, and poor operating economics. The last point was especially important because some members of Congress were perplexed at additional SST funding requests when funding for the curtailed supersonic B-70 bomber program was reaching $1 billion and a B-70 prototype was expected in less than a year.[17]

Interagency cooperation on SST work was not always smooth. Halaby was not pleased with NASA's fiscal 1963 budget request of $21.1 million for SST work, which would mean about 875 new employees for NASA. He was irked enough to want to protest such a request but was dissuaded by Rochte and Dean, who counseled that such an action would be ineffective. The problem of FAA-NASA coordination persisted into the spring of 1962, although the agencies had little choice but to support each other publicly. At Senate FAA appropriations hearings in August 1962 Halaby spoke in favor of the full NASA SST request, and he even went out of his way to praise two basic NASA Mach 3 SST designs, the swing-wing SCAT-15 (SCAT was the NASA acronym for "supersonic commercial air transport") advocated by John Stack and the fixed wing SCAT 4. At the Senate NASA appropriations hearings, also held in August, Webb painted a rather exaggerated harmonious picture: "We [NASA and the FAA] are doing two different things, both of which fit together and we are in very close touch. Our programs have been planned together. We do the basic research and then Halaby picks it up from there."[18]

The White House began exhibiting stronger interest in the status of American SST development, though such concern was not always to the FAA's advantage. The President's Science Advisory Committee, which was chaired by the president's science adviser, Jerome B. Wiesner, became more involved. Both Wiesner and Nicholas E. Golovin, a member of his staff, had expressed support for SST development. Golovin stressed the need for the United States to maintain "the image of technological supremacy in commercial aviation" and urged a state-of-the-art Mach 3 SST. But Golovin also doubted that the FAA possessed the ability or the resources to run an SST program effectively: "The FAA has itself no major technical laboratory centers capable of attracting and keeping top-notch technical competence in the principal relevant technology." Emphasizing the enormous technological complexity involved in SST development, Golovin favored the "application of more of the techniques found successful in forcing progress in space technology" and the creation of a separate organization to manage an American SST program.[19]

Given such skepticism over the FAA's managerial ability, it is not surprising that Halaby and his aides in 1962 tried to strengthen the FAA's administrative control of the SST effort. In March 1962 Halaby suggested using a "systems" approach for the SST, which was then in vogue due particularly to the enormous impact of McNamara's management of the Pentagon. Halaby's promotion of systems techniques and other sophisticated managerial methods eventually resulted in late

July 1962 in a detailed and integrated "techno-economic analyses" plan and a new systems analysis unit (with a program manager) in the SST Office.[20]

But questions about the FAA's managerial ability still remained, both outside and within the FAA, and this nagging doubt continued to dog Halaby as he attempted to promote his program. Within the aviation industry many were uncomfortable with the prospect of the FAA as SST project manager. In mid-June 1962 William M. Allen, president of Boeing and a highly respected figure in the industry, candidly reflected this opinion. He wrote to Roswell L. Gilpatrick, deputy secretary of defense, that the current SST multiagency structure under FAA leadership was "unwieldy, complex, and lacks effectiveness," and he called for a "vigorous, aggressive [SST] management organization." Allen concluded, "The FAA is primarily a regulatory agency and lacks the personnel, skills and experience to undertake management of a program with such magnitude as the SST." Allen offered three more satisfactory alternatives: a separate Defense Department SST research and engineering unit; an "air agency" with the power to develop, purchase, finance, and lease SSTs; or a joint industry-government corporation like COMSAT, which had been established for satellite communications.[21]

Such high-level dissatisfaction with SST management culminated in a meeting of the National Aeronautics and Space Council in early August 1962. Attending were Vice-President Lyndon B. Johnson (the council chairman), Secretary of State Dean Rusk, Defense Secretary McNamara, Atomic Energy Commission chairman Glenn T. Seaborg, NASA administrator Webb, Halaby, and the council's executive secretary E. C. Welsh.

Halaby stressed the SST's technical and economic feasibility and the threat of possible Anglo-French SST collaboration. He even mentioned that the Russians might be "first with the worst," with a Mach 2 short-range aluminum vehicle. (Soviet officials were predicting airline service in 1965.) But McNamara, who already knew that several people were worried about the FAA's overall competence, expressed little faith in the FAA data, which he said were not based on hard evidence. He thought that the FAA estimates on development cost and aircraft purchase price were too conservative and, not for the last time, urged more studies. Voicing views he would cling to for years to come, he argued that the SST had to be justified commercially before development could take place and that the SST needed to be viewed as a straightforward business venture.

McNamara's view prevailed, and the FAA agreed to support outside cost and market analyses before any further action took place.[22] McNamara's enormous influence over SST affairs was thus demonstrated at an early date: he had forced a closer inquiry into nontechnical SST matters and had effectively derailed Halaby's attempt to accelerate SST development.

After this meeting the FAA stood no chance of obtaining a presidential decision on developing an SST before completion of the current SST research and feasibility study period, at the end of fiscal 1963 (June 30, 1963). Halaby was also forced to justify the SST on economic as well as technical or national prestige terms. Two months later the FAA awarded contracts for SST cost and market studies to the Rand Corporation and to the Stanford Research Institute (SRI), respectively. Halaby retreated, and within a few days after the meeting he had eliminated most of his expansive rhetoric in congressional hearings.[23]

Halaby also encountered difficulties obtaining the full $25 million fiscal 1963 request. In almost an exact replay of the fiscal 1962 funding debate, the House cut the original request, while the Senate supported Halaby's request. Eventually they compromised at $20 million for fiscal 1963. Thus for fiscal 1962 and 1963 Halaby secured a total of $31 million for SST feasibility studies.

The SST study effort had widened substantially by the end of 1962. Thirty-seven contractors were awarded one or more contracts for SST analyses on technology and engineering, economics, the sonic boom, and the integration of data from various fields.[24]

Throughout Halaby exhibited an essential characteristic of an effective champion—persistence—and he forcefully continued his campaign. STAG was especially crucial for Halaby at this time for establishing credibility and mobilizing support in industry and government. The initial report of this advisory group—consisting of very influential individuals, most of whom had supported SST development from the beginning—was due by the end of 1962.[25] Halaby pushed for a favorable analysis. Indeed he criticized the proposed SST technical requirements in STAG's first draft report as too high and rigid to be "realistic." He further charged that STAG had ignored certain political, social, and economic benefits in such areas as national prestige, transportation, communication, balance of payments, employment, technological innovation, and national defense. For good measure he added that foreign SST competition was inadequately discussed and that the infeasibility of converting the B-70 bomber into a commercial aircraft should be covered.[26] (The FAA's consultants, Rand and SRI, later reviewed a revised draft, and similarly advocated a less restrictive document. Behaving like FAA strategists, they seemed to call for a STAG report that would, in effect, fully and credibly support the FAA's position.)[27]

Halaby's tenaciousness paid off. The final version of STAG's initial report, which was transmitted to the FAA in mid-December 1962, was all he could have desired. "Passing the speed of sound," the report declared, "will be one of the last major breakthroughs for commercial air transportation in the earth's atmosphere." To STAG, the SST would "represent not only a vehicle to foster economic growth of the nation, but also a vehicle to demonstrate the technological and scientific leadership of the United States," and it would be "vital to the national interest." The report strongly recommended that the United States aim for a Mach 3.5 steel-titanium-based aircraft, at least "as a planning goal." It called for a program that would cost an estimated $1 billion, of which 90 percent would be borne by government and 10 percent by industry.

To Halaby, only the proposed SST management structure left something to be desired. STAG recommended that the SST program director be a presidential appointee to be named in early 1963 and that this director head an independent organization with a board of directors chosen from the FAA, the CAB, NASA, the Defense Department, the Bureau of the Budget, and the public. But STAG also advised that the FAA administrator "should continue to play a prominent role in the program."[28]

The report was not universally applauded. In particular, reflecting McNamara's cautious and commercially oriented views, the Defense Department's liaison with the FAA on SST matters urged a less hurried approach, more comprehensive research, and further consideration of organizational alternatives before establishing an SST management structure.[29]

Still, armed with the report, Halaby pressed vigorously for a presidential commitment for SST development in the fiscal 1964 budget message, which the president would send to Congress in January 1963. Halaby enthusiastically informed Kennedy that STAG, which Halaby characterized as "a distinguished and hard-headed group of aviation industry people," had proposed "an all-out, immediate effort." He told the president in late December 1962 that "the President's Budget Message should state clearly the intention of this country to proceed with the Supersonic Transport development," and he mentioned that he had the support of Gilpatrick, Webb, Magnuson, and other key Capitol Hill figures.

Here, however, Halaby was largely unsuccessful. His demands for a strong public commitment in the budget message were opposed by the powerful Bureau of the Budget. Like McNamara, the bureau was skeptical of Halaby's strong pro-SST claims and urged a delay until a "thorough and objective analysis of its economic and social implications" was available with the completion particularly of the Rand and SRI cost and market studies.

The president followed the course favored by the bureau and McNamara and rejected Halaby's demands in favor of waiting for the economic evaluations. Nevertheless Kennedy directed Halaby, in cooperation with various government agencies, to "take the lead" in drafting a report on all relevant SST research results, which would include recommendations for further action.[30]

Industry Speaks

The private sector's impact and views on SST development were ambivalent. Although the manufacturers and most of the airlines publicly supported the SST, they expressed a good deal of skepticism behind closed doors. Neither the airlines nor the manufacturers desired to invest significantly in SST development, even under vigorous prodding from the government.

The attitude of the domestic air carriers toward the SST was especially confused and would remain so throughout the SST conflict. Airline presidents vigorously supported SST development at congressional hearings and in public letters, but in private they were not particularly worried about Anglo-French SST efforts and, if anything, wanted to slow down American SST work.

Airlines select aircraft primarily on the basis of economic and long-term strategic considerations. A key backdrop for any airline SST decision was that in the early 1960s the airlines were adding expensive subsonic jets to their fleets and they would later face the financial burden of buying subsonic jumbo jets.

Moreover the airlines generally doubted the SST's economic performance. At an International Air Transport Association symposium in April 1961 many airlines expressed such concerns, and even the manufacturers acknowledged the SST's uncertain economics. One engineer reportedly said, "We think we know how to build them, but we haven't the faintest idea about seat-mile cost."[31] At a meeting in June 1961 the airline operators bluntly told Halaby that they were unenthusiastic about the SST as a near-term economic prospect because of their heavy investment in new subsonic jet fleets. Although the airlines wanted a strong voice in determining the SST's design features, the industry basically considered the SST a matter of national prestige rather than an attractive or needed commercial venture.[32] At a public conference in September two leading airline executives—one from American Airlines and the other from BOAC—criticized crash SST development. Interestingly the BOAC official publicly questioned building the British-backed Mach 2.2 aircraft even if its technical feasibility were proven and suggested that the entire goal of developing an SST needed further thought. The general mood of the airlines was summed up at the end of the year by the director of International Air Transport Association: "Any government which decides an SST is necessary should face the full consequences of that decision early in the game. If they want prestige, they must be prepared to pay for it. There will not be enough airlines and enough passengers to foot the bill."[33]

The sonic boom also worried the airlines, and the issue was raised at the April 1961 symposium. Several air carriers, the symposium report warned, "emphasized the vital necessity of obtaining this information [on sonic booms] well in advance if crippling economic penalties were to be avoided." Bo Lundberg, director-general of the Aeronautical Research Institute of Sweden, and even at this early date one of the most persistent critics of the SST, detailed the enormous array of problems associated with the sonic boom, including the difficulty in determining the sonic boom's public acceptability, the inequity of prohibiting supersonic flight over "densely inhabited areas" while permitting it over "sparsely populated districts," and the legal problems related to personal and property damage by sonic booms.[34]

Technical doubts were growing. In April 1961 a number of papers presented at the National Aeronautics Meeting of the Society of Automotive Engineers "went deeper into the subject [of SST technical feasibility] than many earlier efforts . . . and came up with little encouragement and lots of problems." The projected high engine cost, the lack of data on material life span, the inability to transform military airframes and power plants into commercial SSTs or to reduce the intensity of the sonic boom, and many other issues were "now beginning to darken the first rosy glow of optimism."[35]

A STAG meeting in July 1962 with representatives from United, American, Eastern, Pan American, and TWA provided a candid glimpse of airline opinion on the SST. All of the airlines emphasized the need for much more study on safety, economic feasibility, and especially the sonic boom. According to the meeting's minutes, American saw "no urgency for the SST today" and thought it could "wait a long time before buying" an Anglo-French Mach 2 aircraft. United was skeptical of the SST's economics. It believed that the SST "should come along as a natural evolution rather than forcing it" and was not very concerned with foreign SST competition. TWA also commented that there was "no urgency from the public needs standpoint" and also stressed that the SST must be "economically competitive" with "an economical tourist class [subsonic] aircraft."

The U.S. international carriers, however, manifested an ambivalent attitude that continued throughout much of the SST program. They did not entirely ignore the possible effect of a European-built SST. TWA, although preferring an American vehicle, would consider purchasing an Anglo-French SST if it were safe and economical. Pan American expressed respect for British and French technical capabilities. As the minutes reported, "Pan American indicated they would stay on the line with protective orders for as few of the foreign supersonic transports as possible and place orders within three years for the better U.S. Mach 3 aircraft if it were more economical."

The engine and airframe manufacturers appeared more enthusiastic than the airlines at this time. Their views were also visible at STAG's July 1962 meeting, which included representatives from General Electric, Pratt & Whitney, Curtiss-Wright, Boeing, General Dynamics,

Douglas, Lockheed, and North American. Generally the firms favored an advanced Mach 2.5–3-range aircraft, although almost all expressed great worry about the SST's economics, development costs, and financing.

Among the engine firms, General Electric believed that a Mach 3 SST prototype could fly in 1967 if the basic design and go-ahead decisions were made in 1963, although the introduction of the SST should take place only "when the engine and airframe are reliable and the airplane is economical." The firm confidently argued that the mere appearance of a flying advanced Mach 3 American SST prototype in 1967 would "forestall" foreign SST sales. Pratt & Whitney and Curtiss-Wright also argued in favor of a Mach 3 SST, with Curtiss-Wright calling the idea of first building a Mach 2–2.3 aircraft a "weird concept."

But these firms were considerably less enthusiastic about financing SST engine development, with cost estimates ranging between $150 million and $200 million. Curtiss-Wright was willing to consider participation in an industry-government cost-sharing arrangement. General Electric, however, "was not very enthusiastic about cost-sharing." Similarly Pratt & Whitney said it could not contribute large funds, and it recommended government financing through the development stage with a provision for royalty payments to the government on engine sales.

The five airframe manufacturers exhibited less consensus. Boeing described its initial SST work, which was aimed at developing a Mach 2.5 swing-wing prototype having a transcontinental range of about 2,500 nautical miles with a payload of 125 passengers. The company was optimistic about curbing the sonic boom and radiation problems and saw foreign SST competition as "a serious threat to U.S. prestige." Lockheed, on the other hand, felt that "early introduction of an SST by the British or French would not seriously affect the U.S. market," although it too supported development of an advanced Mach 3-range aircraft. Douglas considered this competition an extremely important problem but unlike Boeing and Lockheed favored an aluminum Mach 2 vehicle, which would require a shorter development period. North American emphasized its own experience as manufacturer of the Mach 3 B-70 bomber.

All airframe manufacturers mentioned the necessity of good economic performance. Roger Lewis, president of General Dynamics, reasoned that the SST effort from the very beginning should aim at acceptable seat-mile costs. "The SST is not a Sputnik," he declared. Not surprisingly he also urged taking advantage of his firm's swing-wing TFX fighter-bomber. Boeing, however, estimated the SST's direct operating costs at about 12 percent higher than those of subsonic jets. Using by now classic arguments for government SST financial assistance, the company noted that foreign SST competition was almost completely publicly funded and stressed the great risks involved in SST development because no military predecessor for an American SST really existed. Lockheed and Douglas also saw the need for government funding, although both felt that the government should recoup its investment through royalties.

Regarding management of an SST program, Boeing favored establishing an independent "air agency," General Dynamics backed air force supervision, and Douglas recommended splitting the responsibility among the air force, an interagency committee, and an independent SST "Management Agency." Only Lockheed and North American seemed generally satisfied with FAA direction.[36]

Industry approval of the SST, therefore, was not total. Both the airlines and the manufacturers would give Halaby public support, with the manufacturers being more enthusiastic in private. But all were worried about costs. No firm was willing to take a large financial risk. The source of leadership and funding for an American SST program clearly would have to be the government, although many in industry remained unconvinced that the FAA should direct an SST development effort.

Europe Speaks

From the beginning Halaby had to deal with foreign pressures for SST collaboration, and, like Quesada before him, he consistently fought off such notions. In February 1961 the British once again sent out feelers for collaboration on the SST. The British ambassador to the United States argued that there was room for only one SST and urged that the British and Americans join hands in SST development. A month later, Peter Thorneycroft, British minister for aviation, proposed that the two nations undertake a feasibility study of a Mach 2.2 aircraft, anxiously adding that he would send a team to the United States immediately for preliminary discussions.

But Halaby had no intention of becoming partners with the British. He told the House Appropriations Committee in April 1961, "We want to be there ahead of potential competitors." Because Congress had not yet approved funding for the two-year SST study effort, however, Halaby merely evaded Thorneycroft's request.[37]

The British continued to press during late 1961. In November Thorneycroft met with FAA officials in Washington. He again pushed for a joint development program, and flatly warned that he would recommend an Anglo-French SST project if the Americans would not join the British.

But Thorneycroft's threat failed to budge Halaby, who offered only to consider exchanging SST information on such "environmental aspects" as the sonic boom problem, air-traffic-control systems, navigational aids, air-worthiness standards, and airfield design. Halaby also frankly mentioned the American commitment to an advanced Mach 3.5 SST. Thorneycroft could do little but accept the crumbs offered, and he proposed to exchange information "on the fringe areas and continue thinking of ways to build one U.K./U.S. supersonic transport."[38]

Like Thorneycroft, Georges Hereil, president of France's Sud-Aviation, also publicly favored collaboration with the United States. He suggested that the Europeans concentrate on a Mach 2.2 "transcontinental aircraft" while the United States focus on a Mach 3.5 "transoceanic" vehicle. In the SST's case, he believed, "private enterprise must give way to an interstate organization."[39]

The Americans never seriously considered Thorneycroft's and Hereil's proposals. If anything, American resistance to international SST collaboration intensified in 1962. Reflecting American opinion, John Stack of NASA told two high-level SST advisory groups in early January 1962 that foreign SST technical information would be of no value to the United States.[40]

Anglo-French collaboration, in effect, became the only real SST option for the Europeans, and the Americans were quite aware that U.S. aloofness was a key factor in encouraging Anglo-French SST cooperation. The French and the British separately had favored an SST with about the same speed, about Mach 2.2. Public indications of possible joint efforts between the British Aircraft Corporation and Sud-Aviation to build a Mach 2.2 aircraft surfaced as early as November 1961. In March 1962, with expectations high, top British and French teams met in London and predicted a flying transatlantic SST by 1970.[41]

The Americans obtained more detailed information from a delegation of Sud-Aviation officials who visited Boeing in Seattle in early March 1962. The French claimed to be visiting Boeing because they believed that ultimately the United States could not "afford to stay out of this SST project" and that they considered Boeing the most likely of being selected to participate in any future U.S.-European SST joint venture. A senior Boeing executive told Boeing's president William Allen that the French had decided "for prestige purposes" to build a Mach 2.2 SST and anticipated a market of seventy aircraft; the French government had already earmarked $220 million, and deliveries were expected in 1967; the British Hawker-Siddeley group was already developing an engine, and the British Aircraft Corporation would eventually build a transatlantic version of the French model; ultimately there would be two final assembly lines, one at the British Aircraft Corporation and one at Sud-Aviation.

The Seattle meeting obviously disturbed the Americans, at least momentarily. Boeing realized that American domestic carriers would constitute a major market for the Anglo-French aircraft, and Pan American had already informed Boeing that the airline would have "no choice but to go ahead and purchase some of these aircraft." A Boeing official warned Allen, "This means to me that the race for the supersonic transport has started and we are going to be faced with competition in the United States from foreign manufacturers. . . . This overall situation requires serious consideration on our part at an early date." Copies of a Boeing report on the meeting were sent in mid-April to Roswell Gilpatrick at the Defense Department and to Halaby. The Boeing cover letter again warned Halaby of the "foreign threat" to our "air transport image" and our "national prestige" and called for "much more vigorous" SST effort than was then taking place.[42]

There were even a few calls for a new look at American-European collaboration. In July 1962 a planning officer within the FAA in a detailed, well-argued, and forceful brief suggested a multilateral consortium among the Western nations as "the most economic manner" for placing an SST in service while satisfying American goals. Although acknowledging "jealousies" among competing nations, the officer reasoned, "The West will preserve its prestige vis-à-vis the Soviet Bloc; uneconomic competition for a very small market among the Western Allies would be reduced or precluded; the claim on U.S. public funds will be substantially reduced; the strain on the U.S. private aircraft industry will be lightened *while they would still achieve the largest share of the market* since U.S. airlines would presumably order the largest number of aircraft; all Western participating countries would share in the advanced technology, and there would be a definite political gain in having the key Western nations work closely together on a program of this magnitude."[43]

But any move toward such collaboration ran counter to the views of almost all in the FAA and other groups, and by late summer meaningful consideration of international collaboration had all but vanished. Reflecting prevalent opinion, Raymond B. Maloy, the director of the FAA's International Aviation Service, consistently opposed any substantive SST interaction with the British and French, and he wanted any meeting "limited to an 'exchange of views' on areas of common interest."[44] Maloy believed that the prospective French SST, a short- to medium-haul aircraft, was not competitive with the planned American aircraft and that the design range of the British version, which was directed at the transatlantic market, was "not adequate in the same way that the Comet was not adequate as a transatlantic airplane." He further stressed the traditional Anglo-French mutual distrust and current disagreements. "All the French want from the U.K.," Maloy concluded, "is an engine."

Instead of collaboration, Maloy pressed for a solely American SST effort that would proceed at its "own pace" and aim toward an advanced Mach 3 SST with "true transatlantic range." He reasoned, "If the U.S. is to make the big Mach 3 leap, it can afford to be a little late, but it cannot afford to be even a little wrong technically or economically."[45]

The Americans continued to monitor European SST activities and kept well informed, however. STAG met with the representatives of British and French aerospace manufacturers in July 1962, when the contrasting priorities in SST design and timing were quite evident. Moreover apparent disagreements and lack of coordination between the British and the French merely demonstrated to many American officials how difficult it would be for all three nations to collaborate.[46] When Halaby met with Sud-Aviation officials in France in early September 1962, the Anglo-French plans for organizing and financing a joint SST project were laid out in considerable detail. The French still anticipated passenger service in 1970.[47]

Three months later, in late November 1962, Britain and France dramatically announced a formal agreement to develop an aluminum-based Mach 2.2 SST, called the Concorde. An actual rival SST program now existed. The Anglo-French agreement conformed to the information already provided Halaby. The development cost of the Concorde project was estimated at $448 million, with Britain and France sharing costs evenly. Britain would tend to focus on engine development and France on the airframe. One prototype would be built in each country.[48]

Concorde officials quickly began to press their advantage of being first with an SST program. In mid-December 1962 Sir George Edwards of the British Aircraft Corporation and deputy chairman of the Concorde project requested a meeting with Halaby. According to Halaby, they had a "long, candid and constructive conversation, most of which was about the supersonic transport." Edwards exuded optimism. He appeared to go out of his way to stress that Anglo-French working arrangements and designs were settled and, in Halaby's words, that "there was an even tighter arrangement than before." He told Halaby his sales managers were like "two unleashed tigers eager to get out and start selling the airplane," forcing him to hold them back until he was "more confident of his numbers." Edwards mentioned he had recently met with Pan American's Juan Trippe and others and had told them "straight away that I won't be pushed into revising my design to please everyone who wants to get in line for an order." He looked forward to initial passenger service by January 1971 and minimized the Concorde's engine noise and sonic boom problems.[49]

The establishment of the Concorde project—in effect, creating the specter of a credible foreign threat—was Halaby's ace in the hole at the end of 1962. He had already used this line of reasoning in testimony on Capitol Hill earlier in the year: "We are in a life and death struggle, so far as commercial airframe manufacturers are concerned, with the British, French, and Russians who would seek to sell an airplane which could cross the Atlantic and cross other parts of the Continent at anywhere from 2,000 to 3,000 miles per hour. To the winner, the guy who produces the first commercially profitable, safe, and efficient aircraft, goes a $3 to $4 billion market.''[50] Later in the year he argued similarly that the development of an American Mach 3 SST would be as important to national prestige as being the first to land on the moon.[51]

By November 1962 the dire implications of the new Anglo-French project was Halaby's main lobbying theme. In a report to President Kennedy, Halaby portrayed a successful Concorde as forcing the United States to ''relinquish world civil transport leadership''; over 50,000 jobs would be lost and the United States could be forced to depend on foreign sources for supersonic military aircraft. Halaby clinched his appeal by warning that ''conceivably'' an American president would fly in a foreign aircraft someday.

Halaby offered an effective antidote to the Concorde: an American SST. ''Research and development work under my leadership,'' he declared, ''has taken the British-French Project into account,'' and he stressed that a steel-based Mach 3 SST would be safer, more durable at high temperatures, and more efficient than the Anglo-French design. According to Halaby, the president had three SST options for fiscal year 1964: ''yield'' to the British and French, provide $100 million in the fiscal 1964 contingency fund for a prototype Mach 3 SST to be used as soon as market and cost factors were confirmed in the spring by the feasibility studies, or ask Congress immediately for authorization and appropriation to build a Mach 3 prototype at an estimated total cost of $750 million, with a first-year requirement of $100 million.[52]

Halaby believed strongly that the Concorde announcement was a crucial factor in mobilizing support within the administration for an American SST program. As he recalled years later, ''When DeGaulle embraced the joint Concorde project, it seemed to trigger competitiveness in John Fitzgerald Kennedy. In fact, I think JFK associated the Concorde most with DeGaulle; on more than one occasion, he said, 'We'll beat that bastard DeGaulle' . . . every time I saw the President, from the day DeGaulle made his announcement, he would press me on how our [SST] studies were going—and how the British and French were doing.''[53]

By early 1963 Halaby had undergone significant and frustrating ordeals as an SST champion. Within the administration, though successfully shepherding through a program of SST feasibility studies and being designated the lead person to report to Kennedy on SST work, worry about the FAA and the SST was building in key quarters. In industry public support was counterbalanced at various levels by some private ambiguity and doubt. But the Concorde threat provided Halaby with a new and powerful opportunity to regain the momentum in pursuing his ultimate goal: an SST program under FAA control.

4 Emergence of a Program

In early 1963 Najeeb Halaby began a critical phase of his promotional campaign. The next few months would tell whether he would possess a real development program. Although not totally supported within the administration or industry, he was still in control. Moreover, with President Kennedy assigning him prime responsibility for evaluating "all the probable benefits to government and the national economy" resulting from an American SST program, Halaby maintained his influence. The FAA remained the preeminent domestic institution for American SST affairs, even though Kennedy also directed that the Budget Bureau director and the chairman of the Council of Economic Advisers participate in the extended SST evaluation.[1] The SST was still "on schedule," as an aviation industry journal observed.[2]

Kennedy then created a cabinet-level committee to review the FAA's recommendations on the SST. This committee eventually included members from the Departments of Defense, Commerce, Labor, and the Treasury, as well as from NASA, the FAA, the CAB, and the Office of Science and Technology (OST). In February, to Halaby's great delight, Kennedy named Vice-President Lyndon B. Johnson as committee chairman. Halaby had already told Johnson, "I can't tell you how pleased I am to see you resolving and dissolving some of the resistance and opposition-without-facts to the [SST] Program. As you know, I have been leading this one all alone with many competitors for potential funds."

Halaby was encouraged. After meeting with Kennedy in late January, he found the president "much more nearly in favor of going ahead with the procurement of an SST prototype. One of the factors is the desire to outdo de Gaulle." Energized, Halaby demanded "mobilization of all the talent we can find." He even suggested placing a "task force commander" in charge of a massive three-month, seven-day-per-week coordinated effort for producing a proposal that could be reviewed positively by the Johnson committee in April 1963. It was time, in Halaby's words, to move from a "Gulf Stream level" to a "supersonic level."[3]

But, like before, Halaby soon confronted increasing controversy over SST economics. The Stanford Research Institute (SRI), which had been examining SST economics and markets, concluded that a Mach 3 SST would yield appreciably lower annual profits than would a comparable Mach 2 aluminum version. Moreover although a Mach 3

SST would be most profitable if it charged only first-class fares and had high-density seating, this fare policy might also cut the SST market by as much as two-thirds. SRI found that all SST versions would have considerably higher direct operating costs per available seat-mile than those of subsonic jets. But because of the SST's allegedly positive effect on employment and balance of payments, SRI still strongly favored developing such an aircraft.[4]

The SRI study provided a starting point for an expanded discussion of the SST's economic feasibility and also led to the first of a series of conflicting opinions on SST economics that would plague the program practically throughout its entire existence. In mid-February 1963 E. C. Welsh, National Aeronautics and Space Council executive secretary, asked the Council of Economic Advisers and the Department of Commerce to review the SRI work as well as Rand Corporation SST cost studies, which were also becoming available.

A month later Walter W. Heller, chairman of the Council of Economic Advisers, sent Johnson a severe critique of the SRI study: sources of air-traffic-growth projections had not been indicated, and much more attention should have been paid to the effect on SST demand of alternative modes of air transportation. He suggested that SST demand would be highly sensitive to the fares charged, and he urged a detailed evaluation of the Anglo-French Concorde program. Objecting to SRI's "overly narrow framework," his own examination was extremely comprehensive. He called for a broader and more detailed look into the SST and warned against plunging ahead. Heller also suggested that an independent group of experts evaluate FAA sonic boom limits for the SST's designs. He wanted this group to include experts on human reaction—physicians and psychologists—in addition to structural engineers. He even urged performing a systematic evaluation of all alternative uses for government funds of the magnitude envisioned for the SST, using the criteria of international prestige, technological leadership, and more efficient transportation.[5]

Heller's concern about the sonic boom was not occurring in a vacuum. At that time research on the subject was taking place at Edwards Air Force Base in California and elsewhere, and the National Opinion Research Center at the University of Chicago had been hired to interview one thousand people about effects of the sonic boom. In fact, a remarkably perceptive air force document in March 1963 called attention to the "growing number of complaints and damage claims" due to sonic booms from military aircraft and then declared that the effect of sonic booms on the general public "remains an open question." It asked if people could "learn to tolerate sonic booms in the same way that they now tolerate railroad and traffic noises[.] Should they?" It concluded, "The sonic boom problem cannot be ignored or treated lightly. It is severe and controversial now and will be magnified by future supersonic aircraft development."[6]

The Commerce Department review was quite different from Heller's. Clarence D. Martin, Jr., the undersecretary for transportation in the Department of Commerce, was more optimistic and considered mostly economic considerations. But he also favored ranking alternatives and weighing "intangibles" and stressed the need for studies of market conditions, including Concorde competition, potential air traffic patterns, and effects on the country's economy.[7]

Halaby, however, was not without allies. Welsh, who was something of an SST advocate, did not appear particularly impressed with either response, and he especially disagreed with Heller. Welsh wanted to rely on current sonic boom studies, and he considered national prestige an important issue. He told Halaby, "I would assume that the technological advantages of leadership in the aviation field would weigh heavily in the SST decision," He found it "disturbing" that "in an area in which this nation had been the acknowledged leader for several decades, there is any uncertainty as to whether we should continue to compete in the future." Welsh stated categorically that the vice-president should present to the president a clear "go" or "no-go" recommendation.[8]

Meanwhile the SST Office within the FAA was upgraded and enlarged. Its SST research program grew in number and variety with approximately fifty-five contracts awarded to twenty-eight contractors in fiscal 1962 and 1963. It was also under tremendous pressure to produce data and conclusions in time for the Johnson committee SST review, which was expected in May 1963.[9]

In February 1963 Colonel Dale D. Davis replaced Rochte as the SST program manager. In late March 1963 Halaby gave the SST Office an even more visible bureaucratic profile. He directed his assistant administrator for appraisal, Gordon M. Bain, to assume overall responsibility for SST activities. Bain had been a CAB member for many years and later had served as a high-level executive with Northwest Airlines. Halaby told Bain that the SST effort "is now at a stage which requires that it be given the highest possible priority" and that "all of our resources [must] be mobilized under a single individual." During the five-month period ending in May 1963 the SST Office grew from eight to thirteen people.[10] Bain, a tough and committed administrator, would run the day-to-day SST activities and would play an extremely influential SST policy-making role in the FAA over the next two years.

But even within the FAA this vigorous promotion and upgrading did not go unopposed. Frank E. Loy, the director of the FAA's Office of Policy Development, rather courageously (though without real impact) contrasted the "cautiously optimistic" airline attitude toward commercial jets in the 1950s with the less-than-enthusiastic mood of the manufacturers and airlines toward developing the SST. Practically alone in his views, he forlornly told Halaby in early May 1963, "I am fully aware that lately my principal contribution seems to have been to say 'no' to various proposals you make. . . . Nevertheless, it seems worth saying. Perhaps my highest and best use is somewhere like the Bureau of the Budget where saying 'no' is virtuous."[11]

A similar note of caution was also expressed by the director of FAA information services, who in late February 1963 recommended that Halaby not give a particularly grandiose draft speech entitled "Beyond the Supersonic Transport." He advised Halaby to "shelve all supersonic talk—much less, beyond supersonic talk—until this [SST] decision is made or until a decision is at least close at hand."[12]

But Halaby, ever the committed champion, cast aside such internal expressions of caution and doubt. As he had in late 1962 Halaby again looked forward to a highly positive supplemental report, expected in May 1963, by his high-level SST advisory group, STAG. Under Halaby's prodding, STAG in December 1962 had produced a very optimistic first report. An observer told Halaby that he found this document a "rather convinced, almost impassioned, polemic." Halaby agreed and called STAG's case "a powerful one."[13] The supplemental STAG report, which came out in May 1963, was even better from Halaby's perspective. STAG now not only strongly favored American SST development and proceeding with the design competition, it also no longer backed an independent SST public corporation. Instead it strongly endorsed the FAA as the agency to direct SST development because of the FAA's "demonstrated ability to advocate and concentrate on" an SST program.[14]

Halaby also continued to use the existence of the Concorde program to his advantage. His position was strengthened in this regard by a detailed and extremely favorable report on the Concorde by FAA assistant administrator George Prill in April 1963. According to Prill, who gathered his information from firsthand observation and a number of European publications, the outlook for the Concorde was strong; the project was vigorously supported by both the French and British governments; the developers viewed the European state-owned airlines as "captive customers"; moreover if Pan American ordered the Concorde (as it was about to do), other American-flag air carriers would certainly follow suit; the British and French minimized the sonic boom problem and were in complete agreement about design specifications and development methods; two prototypes would be ready to fly by early 1967, and airline service was due by 1970. Prill's single negative comment about the Concorde was that development did not fully utilize the much-touted systems approach espoused by the Americans.[15]

Halaby also felt that Vice-President Johnson was in his corner, which was fortunate since by late spring of 1963 the center of SST activity had shifted temporarily from the FAA to Johnson.

Johnson meanwhile was hearing from a number of other vigorous SST advocates, including union leaders and airline and manufacturer executives. For example, Robert F. Six, president of Continental Airlines, in addition to stressing the need for government funding and for SST operating costs comparable to those of subsonic jets, argued that a short lag—two or three years—between introduction of a European Mach 2.2 transport and an American Mach 3 SST "would pose little threat" to the United States. "If, however," Six declared, "the Mach 3 appears to be five or more years behind the European model, then I believe that all carriers will meet full system requirements with the Mach 2.2 transport." At the encouragement of his aide Bobby Baker, Johnson met with Lee Atwood, president of North American Aviation, which had developed the supersonic B-70 bomber. Atwood presented an informational document categorically stating that "North American Aviation is the one organization with the necessary knowledge and experience to develop an American SST."[16]

Johnson received similar pro-SST advice from his immediate staff.

In late April 1963 his influential military aide, Howard L. Burriss, came out quite strongly in favor of developing an SST, with arguments that read like a brief from SST boosters in the aerospace industry. Burriss declared that any "lingering doubts" about the SST "should not alter nor obscure the *inalterable necessity for the U.S. to proceed* energetically with the development without delay, not only for the purposes of national prestige and defense but for the American economy as well." Like most other SST advocates at this time, Burriss saw the SST program as a more or less conventional aerospace development effort that primarily involved "straightforward technical direction and coordination." To Burriss, SST operating costs would present "no problems since these costs will be comparable to those of a 707 or DC-8," with minor increases "absorbed in fare increases and through anticipated growth in traffic density." Although he acknowledged supersonic overland flights as "somewhat troublesome," available data suggested that "sound problems can be held within acceptable limits," with trans-oceanic SST flights presenting "practically no problems." Opposed to the "myriad of committees and organizations presently participating in SST studies and investigation," he recommended merging all SST activities into a "more cohesive unit." Halaby could take exception to only one point: Burriss did not favor the FAA's directing SST development and instead recommended placing the program in the Defense Department or NASA.[17]

Halaby actively attempted to influence the Johnson committee as May approached. In response to charges by Senator Estes Kefauver (D-Tennessee) that STAG's membership was not broad enough and was too closely identified with the aerospace and aviation industries, Halaby recommended to Johnson that "some benefit" might be derived "from having a group of outstanding individuals not associated with the aviation industry giving us the advantage of their opinion on the final report before it is submitted to the Cabinet Committee." He even sent Johnson a list of such people, though by May the time pressure and decrease in criticism lessened the need for a new advisory board and none was formed. Meanwhile on April 26 the FAA presented to the Johnson committee a highly detailed progress report.[18]

In mid-May 1963 Halaby sent Johnson a final strongly favorable FAA report that reiterated a number of standard pro-SST arguments, such as foreign SST competition and the possible loss of national prestige, terming the SST "the next inevitable advance in commercial aviation." It was more specific on the issue of government-industry cost sharing than previous statements, recommending that the airframe and engine manufacturers "underwrite" 20 percent of the development cost "in accordance with a specified formula" and that the airlines make advance royalty payments ranging from $200,000 to $500,000 per aircraft. Also the report was less technically narrow than earlier documents, discussing the costs and benefits of both a steel-titanium Mach 3 and a less advanced aluminum-based Mach 2.2 SST. Moreover Halaby was sophisticated enough not to present a blatantly pro-FAA discussion. In fact, the report suggested that a supersonic transport development authority be established in either the FAA or NASA, with a director appointed by the president and confirmed by the Senate. However, as seen particularly in its handling of the relatively controversial areas of the sonic boom and economic feasibility, the report's drafters had written an optimistic introductory section (which basically listed the report's recommendations and would receive the most attention) and a more cautious overall text. This positioning worked in at least one case. Welsh, who was consistently pro-SST, specifically called Johnson's attention to the highly positive introductory section when he transmitted a draft copy to the vice-president.[19]

Still, as Halaby was well aware, not all of the SST opinion that Johnson received was categorically favorable. In fact, the whole review process had triggered a series of negative or skeptical commentaries.

Johnson's close aide George Reedy received the FAA report and urged Johnson to deal with one major area left "unresolved"; choosing an agency to direct the SST program. The savvy Reedy accurately sensed "an obvious preference for the FAA" and anticipated the SST program igniting "an economic scramble" among the manufacturers. He wanted to keep federal agencies removed from such squabbles and therefore backed the creation of a totally independent SST program, with its director reporting directly to the president.[20]

Johnson also directed the member agencies on his committee to review the FAA report. The most negative reaction came from the Department of Commerce, which actively promoted its views. Even before the report was distributed in mid-May, Commerce appeared highly skeptical of the SST's alleged economic advantage. In early May Undersecretary of Commerce Martin informed Johnson that a Commerce Department SST study had found that balance of payments, domestic employment, and defense readiness were not compelling reasons for developing an SST. Another Commerce Department report produced in mid-May was even more unfavorable in the areas of sonic boom and SST engine noise, claiming that public reaction to the SST's sonic boom probably would be quite intense; it was therefore likely that only over-water SST operation would be allowed, and early SSTs should be designed for intercontinental and transoceanic range. Martin and the assistant commerce secretary, J. Herbert Hollomon, lobbied vigorously to have the Commerce findings heard by the key members of the Johnson committee and met with Heller, presidential science adviser Jerome Wiesner, and others, though pointedly avoiding Welsh and other strong SST advocates. Commerce then pressed its attack in May with a series of SST analyses that directly disagreed with the findings of the FAA report. The department portrayed the SST as too focused a venture for government sponsorship and criticized the FAA for offering the SST program only "to counter the competitive threat posed by the Concorde, with little consideration to the potentially more serious threat presented by an advanced subsonic aircraft for cheaper mass transportation."[21]

Further negative findings, though more favorable than Commerce's, came out of the joint review by the Bureau of the Budget and the Council of Economic Advisers. These two agencies believed that "the ultimate measure" of SST success "unlike the space program . . . will be its commercial feasibility" and therefore placed high importance on substantial private industry involvement, recommending that private industry contribute a minimum of 25 percent of the development cost. The two agencies rejected the pro-SST arguments of national prestige, balance of payments, and employment and instead pressed for "active exploration" of such difficult problems as aircraft design, the sonic boom, and economic issues before authorizing a development program. Even more threatening to Halaby, though recommending that the FAA direct the SST design competition, the review urged that NASA oversee actual SST development because an FAA administrator could well face a conflict-of-interest situation when the FAA had to decide on Concorde certification.[22]

Also countering Halaby's enthusiasm were the Defense Department's views. Reflecting McNamara's own beliefs, Defense favored an

SST program only if there was an "overriding justification"; the agency had no SST military requirement and therefore would not help finance an SST program.[23]

But Halaby had powerful allies also reporting to Johnson. At this time the SST staff member at the Office of Science and Technology, Nicholas Golovin, placed great emphasis on the SST's technological benefits. Golovin wanted an aircraft development program that would be "economically successful *and* an effective response to the British/French technological challenge" and, disagreeing with the Commerce Department, believed that developing a large subsonic jet "would clearly not be an appropriate answer" to the Concorde challenge.[24] Similarly NASA stressed the SST's technological importance and, though urging flexibility, strongly favored developing an SST as quickly as possible.

The Treasury Department vigorously favored beginning an SST program, even to the extent of subsidizing production, and it anticipated balance-of-payments benefits. The State Department agreed with the FAA recommendations and argued that the SST would improve national prestige and public relations for the United States.[25] Finally Welsh, who was Johnson's key aide for aerospace matters, encouraged the vice-president in late May 1963 to "move ahead on an SST program," to undertake an SST design competition, and to recommend a supplemental SST appropriation.[26]

Johnson, the key decision maker for the moment, was fully cognizant of the intense disagreement over the SST. In a revealing telephone conversation he commented to presidential aide Theodore Sorenson in early June 1963 on the huge amount of "jealousy and envy" he had witnessed. He particularly noted that "McNamara is just against—he has no need for it and I think pretty generally is against supersonic stuff." But Johnson was clearly pro-SST. He called the SST a "far-reaching decision" and urged prompt action in announcing a program. Johnson was greatly disturbed about a rumored Pan American decision to order six Concorde aircraft. He warned against studying the SST to death. "My judgment," Johnson argued, "is we ought to keep the leadership in aviation and the only way to keep it is not sit on your fanny and let the British and French build a Mach 2.2 and let us do nothing." He called for a "czar" to run the program. Equating the SST with the space program, Johnson pressed for a visionary and uplifting announcement by the president: "It's a billion dollars. They'll ridicule him [Kennedy] for it. They'll laugh at him, just like they do the space thing. But my friend, if he didn't do it he'd be in worse shape. . . . We're going to have a storm on it, but I think you're going to be proud of it twenty-five years from now."[27]

Given Johnson's own inclinations therefore it was not surprising that at the end of May the Johnson committee generally came out in favor of SST development. Johnson recommended to the president that SST development begin as soon as possible, and he called for a detailed SST design competition, after which a single airframe-engine combination would be selected; competing contractors as well as the airlines would be expected to contribute a "significant portion of development costs."[28]

Halaby was pleased because Johnson favored establishing a new deputy administrator in the FAA to run the design competition and characteristically was poised to take advantage of Johnson's recommendations. Halaby wrote the president on June 3, 1963, and proposed a number of actions, assuming the president approved the general policies outlined by Johnson. Halaby wanted the president to make an SST announcement in a forthcoming speech at the Air Force Academy on June 5, and he even enclosed a possible text on the SST for this speech. He laid out a three-phase timetable for the SST program and confidently informed the president that "the ground has been prepared" in Congress for an SST fiscal 1964 SST funding request of $100 million, the funding level that Johnson had recommended.

Halaby also promoted the FAA's lead role. He told the president that a new and autonomous SST Office would be established under a deputy administrator "whose appointment you would make or approve" and who "would be a man in whom the airlines and the Congress would have confidence." Halaby mentioned that he even had "an outstanding candidate for your consideration."

But Halaby also continued to demonstrate his flexibility, and he did not exhibit naive enthusiasm. Answering some of the outstanding objections, the FAA now recommended a manufacturers' cost-share contribution of 25 percent instead of 20 percent, initial airline royalty payments on each SST ordered, a limit to government SST financing of development costs of no more than $750 million or 75 percent of the total development cost, whichever was lower, and a goal of recovering the government's development contribution, excluding interest, through a royalty payment charge of 1.5 percent on each SST fare sold until total recovery was achieved. Also, Halaby warned Kennedy, "Mr. President—the two biggest risks in this program are that U.S. industry may not be able: (1) to overcome the sonic boom so that it will be tolerable by the population, and (2) to get the cost down to a competitive level." Of course, he quickly added, these risks appeared to be worth taking.[29]

It was the existence of the Concorde, however, that in the end gave Halaby his program. More specifically it was the effective lobbying and manipulations of Pan American Airlines using the Concorde that ultimately pushed Kennedy into announcing the SST program. Pan American was, in effect, the flagship of American air carriers, and it flew almost all international routes, including the crucial transatlantic run. Of all airlines, it was most influential on SST matters. Traditionally Pan American had led the way in ordering new aircraft.

During the spring of 1963 Pan American made sure that the FAA and the White House knew that although the airline was considering ordering a number of Concordes, it preferred to purchase a fleet of larger and faster American Mach 3 SSTs. Juan Trippe, president of Pan American, visited a number of high-level officials in Washington, including Halaby, Alan Boyd (CAB chairman), and Secretary of the Treasury C. Douglas Dillon, and told them that he intended to place a "protective order" for six Concordes.

Trippe's actions had their desired effect. As he demonstrated in his telephone conversation with Sorenson, Johnson was very worried about the impact of Pan American's Concorde decision, and, with Johnson's recommendation and the knowledge of Trippe's intentions, Kennedy decided to announce an SST program.

Even after Kennedy made his decision the actual timing of a Pan American Concorde announcement concerned Kennedy and Halaby. On June 1 the president called Halaby and told him that he did not want Pan American to announce its Concorde decision before the president had publicly declared an American SST program. The next day Halaby telephoned Trippe and informed him that because the president would soon announce his SST decision, a Pan American Concorde order before Kennedy's SST announcement would greatly embarrass the president. But although Halaby thought he had Trippe's agreement, his efforts failed. On June 4 at 5 P.M., one day before Kennedy's declaration, Pan American announced its order for six Concordes. Halaby was shocked.[30]

Still Halaby would have a program. By the end of May he and Johnson had even obtained McNamara's limited but crucial support for beginning SST development.[31] Therefore, with Johnson's strong recommendations, the acquiescence of McNamara, and the Pan American Concorde decision, Halaby was finally able to marshal enough forces to convince the president to announce formally an SST program at the Air Force Academy on June 5, 1963.

Halaby was buoyed by the generally positive and vigorous thrust of Kennedy's SST commitment. The president declared, "Neither the economics nor the politics of international air competition permits us to stand still in this area. . . . But if we can build the best operational plane of this type—and I believe we can—then the Congress and the country should be prepared to invest funds and effort necessary to maintain this Nation's lead in long-range aircraft, a lead we have held since the end of the Second World War, a lead we should make every responsible effort to maintain. . . . This commitment, I believe, is essential to a strong and forward-looking nation, and indicates the future of the manned aircraft as we move into a missile age as well."

On the other hand, the rather grandiose and enthusiastic ideas expressed by Johnson and the original FAA draft speech were essentially scrapped. The president also eliminated all references to funding amounts, the alleged similarity to the space program, and the designation of the agency to run the program. Moreover the president explicitly called attention to the need for good economics and significant private risk capital: "If these initial phases do not produce an aircraft capable of transporting people and goods safely, swiftly, and at prices the traveler can afford and the airlines find profitable, we shall not go further."[32]

In June 1963 Halaby was a victorious champion. After confronting countless delays and strong internal opposition, a presidential decision to move ahead with the SST had finally been given, with the FAA in all probability to be designated as the leader of development. The pro-SST forces were clearly in control. It looked as if the initial debates and feasibility studies had presumably ended and a milestone or point of no return had been reached.

But a number of technological and economic uncertainties remained, and the president's commitment was not absolute. The FAA's power was still limited. It remained to be seen whether the agency, even with Halaby's promotional skills and persistence as a champion, could shepherd the SST effort through to development.

II FRAGMENTATION

5 Power Shift

Creating a Program

President Kennedy's formal announcement of an American SST program in June 1963 marked the high point of Halaby's campaign as SST champion. Although he had proven extremely effective in mobilizing sufficient support to inaugurate a program, the seeds of doubt—in such areas as the economic performance of the SST, the sonic boom, and the FAA's management capability—had also been planted. Halaby's challenge was to maintain the momentum created in the wake of Kennedy's announcement and to block the growth of negative opinion. Halaby was more than ready to act. He had the May 1963 FAA report "reworked in line with" Johnson's recommendations to the president and submitted to the White House at least two draft presidential messages to Congress on the SST.

Kennedy's actual message, sent to Congress in mid-June 1963, however, was more cautious than the FAA's drafts. Although Kennedy appeared to be a vigorous SST supporter, characterizing the SST as "the logical next development of a commercial aircraft," he proposed that the manufacturers provide at least 25 percent of the development cost and that the airlines make royalty payments. He repeated that the government would invest no more than $750 million and would not subsidize production, selling prices, or operating costs. He did not emphasize the FAA's leadership role, merely stating that the "United States, through the Federal Aviation Agency, must proceed at once with a program of assistance to industry . . . in order to permit" government participation in an SST program.[1] While in May the FAA had called for $73 million and Johnson had recommended $100 million, Kennedy, taking his cue from Budget Bureau director Kermit Gordon, requested $60 million.[2]

But Halaby, riding high on Kennedy's SST announcement, continued to play the role of an active champion. For example, a June FAA report vigorously promoted the agency's lead role in running the SST effort.[3] Halaby also released the pro-SST report by STAG. On Capitol Hill he stressed the commercial nature of the SST program, the FAA's careful management, and Kennedy's policy to "postpone, terminate, or substantially redirect this Program whenever it fails to achieve its objective."[4] In public Halaby dwelled on the SST's intangible benefits. He termed the SST "perhaps the greatest adventure civil aviation had known." "The opportunity isn't just in profit," he stated, "the opportunity is in adding to the Nation's prestige, in adding to the Nation's industrial capacity and strength, in writing another brilliant chapter of U.S. aviation." In his words, SST development presented "a challenge to free-enterprise American industry to show it can compete [with] and beat [the] nationalized efforts of the French-British and, perhaps, the Soviet Union."[5]

Halaby also moved to counter skepticism over the FAA's managerial capability. By late May he had already informed Vice-President Lyndon Johnson that he intended to appoint Gordon Bain as director of the SST program. Bain was already running the SST Office. Johnson agreed with Halaby's choice, and in late July 1963 the FAA formally named Bain deputy administrator for SST development. In a draft memorandum to the president Halaby confidently informed Kennedy that this choice had been greeted in government and industry with "support and some enthusiasm"—and then crossed out the word "some."[6]

But within his own agency creating a strong and independent SST unit presented difficulties. In mid-June Alan Dean told Halaby that although he generally agreed with the notion of a strong SST director with a large amount of autonomy, he favored keeping the SST Office's "administrative support staff" small and using existing FAA resources for the SST whenever possible. Halaby took this advice. The number of formal positions in the SST Office (now renamed the Office of Supersonic Transport Development) grew from about fifteen people in May 1963 to only about twenty-three by the beginning of January 1964.[7]

Halaby and Bain faced constant and even growing concern over the FAA's managerial ability. In mid-August 1963, for example, Bain learned that Senator A. S. "Mike" Monroney (D-Oklahoma), the powerful chairman of the Senate Subcommittee on Aviation, questioned the FAA's capacity to manage the SST program and was dissuaded from advocating the transferral of SST control to NASA or the air force only by the argument that NASA was too "research minded" and that the air force would design a commercial SST overly suited to military needs.[8]

Bain's management challenges were huge as he attempted to counter such criticism. He acted on a diverse number of fronts. He supplied supporting information for a supplemental SST budget request of $60 million for fiscal 1964. By the end of August, after numerous internal discussions and briefing papers, he had also determined the kind of organizational structure that he wanted, had specified critical technical areas, and had even worked out desired control and information systems. Still he had not yet begun full-scale recruiting.[9]

The interorganizational and externally oriented demands were just as, and perhaps even more, critical for managing the SST program successfully. Halaby and Bain knew that they had to promote continued support for the SST in the White House, Congress, other agencies, and industry. They therefore made it a point to keep the White House informed of the program's status in the House and Senate and, working closely with the pro-SST congressional forces, provided a flow of SST information to Capitol Hill.

Congressional doubts on whether the FAA possessed the necessary authorization and legal status to direct the SST program proved especially time-consuming and troublesome. Monroney initially was insistent on this point. But Halaby clearly wanted to avoid new authorizing legislation, and, fortunately for him, the whole matter proved to be something of a red herring. Discussion rapidly diminished, and by late 1963 Bain was finally free to state categorically in a letter to an influential congressman that the FAA did not require new authority to conduct the SST program.[10] Still the underlying concern that caused Monroney to make such an issue of authorization—doubts about the FAA's managerial ability—continued to be voiced.

Senator Estes Kefauver, chairman of the Senate Subcommittee on Antitrust and Monopoly, presented the FAA with another potential hazard. Throughout the spring and into the summer of 1963 Kefauver insisted on expanding STAG because of the alleged absence of public-interest representation. Halaby learned in June that Kefauver actually wished to hold antitrust hearings on the SST, which would have been especially dangerous because of a possible jurisdictional dispute between Kefauver's Subcommittee on Antitrust and Monopoly and the Subcommittee on Independent Offices, which was chaired by Halaby's key Senate ally, Warren Magnuson. Halaby immediately alerted Magnuson, but the danger was averted when Kefauver died suddenly two months later.[11]

An example of interagency difficulties Halaby and Bain faced was a squabble with NASA in the summer and early fall of 1963. NASA became incensed when Halaby suddenly stopped awarding SST contracts to make certain that further contracts were consistent with the president's new guidelines. NASA complained strongly about this "unilateral" action and, stressing its own SST work, publicly hinted that it was considering acquiring a second or third B-70 bomber prototype for a proposed SST research program of its own. NASA continued to tout its SST activities with, in Bain's words, a "deluxe press conference" and limited lobbying efforts for control of the SST program. It also released its own evaluations of various SST designs, including some developed by NASA. In October 1963 NASA administrator James Webb testified in favor of NASA's SST work with, as one FAA official put it, "Cassius Clay-like modesty." NASA's actions astounded Bain, and he pointedly told Halaby, "This nonsense should be stopped by the President." He also dismissed the NASA designs as "impractical" and argued that they were based on "older" ideas developed for the air force B-70 bomber.[12]

This flurry over SST control, however, was actually something of a political and bureaucratic shadow play. In fact, NASA, concerned basically over the principle of the seemingly unilateral actions taken by the FAA, wanted to protect its own dominance in aerospace research and development; it had little real interest in running the SST program. As early as July representatives of the FAA, NASA, and the Defense Department had begun meeting to clarify areas of SST responsibility. At the end of 1963 Halaby and Webb signed a final "memorandum of understanding," which confirmed the FAA's responsibility for directing the SST program and which delegated to NASA a technical support role.[13] Still this FAA-NASA episode illustrated how the SST effort was being manipulated for political and bureaucratic ends.

Halaby and Bain continued to use the existence of the Concorde program to promote the American effort. The Concorde had been a crucial catalyst for Kennedy's SST decision, especially after Pan American's public announcement to order the aircraft. Halaby saw the Concorde effort only as a competitive spur and, like before, rejected new Anglo-French approaches for a tripartite SST development consortium. The British minister of aviation at this time acknowledged that there was "zero [American] interest in SST cooperation."[14] Instead with the Concorde program apparently making progress during the summer and fall of 1963, with potential orders from Pan American, BOAC, Air France, and Continental, Halaby urged FAA participants in a mid-November 1963 meeting in London to demonstrate "our best leading edge."[15]

Halaby also had another potential weapon: Soviet SST efforts. He was receiving fragmentary information on this subject at the time, but because the Russians appeared to be about five years behind the Anglo-French program, he did not for the moment portray the Russian plane in the Concorde's league as a significant challenge to American hegemony.[16]

The American-flag air carriers were key participants in the American SST effort from the beginning and a major constituency for Halaby and Bain to cultivate. The airlines' attitude continued to remain ambiguous, though in public they strongly backed a U.S. effort. The president of Continental, Robert F. Six, confidently told the press, "Actually, addition of supersonic equipment will be far less of a financial burden to Continental than was the initial move from piston aircraft to subsonic jets in 1959." American Airlines president C. R. Smith also favored rapid development.[17]

To demonstrate further air carrier confidence in the American effort in the face of the fact that American-flag airlines had ordered Concordes, Halaby successfully pushed the idea with top airline executives, the president, key members of the House and the Senate, and Budget Bureau director Kermit Gordon that airlines be permitted to place immediate orders for the American SST.[18] Halaby then obtained agreements from American and TWA to place orders. (Senator Magnuson applauded TWA's decision and was not at all disturbed by grumbling from Pan American about procedural difficulties, angrily recalling, "They were the first to buy the British-French Concorde.")[19] Later other airlines, including Pan American, also agreed to participate. "A few weeks ago," Halaby observed in October 1963 with satisfaction, "it appeared that not only was the United States far behind in this competitive effort but that our European friends were getting all of the orders and it was a hopeless situation. . . . This week we are confronted by an equally difficult problem—one of abundance."[20] By mid-November he had established a delivery position policy with the following priority scheme: American-flag carriers, Atlantic; foreign-flag carriers, Atlantic; American-flag carriers, Pacific; foreign-flag carriers, Pacific; and American domestic carriers.[21]

Although not everyone was happy over the exact allocations—the presidents of Northwest and American complained to Bain and Halaby—by early December 1963 the FAA had received deposits for forty-five planes, and no "unfavorable comments" had come from Capitol Hill.[22] The FAA clearly viewed these airline orders as politically significant expressions of support and later observed that they "proved to be one of the major factors that encouraged the [Senate Aviation Subcommittee] to authorize further development of the [SST] Development Program."[23]

The Early Cost-Sharing Debate

Halaby had achieved victories and accomplishments, including developing a management system, maintaining congressional relations, and winning airline acceptance on advanced orders. But a growing controversy over cost sharing led to an ultimately fatal decline in his effectiveness as a champion. Cost sharing was a critical issue because the arguments surrounding it represented a proxy debate on the SST's economic prospects and on the associated question of how much risk capital the manufacturers were willing to sink into SST development. It was over the cost-sharing issue that Robert McNamara, Kermit Gordon, Walter Heller (chairman of the Council of Economic Advisers), and representatives of the Commerce Department clashed most strongly with the ardent SST advocates in May 1963 when the Johnson committee had reviewed SST options. McNamara was especially insistent on Kennedy's 25 percent industry contribution requirement (Kennedy also limited the government's SST contribution to $750 million), and McNamara warned Halaby that if the manufacturers rejected this percentage, he would "reconsider" his support of the SST.[24]

If cost sharing was a true test of industry's view on the SST, then their real attitude was one of significant doubts about the program as a commercial venture. It seemed as if as soon as Kennedy set out his cost-sharing formula, manufacturers uniformly took a stand against the 25 percent cost share, and industry pressed vigorously for increased government funding—up to 90 percent of the SST development cost.

Bain informed Halaby in late June that all industry representatives doubted that they could meet the 25 percent participation requirement.[25] By July bitter industry complaints about cost sharing had reached high-level government officials. Courtlandt S. Gross, president of Lockheed, told Halaby that the provisions presented "a most difficult obstacle," terming them "unwise from a prudent financial standpoint." William Allen of Boeing expressed similar concerns.

The manufacturers' objections intensified in August, and these firms mounted an energetic campaign in the FAA, in Congress, and in other parts of the administration to reduce their cost-sharing burden. Gross reiterated Lockheed's fears even, he said, at the risk of being disqualified from the design competition. FAA officials met with Allen, and, according to FAA notes, Allen stated that "we want to see it [the SST] go but don't see how we can do it on the present basis." Donald Douglas of Douglas Aircraft echoed such beliefs.[26] Bain acknowledged that the cost-sharing ratio was the manufacturers' "single greatest problem" with the FAA SST guidelines for the initial design competition, though he still insisted that lowering industry cost sharing would only discourage the manufacturers from assuming a "legitimate risk."[27] Meanwhile industry complaints reached Vice-President Johnson, McNamara, and CIA director John McCone, who possessed close ties with the business world. Halaby was compelled to warn President Kennedy that the manufacturers "will exert every pressure they can to have you reverse the cost sharing requirements." The manufacturers, Halaby declared, "simply will not recognize that this is a commercial transport for which they must risk their own capital to achieve a market which could reach $5 billion."[28]

Then in late August Halaby received a serious shock: Douglas Aircraft dropped out of the SST competition. Reflecting prevalent opinion, the *New York Times* observed that "whatever else it means, the Douglas move was a reflection of widespread industry caution about plunging into commercial supersonic flight."[29] Ultimately six firms—Boeing, Lockheed, North American Aviation, Pratt & Whitney, General Electric, and Curtiss-Wright—chose to participate in the initial SST design competition; in addition to Douglas, General Dynamics and McDonnell also declined.[30] The final request for design proposals still firmly followed the cost-sharing guidelines Kennedy laid down.[31]

Most of the key points of contention over SST economics and cost sharing surfaced publicly in October 1963 at hearings on the SST by the Senate Aviation Subcommittee. Setting the tone, the first witness, Alan S. Boyd, CAB chairman, was not particularly enthusiastic over the SST's economic prospects. He termed the Concorde "a loser," cast doubt on the American SST's performance, and questioned the validity of favorable FAA economic analyses. A certain amount of bureaucratic positioning was also behind Boyd's conclusion: he wanted his agency to play a dominant role in analyzing the SST's economics. Similarly the Department of Commerce continued to be strongly in favor of Kennedy's cost-sharing formula. And like the CAB, it emphasized its own ability to perform the economic analyses.

The manufacturers remained adamantly against the 25 percent requirement. J. L. Atwood of North American Aviation, Courtland Gross of Lockheed, William Allen of Boeing, and all representatives of the engine manufacturers believed that the current cost-sharing formula would result in large and unacceptable financial risks to their firms. Allen declared, "If the [SST] program is considered to be in the best interest of the country, and if it must proceed upon the schedule proposed, Government must be prepared to render greater financial assistance than presently proposed."

The constant lobbying and complaining by the manufacturers were having results. Senator Monroney appeared quite convinced by the industry's arguments. He told Halaby on the final day of hearings, "I can't help but feel that they [the manufacturers' concerns about financial risk] are genuine and real." Halaby himself admitted that the cost-sharing ratio was not ideal.[32] He informed Kennedy at this time that although he backed the current formula, he also would "consider, after careful review of the situation, alternatives that protect the public interest."[33]

The great blow to Halaby to grow out of increasing industry unease over cost sharing was Kennedy's decision in August 1963 to call in a respected outside adviser to deal with the whole issue of industry-government SST financing. This action triggered a chain of events that ultimately removed the FAA and Halaby from the power center of the SST effort.

Kennedy first broached the notion of an outside adviser with Halaby in July 1963. Although Halaby outwardly approved the idea and submitted to the president a list of candidates, he was clearly worried about "being bypassed" at the very beginning of the SST program. He and others in the FAA believed the interjection of an outsider originated with McNamara. In late July Halaby expressed his concerns in a memorandum to the president.[34] Bain was even more bitter. He later ruefully commented to Halaby, "It is doubtful that a financial adviser could make any suggestion with respect to the program which has not already been considered many times."[35]

But in mid-August Kennedy went ahead and made one of the most far-reaching decisions of the entire SST program: he appointed Eugene Black, former chairman of the International Finance Corporation and former director of the World Bank, as special adviser to the president for reviewing the financial aspects of the SST program. Later Stanley de J. Osborne, chairman of the Olin Mathieson Corporation, was appointed as Black's deputy. Moreover Kennedy gave Black and Osborne a wide mandate, directing them to take into account the views of all interested parties in both the private and public sectors and asking them to review the financial capabilities of private industry that could be brought to bear in developing the SST.[36] Critical decision-making power was no longer solely under Halaby's control.

A New President

The assassination of President Kennedy on November 22, 1963 was another severe setback for Halaby and his SST ambitions. Halaby became a great admirer of John Kennedy. He nostalgically recalled years later that Kennedy "gave me more than just *carte blanche* in running the FAA since he had made me his 'principal aviation adviser.'" Halaby appreciated "JFK's personal thoughtfulness," his capacity for "fun," the creativity and "first class" quality of the White House staff, his own free access to the White House, and the general *esprit de corps* of the Kennedy administration. "I had grown, really, to kind of love Kennedy, and adulate him," Halaby later admitted to an interviewer. Halaby spoke of the exciting "social whirl" of the Kennedy years, the parties, the small intimate dinners at the White House, and the "growing personal friendship" between Kennedy and himself. "Most of us," Halaby reflected, "didn't just like and/or respect JFK, we also tended to idolize him."

The new president, Lyndon B. Johnson, had been a consistent and enthusiastic SST supporter. But the context had totally changed. Halaby suddenly lost the close contact and easy admission to the White House that he had previously enjoyed. Halaby, after all, was now a Kennedy appointee in a Johnson administration. Halaby also resented the increased overt political pressure of the Johnson presidency. Moreover he personally did not like Johnson and found him "kind of disgusting to me in his language, in his vanity, in his egotism."[37]

Weakened as a champion in a number of ways, Halaby was left to confront the Black-Osborne report, which was submitted to Johnson on December 19, 1963. The document marked a turning point in the whole SST program. It gave expression to the festering doubts about the FAA's SST role and clearly went beyond its original charge. As expected, the report strongly agreed with industry that the current cost-sharing formula was overly burdensome and recommended instead 10 percent industry cost sharing and 90 percent for the government and easier overrun provisions. But more significantly, the report, unlike the FAA, saw no compelling need for building an American SST as quickly as possible in order to compete with the Concorde and actually warned against a crash program. Finally Black and Osborne even recommended that the SST program be taken away from the FAA altogether, and they came out in favor of creating an "independent authority" to manage the project.[38]

Halaby acted immediately in an attempt to protect his enterprise. In a Christmas Eve memorandum to Johnson he pleaded not to "break faith" with Congress, which had just voted $60 million for an SST program, or to change the SST ground rules suddenly. He also urged the president to withhold the report until the manufacturers' proposals for the initial design competition were received in mid-January 1964. He nervously added that he was "most anxious" to see Johnson to "clear up the points" that he had expressed regarding the Black-Osborne recommendations.[39]

Halaby's anxiety was certainly justified. Johnson quickly asked McNamara, who had never been particularly favorable about the SST, to review the report, and by early 1964 both McNamara and Kermit Gordon had urged the president to prepare legislation for establishing an independent SST authority under an SST administrator who would also serve as a special presidential assistant; they suggested issuing a temporary executive order to create this new entity. McNamara told Johnson that choosing a "director with sufficient stature and imagination" was crucial, and he recommended five high-level private executives for this post.

Of even more ultimate significance, McNamara and Gordon also suggested establishing an SST "advisory board" composed of the defense secretary, the FAA administrator, the CAB chairman, Black, Osborne, the treasury secretary, the CIA director, and possibly the secretary of commerce. They also wanted to limit severely the FAA's managerial role, if not eliminate it completely.[40]

As SST champion, Halaby had never faced a problem as serious as the threat represented by the Black-Osborne report and the resulting McNamara-Gordon proposals. Halaby was determined to maintain FAA authority, and he called for intense staff work within his agency to generate material supporting continued FAA control.[41] He also met with the president at the LBJ ranch in Texas. Trying to create close ties with the president, Halaby later acknowledged to Johnson that the SST program must be a "President Johnson Program" and "must provide for a more realistic, flexible, and cooperative participation by industry, a maximum practicable contribution . . . and more protection for you from Congressional and public controversy and criticism." He assured Johnson, "You have wisely set me thinking anew." He claimed his ranch visit was inspiring and thanked the president "from the heart"; "You took us to the top of the mountain!" Halaby even hinted that he was developing a "better formula."[42]

Johnson displayed little reaction to Halaby's actions. He simply thanked Halaby and told him that he was "very happy with the energetic way" Halaby was tackling the SST's problems.[43]

Halaby continued to try to prevent further deterioration. He was particularly exasperated with McNamara, who was becoming increasingly powerful in SST matters. Halaby felt that not only had McNamara proposed the Black-Osborne review, but now the Defense Department wanted to oversee the SST contractors. The FAA in mid-February 1964 submitted its own optimistic report to the president to counter the Black-Osborne and McNamara-Gordon initiatives. Halaby also urged Johnson to defer action on cost sharing until January 1965 on grounds of timing ("after the election") and avoiding increased congressional criticism; in January major congressional or executive action would be needed in any case. Finally he complained that a "rather high pressure campaign" was being mounted by the "military-industrial complex" with the aim of transferring "all the risks of the program to the taxpayer."[44]

While Halaby was trying the soft sell, he allowed Bain the hard sell. At the end of February 1964 Bain sent a detailed and blunt review of the Black-Osborne report to the president. Noting that the report dealt "somewhat cursorily" with SST financing and instead "devoted major attention to the substantive aspects of the Program," Bain refuted the major Black-Osborne findings point by point. He implied that the report would have avoided some "misconceptions" if Black and Osborne had found the time to discuss the program in detail with the FAA, and he particularly objected to removing the SST program from the FAA.[45] Undercutting the impact of Bain's review, however, Halaby quickly informed Gordon that Bain's views were essentially his own.[46]

Meanwhile the Black-Osborne report, the findings of which were gradually being leaked to the press, was continuing to damage the FAA, in public as well as within the administration. In February anti-SST views appeared in the *New York Times* and the *Saturday Evening Post*, and the *Wall Street Journal* and *Aviation Daily* mentioned key aspects of the Black-Osborne report.[47] Recommending an FAA media response, the FAA director of information services advised Halaby to assume "the coolest possible attitude toward this flurry of interest" in the report, hoping that "we are now at the bottom of the wave."[48]

By early March 1964 it was clear that Halaby could not withstand the deluge of criticism ignited by the Black-Osborne report and subsequent proposals. He was forced to compromise in order to salvage FAA operational and managerial control over the SST program. He fought off attempts to create an independent SST authority, therefore, by accepting another major component of the McNamara-Gordon plan: a high-level advisory board.[49]

Halaby was helped because Kermit Gordon, who was placed in charge of an administration-wide review of the Black-Osborne report, had himself moved away from endorsing both an independent SST authority and an advisory board. In February he had recommended letting the FAA manage the SST program, while creating an SST advisory board headed by McNamara. Gordon reasoned that any wholesale shift in the SST management structure would cause critical delays and that only legislative action could relieve the FAA administrator of SST contractual obligations. Gordon's staff then drafted an executive order for establishing an SST advisory board.[50]

Gordon was backed by the rest of the administration. He already had received statements on the Black-Osborne report from the secretaries of state, commerce, and treasury, from the administrators of NASA and the FAA, from the chairmen of the Council of Economic Advisers, the CAB, the Atomic Energy Commission, and the National Aeronautics and Space Council, and from the director of the Office of Science and Technology. The agency comments focused primarily on management, financing, and timing. With the possible exception of Treasury Secretary C. Douglas Dillon, none of the officials surveyed appeared to favor an independent SST authority, although not all considered the FAA the first choice to direct the SST program. (Significantly McNamara did not participate in this interagency review because, as Gordon acknowledged to Johnson, "we understand that he has already made his views known to you on the principal Black-Osborne recommendations.")[51]

By March the FAA's fortunes had taken a turn for the better. The most dangerous suggestions of the Black-Osborne report and the McNamara-Gordon plan had not been accepted. Although an SST advisory board would clearly be established, the FAA would retain managerial control.

The report's impact was further weakened by a presidential announcement of the existence of the advanced 2,000-mile-per-hour experimental military aircraft, the A-11, which was being developed by Lockheed. *Aviation Week and Space Technology* commented, "It appeared that the authors of the [Black-Osborne] report had not been briefed on the USAF Mach 3 A-11 aircraft project, . . . and thus were forced to base their conclusions and findings on the false assumption that much work remained to be done in the field of heat-resistant metal fabrication."[52]

Meanwhile during the first quarter of 1964 the FAA successfully supervised the SST initial design competition. Over two hundred government experts from various agencies participated in the government's evaluation of the manufacturers' design proposals, and ten airlines also undertook independent evaluations. Boeing, Lockheed, and North

American submitted airframe designs, and General Electric, Pratt & Whitney, and Curtiss-Wright submitted proposals for the engine. Lockheed and North American proposed a fixed-wing aircraft, while Boeing emphasized swing-wing designs based on models developed in an earlier fighter-bomber competition.[53]

Bain presented his summary findings and recommendations to Halaby by early April. In what would be a consistent FAA viewpoint, Bain's review clearly favored the Boeing-General Electric design over all others. He told Halaby that according to the government evaluation team, only the Boeing proposal incorporating the General Electric engine "offers a reasonable expectation of success in the time period of the Development Program." Eight of the ten airlines also recommended the Boeing aircraft, and seven of the ten airlines "regarded the GE engine . . . as having sufficient potential to warrant development." Still, unlike the FAA, the airlines generally urged developing two competitive SST prototypes to hedge their design options.[54]

Halaby was clearly impatient to move on to the next stage of the SST program. He was hoping to make the SST award on May 1, 1964, based on the recent design competition results, as the FAA had originally planned. But by the end of March he still had not received a clear go-ahead from the White House.[55]

The key reason for this crucial inaction during March was that a radical shift of power was taking place. Momentum kept building for establishing an SST advisory board, which Gordon had formally recommended to the president in late March.[56] Myer Feldman, a key presidential aide for SST matters at this time, also supported the idea,[57] and the *New York Times* in mid-March reported that an SST "board of directors" was under serious consideration.[58] Halaby had to acquiesce, though he told Feldman that he disagreed with the "use of [the] Advisory Committee in the *selection* process." But Gordon was clearly anxious to have this body functioning by the first week in April so that it could "participate fully" in the ongoing evaluation.[59]

The president agreed, and on April 1, 1964 he signed an executive order that established the President's Advisory Committee on Supersonic Transport (PAC). The PAC was initially composed of the secretaries of defense, commerce, and the treasury and the administrators of NASA and the FAA and would soon include Black, Osborne, and the CIA director. The president designated McNamara as PAC chairman.[60]

The PAC's authority was extensive. Johnson directed the committee to "study, . . . advise, and make recommendations to the President with regard to all aspects of the supersonic program. The Committee shall devote particular attention to the financial aspects of the program and shall maintain close coordination with the Director of the Bureau of the Budget in this regard."[61]

The creation of the PAC was an important milestone in the SST program. The lines of power had changed. The FAA had not lost as much authority as it might have if all of the Black-Osborne recommendations had been implemented. Still a new and powerful body now assumed control over the major program decisions, debate had been elevated to an interagency level, and McNamara—a known SST skeptic—emerged as the most powerful figure in the SST program. The choice of McNamara as PAC chairman and the recent unveiling of the Lockheed A-11 program indicated to some that the White House would now rely on the Defense Department for achieving tighter program control.[62] But the PAC also provided an opportunity for other agencies to become involved, and it was uncertain whether the committee would prove to be a force for stabilization or fragmentation. Only one thing was clear: the committed SST champion, Halaby, was no longer the most powerful direct participant in SST decision making. A major power shift had occurred.

6 Sonic Boom One: The Seeds of Doubt

Compared to such controversial issues as program control, bureaucratic conflict, and even SST economic performance, the discussions about the sonic boom were initially on the periphery of SST decision making. But the sonic boom grew as an issue during the 1960s, becoming one of the critical forces that undermined the entire SST program. Moreover the sonic boom subdrama documents the rise of technology assessment—evaluating technology using a broad range of criteria, including social and environmental considerations—and this new movement's growing influence in affecting the direction of major decisions by government and industry during the late 1960s and early 1970s.

What is a sonic boom? A pebble thrown into a pool activates waves that move outward in ever-widening concentric circles. A wave is simply a disturbance that moves from one place to another at a characteristic and measurable speed. Similarly a sharp disturbance in air creates a wave that is perceived as sound when it reaches the ears. Such waves are called sound waves. The human ear perceives sound by detecting changes in air pressure produced by the object making the sound. Such changes of pressure move like waves through the air at the speed of sound. (The speed of sound depends on temperature and altitude. It is about 760 mph at sea level at 59°F but is only about 660 mph at 65,000 feet altitude, where the temperature is −70°F.)

Sound waves emitted from a stationary object, such as a bell, move out uniformly in all directions. Sound waves moving out ahead of an airplane that is flying at subsonic speeds are closer together since the source of the sound is also moving forward. When an airplane travels faster than the speed of sound, the aircraft structure collides with the air in front of it. The air is compressed and waves are created, essentially expanding on top of one another. Such a coalescing of waves forms a shock wave, which moves the air like a conical sound wave, only somewhat faster. A shock wave has been described as a "moving wall of compressed air." It causes a sudden change in air pressure on the ground that is perceived as a thunderous noise, a phenomenon known as a sonic boom. The magnitude of the shock waves produced by a supersonic aircraft is affected by such factors as the aircraft's weight, size, shape, speed, altitude, attitude (the angle of attack), weather conditions, and the terrain below the aircraft. Weight and altitude are especially important. Generally the larger the aircraft's weight, the stronger the shock wave; the higher the altitude, the weaker the shock wave since there is less air to compress at higher altitudes.

The severity of the sonic boom is usually measured by the sudden rise in pressure, called overpressure, and expressed in pounds per square foot (psf). Nominal overpressure is the calculated value of overpressure expected, excluding modifying effects such as atmospheric conditions. Superbooms, a sonic boom at a given location with twice or more the nominal overpressure, can also occur as a result of such factors as transition (accelerating through the speed of sound), atmospheric conditions, the reflection effects of hard surfaces, curved valleys, or massive buildings, and the maneuvering of an aircraft during supersonic flight.

A sonic boom does not occur only when an aircraft breaks the sound barrier. On the contrary, as long as an airplane is flying supersonically, a shock wave spreads out behind it in a conical pattern. It was estimated that an American SST flying at an altitude of 60,000 to 70,000 feet would generate a cone's edge at ground level covering an area about 60 miles wide; a supersonic flight, therefore, actually creates a "bang-zone" or sonic boom carpet (having as its width the length of the cone's edge at ground level) that extends for the length of the supersonic flight path.[1]

The sonic boom, of course, was never entirely neglected by the American government. By 1964 various U.S. government agencies had already conducted sonic boom tests: NASA tests at Wallops Island, Virginia, in 1958 and 1960; joint NASA-air force-FAA tests at Nellis Air Force Base in 1960 and 1961 and over St. Louis in 1961 and 1962; and joint NASA-air force-navy tests again at Wallops Island in 1962. Also some initial opposition to the SST in the early 1960s, particularly by Bo Lundberg, director-general of the Aeronautical Research Institute of Sweden, was based primarily on sonic boom hazards.

The year 1964, however, marked the beginning of the sonic boom as a significant influence on the evolution of the entire SST conflict. In early February 1964 the most extensive sonic boom tests yet undertaken began over Oklahoma City. Scheduled to end in late July 1964, these were aimed primarily at evaluating the sonic boom's effect on physical structures and on public attitude and at helping to develop insurance damage standards and some means of predicting sonic boom overpressures under nonstandard meteorological conditions. Sonic boom intensity was limited to 1.0 to 1.5 psf for the first twelve weeks and then increased to 1.5 to 2.0 psf for the final fourteen weeks. There were twenty-eight sonic booms during the first week and thereafter eight sonic booms per day, seven days a week. In addition, the National Opinion Research Center of the University of Chicago conducted detailed public opinion surveys at various times during the test period.[2]

The Oklahoma City tests immediately became controversial. Within the administration, the Bureau of the Budget attacked the FAA's management of the public opinion poll and survey format, maintaining that the FAA's vigorous public relations effort and the fact that the tests included no night sonic booms could "bias" the survey's results.[3] The FAA itself acknowledged in mid-May certain difficulties due to poor management, lack of internal coordination, and poor control over its public opinion contractor.[4]

In addition, the FAA faced the emergence of a public outcry against the tests. As early as February 1964 one FAA representative in Oklahoma City had urged a much shorter test period in the face of growing protest, especially over property damage.[5] Local residents were becoming visibly upset. In May the FAA warned a National Academy of Sciences (NAS) review panel about local "high feelings." Although the panel did not seem concerned and found the tests "soundly conceived and capably conducted,"[6] NAS president Frederick Seitz a month later prophesied that the sonic boom problem could turn out to be "the crucial one" in the whole SST effort.[7] Meanwhile the local protests continued[8] and were reported in the national media. The *Saturday Review* in an article entitled "The Era of Supersonic Morality" detailed how the FAA had "hired Air Force jets and jet pilots under contract to break the sound barrier across the heart of the city" without consulting local elected officials.[9]

The FAA grew increasingly concerned. Although Gordon Bain optimistically reported to Halaby in mid-June that there was no indication thus far that the Oklahoma City population would not tolerate sonic booms with overpressures up to 2 psf, he counseled against night booms because Senator Monroney vigorously opposed them and because the FAA had already promised publicly that there would be no night booms.[10] Also both Halaby and the White House continued to receive complaints,[11] and in mid-June even the previously silent mayor of Oklahoma City informed the FAA that local opposition to the tests existed, particularly over the FAA's apparently tough damage claims policy.[12]

In the midst of this growing debate the anti-sonic boom protesters received credible support. Bo Lundberg issued a devastating attack on the SST: the sonic boom made commercial supersonic aviation "infeasible"; even an overpressure as low as 1.5 psf woke up a great many people, including the sick, the elderly, and those dependent on daytime sleep, and caused "partial damage" to structures; and superbooms caused substantial physical damage and created extreme hazards for people.[13]

By early July news of the swelling unease in Oklahoma City had spread. The *Washington Post* mentioned the increasing nervousness of Oklahoma City's political establishment.[14]

After the tests the controversy over damage claims intensified and hurt the FAA's image. A local citizens' sonic boom advisory committee created to review damage claims procedures (originally established by the FAA) expressed dissatisfaction with FAA methods, criticizing the structural investigations, calling for "more imaginative" engineering reports, and urging the settling of all doubtful damage claims with a more lenient "special procedure."[15] The growing protest over claims bothered the powerful Monroney, who had been receiving letters complaining of the FAA's "cavalier manner" of rejecting questionable claims, and he urged Halaby in October simply to pay the doubtful claims.[16] But the protest over damage claims did not abate during the remainder of 1964, and Monroney remained extremely upset with the FAA. He wanted various claims reexamined and insisted that a weekly list of settlements be sent to his office. "Can't *somebody* do something on this headache!!" he demanded angrily. "Get with it!!"[17] The tests had, at least momentarily, weakened the goodwill of a critical SST supporter.

Another blow to the FAA was the National Opinion Research Center's study of the tests, which was issued in December 1964. Although emphasizing many favorable factors, the study found that only 73 percent of those interviewed said that they could live indefinitely with eight sonic booms per day. The FAA objected to the study's allegedly confused and overly academic presentation and demanded substantial revisions for the final draft.[18]

Even more important in the long run than the immediate impact of the tests was that the FAA was no longer the center of decision making concerning the sonic boom. At the suggestion of the President's Advisory Committee (PAC) and over the mild objections of Halaby, the president in May 1964 asked the NAS to evaluate and plan sonic boom tests and to provide overall guidance in this area.[19] The NAS then established the NAS Committee on SST-Sonic Boom, chaired by John R. Dunning, dean of the School of Engineering and Applied Science at Columbia University. Dunning was a full member of the academy and a veteran of the Manhattan Project. One NAS Committee member characterized him as a "shaker and mover." Other initial members included Richard Folsom, a well-known professor of mechanical engineering and president of Rensselaer Polytechnic Institute; William R. Sears, a professor of aeronautical engineering and director of the Center for Applied Mathematics at Cornell University; Angus Campbell, a distinguished social psychologist at the University of Michigan and director of the Institute of Social Research; William Littlewood, vice-president of American Airlines; C. Richard Soderberg, a former professor of mechanical engineering and dean of engineering at MIT; Hallowell Davis, a distinguished physiologist and expert on auditory matters, who was with the Central Institute for the Deaf in St. Louis; Richard H. Tatlow, a well-known civil engineer, chairman of the NAS Building and Research Advisory Board, and president of Abbott, Merkt and Company in New York City; Charles J. Haugh, an insurance expert; and Everett F. Cox, an engineer at Whirlpool Corporation.

At its first meeting on July 29, 1964 the committee was briefed by the air force, NASA, and the FAA. Halaby desired broad advice from the committee yet stressed the positive aspects of the Oklahoma City tests. Apparently 10 to 20 percent of the Oklahoma City population had been disturbed by sonic booms, with 3 percent actually telephoning, suing, or writing protest letters; there had been no complaints from Oklahoma City hospitals or surgeons; and film studies showed "minimal reaction" by individuals, including children. Bain concluded that people already somehow disturbed were likely to be the most disturbed by the sonic boom. At the committee's second meeting on August 26 Bain called for an aggressive public relations program, and the committee was shown an FAA documentary film which the agency claimed clearly demonstrated "that in Oklahoma City active people rarely showed even an awareness that a boom had taken place." Indeed throughout 1964 most of the briefings to the committee on the basic aspects of the SST program (by, for example, the Department of Commerce and the Institute for Defense Analyses) were generally quite positive.

Although the FAA pressed for an early conclusion to the committee's activities, the committee immediately signaled its desire to examine sonic booms in great depth. For example, at its second meeting one member urged that the next test city be less intimately connected with the aviation industry, another suggested testing a wider variety of structures with more automatic equipment, and John Dunning, the committee chairman, recommended using laboratory testing techniques, such as simulating the sonic boom with explosives.[20]

In spite of the committee's general tendency to view the SST favorably, its diverse membership in terms of research disciplines and its general commitment to scientific methodology—placing a high value on thoroughness and caution—meant that for the first time the sonic boom problem would be comprehensively explored. Specialized investigations were established in such areas as aeronautics, legal and insurance problems, financial matters, public relations, and structures.

At this early stage the committee's primary interests were structural effects—damage to buildings and property. A special panel on structures met in September 1964,[21] and the committee subsequently agreed to participate in a series of sonic boom structural tests at the White Sands Missile Range in New Mexico at the end of the year.[22] The committee later asked the NAS Building Research Advisory Board to prepare a complete structural test program and sent a formal statement to the FAA detailing its specific concerns relating to the tests.[23]

But nonstructural matters also emerged. In September Hallowell Davis, a committee member, suggested to Dunning that the annoyance caused by sonic booms and the consequent public reaction would rise sharply when sonic booms became strong enough to cause either auditory or physical discomfort. He urged testing in these areas, and Dunning, who agreed, relayed this recommendation to the FAA.[24]

Initially, however, public response issues were really only public relations concerns. Reflecting this view, in October Kenneth Youel, a

Washington public relations specialist with extensive experience in aircraft noise controversies, and Robert Harper, director of public relations for the School of Engineering and Applied Science at Columbia University, sent the committee a report that called for the increased use of public relations methods for maintaining public support in the face of potentially widespread SST opposition. Youel and Harper recommended establishing a high-level public relations subcommittee. They also suggested specific actions in conjunction with the approaching sonic boom tests: news announcements, a press conference for major newspaper editors, prearranged news releases, and a plan to enlist the cooperation of television and radio networks. In addition, in order to "pave the way" for night boom tests, they urged considerable advance work in the test city, cultivating particularly local civic leaders, and called for a film to show to influential individuals and groups. Finally they proposed a long-range public information program "to dispel false ideas and unbounded fears."[25]

But some committee members objected to such an aggressive pro-SST public relations stance on the grounds that the committee would be behaving like an "interested party." Moreover the committee gradually was acquiring a broader awareness of the human response aspects of the sonic boom. In October the committee heard from Irving Janis, a psychologist at Yale University, that public acceptance of the sonic boom could disappear if people were chronically annoyed and if there were an acute triggering incident, such as a crash. A subcommittee to study psychological aspects of the sonic boom was soon created, and in November the NAS Committee specifically urged the FAA to expand testing and research in this area. The Institute for Defense Analyses also stressed the uncertain but potentially damaging economic impact of a massive number of damage claim lawsuits as a result of the SST's sonic boom. Finally the Department of Commerce indicated at about the same time that the regions benefiting the most from SST operations would actually suffer the least from sonic booms; in fact, it appeared that northern Arizona would experience the greatest number of sonic booms: 166 per day.[26]

The difference in perspective between the NAS Committee and the FAA led to some friction by the end of 1964 after two other NAS bodies, the NAS Committee for Hearing and Bioacoustics and the NAS Building Research Advisory Board, reviewed the White Sands tests.[27] The Committee for Hearing and Bioacoustics reported in November that many of the NAS Committee's recommendations had not been implemented by the FAA in the November-December phase of the White Sands tests.[28] The Building Research and Advisory Board, in an interim report to the NAS Committee, then severely criticized the tests: most of the test structures were not typical of many kinds of construction in the United States and SST-generated sonic booms could have very different characteristics from those generated at the tests. Other inadequacies in both the tests and the instrumentation were highlighted.[29] Conflict broke out once again the following month with the submission of a second advisory board interim report that urged further research and questioned the FAA's assumption that its sonic boom design limit of 1.5–2.0 psf would be acceptable.[30]

The FAA reacted angrily, especially to the first interim report, charging that it was misleading and inadequate, challenging many of the criticisms of FAA planning, and berating the board for not previously discussing the report with the FAA.[31]

On December 2 a dramatic incident at the White Sands test site increased apprehensions about the sonic boom. The FAA had prepared a demonstration for the NAS Committee, the Building Research Advisory Board, and the media. From the beginning the test did not go well. The FAA could not locate a jet transport and had to ferry the group on a propeller-driven DC-6. The guests arrived ill tempered after a long and bumpy flight. At the site they were briefed and then witnessed fifteen overflights of three planes each. Nothing particularly startling happened, and for the reporters at least, the actual tests seemed anticlimactic. But after the scheduled tests were over and most of the visitors had gone inside for a final briefing, an F-104 fighter plane made one last low overflight for the benefit of press photographers. Unfortunately for the FAA, the pilot unintentionally accelerated to a supersonic speed only two hundred feet overhead, creating a massive superboom of about 40 psf. The boom cracked windows and plaster, damaged molding on a test storefront, rocked an instrument trailer, and interrupted broadcasts from the site's radio station. The air force general who was giving the briefing had a glass of water in his hand; just as he was stressing the benign nature of the boom, the superboom thundered over the site and he dropped his glass. Later many of the guests complained of headaches. The media now had something dramatic to report. One of the SST's alleged major flaws had been indelibly implanted in the minds of a large group of reporters.[32]

For the NAS Committee and advisory board visitors, the incident heightened concern about accidental superbooms. Board members argued that it would "be quite fruitless to continue the tests at lower overpressure levels." At the end of the year the board's executive director took the unusual step of informing Dunning about certain issues that "did not seem appropriate" for inclusion in its report, reiterating the board's serious doubts about the test structures, the overall test design, and the instrumentation used. The board wanted a reevaluation of the entire sonic boom test effort. In essence, the leadership of the FAA was being questioned.[33]

A certain ambivalence and even disagreement had emerged in the NAS. On the one hand, there were the highly negative views of the Building Research Advisory Board and of certain NAS Committee members.[34] But more optimism was expressed by other committee members, especially Dunning, who held the powerful position of chairman. Dunning's views were described by Halaby in mid-January 1965. Dunning felt that many adverse sonic boom effects had been exaggerated and that new technology would reduce sonic boom overpressures by as much as half. He favored proceeding with prototype construction.[35] He also saw to it that the NAS Committee's first report,[36] submitted in January 1965, was more hopeful about the SST's sonic boom than an earlier draft had been and that the cover letter from Dunning to Seitz was even more optimistic than the report.[37]

The strongly favorable final report, the committee's first formal statement on the sonic boom, concluded that moving to the next phase of the SST design competition—selecting a prototype design—was "clearly warranted by the evidence," although more aeronautical study was needed. The committee also asserted that the SST's sonic booms would have no significant, direct physiological effects on people and would do "essentially negligible" physical damage to structures, though it admitted that only a narrow range of structures had been tested. With regard to physiological effects, maximum sonic boom overpressure from an SST would "not cause direct injury to the normal human body." On the other hand, the committee took note of startle reactions that could precipitate such incidents as a fall from a ladder, an automobile crash, the slip of a surgeon's knife, or a heart attack, but proving the sonic boom as the sole and direct cause of such incidents would be extremely difficult. The committee did acknowledge that sonic booms would disturb sleep and "would not be desirable" in hospitals. The committee also emphasized the need for sound public relations, following the approach already outlined by Youel and Harper. The report stressed proper press handling. The basic approach, the report also advised, "should be to inform and educate opinion leaders, so they in turn would be able to interpret developments intelligently in their contacts, writings, and speeches, and to develop attitudes sympathetic to the problem."

But psychological response had assumed increasing importance by this time. The committee called for extensive research on psychological impact, noting that human acceptability varied depending on the time the booms occurred, the cumulative number of booms experienced, and the individual's personality. But the report implied that the sonic boom might be more a psychological hazard to those who were "neurotic" or on the verge of mental illness and argued that "people who dislike sonic booms may be more easily and profoundly disturbed than the average." On the other hand, the committee was worried about protests against the sonic boom; even a small number of opponents could be quite effective. The report offered a prophetic warning: the success of the whole SST program might ultimately hinge on the amount of political support the program could muster in the face of such protests.[38]

The committee's increasing concern about the sonic boom's psychological impacts was given further credence by the FAA's public release in February 1965 of its public opinion survey results of the 1964 Oklahoma City tests. Although strongly criticizing the survey, the FAA stressed to the committee at that time the positive findings, stating that "the overwhelming majority [of those interviewed] felt they could learn to live with [the] numbers and kinds of booms experienced during the six-month study." Halaby stated confidently that designing an SST with an acceptable sonic boom was possible and that proceeding with the next phase of the SST program was "clearly warranted." Dunning supported Halaby, praising the survey as helpful in developing an SST design that would minimize the sonic boom and "any resultant adverse public response." In fact, the FAA asked the committee for advice on how to mitigate negative public opinion. Nevertheless, according to the survey, about one-fourth of all people felt they could not learn to accept the booms.[39]

Also, the sonic boom increasingly worried at least part of the committee. Even before the survey results were released, at least one member of the NAS Committee, sociologist Kingsley Davis of the University of California at Berkeley, had begun to express basic doubts about the whole SST program. Davis, who had joined the committee shortly after it had been established, declared in March that the sonic boom could be a key barrier to SST development. Unlike Halaby he found the Oklahoma City public opinion poll "an excellent survey of public reactions to the boom," noting that practically all of the respondents reported some interference from sonic booms and that almost half were "more than a little annoyed." Because night booms had not been used, Davis continued, the major impact of sonic booms on sleep and rest had not been studied; even so, 80 percent of those whose sleep had been disturbed had been more than a little annoyed. Davis was highly critical of the committee's public relations approach, urging a new sonic boom test with a variety of overflights, a separate study on public views of the SST program, and the formation of a committee of social scientists to study public opinion.[40]

In July 1965 the Building Research and Advisory Board issued another negative interim report,[41] and the NAS Committee formally endorsed the board's recommendations.[42] It also issued a new, and this time comparatively neutral, report of its own. Both the board and the NAS Committee came out against the FAA's or any single agency's controlling future tests. The committee now also acknowledged that the hope of finding an answer to the sonic boom problem through design research appeared very low. The report's discussion of psychological response was also pessimistic, finding the public opinion survey "professionally competent and free of bias." The committee even predicted that the sonic boom could be a "sufficient barrier" to SST development if other pro-SST factors were not overriding and that a public relations effort would be useless in the face of large-scale opposition. Finally the committee urged that the next sonic boom test concentrate primarily on public reaction and only secondarily on structural damage.[43] Apparently neither of the two reports by the committee in 1965 was made public at the time.[44]

By the summer of 1965 the NAS Committee's views had changed

markedly. The committee had begun with an optimistic and sympathetic attitude; difficulties resulting from the sonic boom had been perceived as obstacles to be overcome in the natural course of SST development, and emphasis had been placed on physical and structural effects. The problem of psychological response had been seen in terms of public education and media campaigns. Most of all, there had been an implicit belief in the ultimate power of basic design research to alleviate the sonic boom's impact. But gradually doubts emerged in the face of negative data and increasingly diverse and expanding analysis. The NAS Committee became more neutral in its views and began to acknowledge the limits of design research in reducing the sonic boom. The Oklahoma City public opinion survey alerted the committee that public protests over the sonic boom could become widespread and might prevent public acceptance of the SST. By mid-1965 the committee had decided that its basic focus should be on psychological rather than structural factors. Some committee members were also beginning to exhibit distrust of the FAA and to call for independent and multi-institutional investigations concerning sonic booms.

7 Disintegration of a Program

The New Power Center

Until the birth of the President's Advisory Committee on Supersonic Transport (PAC) and the designation of Secretary of Defense Robert McNamara as PAC chairman in the spring of 1964, the history of the SST program since 1961 was largely of a series of successes by Najeeb Halaby as he overcame one obstacle after another in pursuing his goal of an SST program under complete FAA control. With the emergence of the PAC, however, the FAA was no longer preeminent, and Halaby faced a strong new rival for command of the entire undertaking.

McNamara was the antithesis of Halaby in many ways. The FAA administrator was the committed advocate, the defense secretary the skeptic. Halaby was a romantic and almost lyrical at times; McNamara was, at least in public, analytical and cool. Halaby was a trained test pilot and lawyer; McNamara received an MBA from the Harvard Business School and was a member of its faculty for a time. He was one of the managerial "whiz kids" in the army air forces during World War II (when Halaby was testing new aircraft). After the war McNamara rose in the Ford Motor Company to become its president in 1960, though his tenure as Ford's president was quite short.

John Kennedy tapped McNamara to be his defense secretary, and McNamara soon became one of the brightest stars in the whole Kennedy administration. He almost singlehandedly attempted to transform the Pentagon with his associated doctrines of a flexible response in war and cost-effectiveness in weapons system development. During the Kennedy presidency McNamara was something of a champion for an aircraft that seemed to meet his criteria in both areas, the TFX swing-wing tactical fighter. In particular he insisted on "commonality": that the aircraft be developed for both the air force and the navy. He pushed through the commonality concept for the TFX over the severe objections of both armed services. On the grounds of cost-effectiveness in November 1962 McNamara selected the less sophisticated and less technically advanced General Dynamics design over the Boeing TFX model, although the Boeing submission was the unanimous choice of the Pentagon's TFX evaluation group. But McNamara's victory was a Pyrrhic one. The military adeptly used their bureaucratic and political skill to frustrate the commonality objective, and the TFX itself experienced such serious cost overruns and technical difficulties in the middle and late 1960s that the navy version was eventually scrapped and the number of air force aircraft actually produced was drastically reduced. By 1964 McNamara's frustrating TFX experience had reinforced and intensified his general skepticism toward proponents' claims for advanced aircraft, like the SST.

After Lyndon Johnson assumed the presidency in November 1963, McNamara's stock rose even higher than it had been during the Kennedy years. Johnson asked McNamara to undertake a number of diverse missions, including becoming PAC chairman. As Halaby recalled, McNamara in the early Johnson era "at times really acted as Executive Vice-President of the United States due to LBJ's almost total confidence in the man with 'stickum on his hair'!"

McNamara therefore was given control over SST policy determination at a time when he was at the peak of his overall prestige and power in Washington. McNamara was then also extremely self-confident and abrasive. His major failures, such as the Vietnam War and the TFX, were not yet apparent, and Johnson would not sour on him for another two or three years.

The SST, moreover, was a perfect target for McNamara. It had serious substantive technical, economic, and environmental problems. It was ostensibly to be a commercial aircraft, which meant that McNamara could apply the sophisticated profitability and return-on-investment measures that he had learned at the Harvard Business School and had used as a corporate executive. Finally the SST would never measure up as a cost-effective aircraft in McNamara's eyes. He continually reiterated that the Defense Department did not require an SST for military missions. Halaby, even years later, complained that although "many of the generals and admirals thought it [the SST] was a good idea . . . we [the FAA] had to go all alone." According to Halaby, McNamara also did not want any "diversion" of Pentagon funds; he did not want to use any of his "brownie points with the Congress." In any event, in spite of his farflung and diverse responsibilities, McNamara allocated a surprisingly large amount of time and resources to the SST program. He used his substantial clout and talent to reevaluate and to reshape the whole program. Halaby clearly faced a powerful and perhaps fatal adversary.[1]

The FAA's objective at this time, the spring of 1964, was reflected by the findings of the recent interagency government evaluation of the initial competition under Gordon Bain's direction: to award design contracts to Boeing and General Electric for their swing-wing SST model. Bain also supported the spirit, if not the exact formula, of Kennedy's original 25 percent industry cost-sharing requirement. He even declared that if the manufacturers were "unwilling to accept even normal commercial risks," there was no longer "a clear case for continuing" the SST program.[2]

But this rapid-development schedule for the SST design competition drew fire from McNamara's office even before the PAC first met in mid-April 1964. In what was to be the first of several highly negative Defense Department evaluations on the FAA's SST analytical work, McNamara's influential, bright, and aggressive aide and key SST staff member, Joseph A. Califano, severely criticized Bain's recommendations and the entire interagency evaluation. Califano sent a devastating memorandum to all PAC members. This critique from Califano's office questioned the selection of only the Boeing-General Electric design, pointing out that Lockheed and North American ranked higher in a number of technical areas. It favored a thorough reexamination of the economics of the Lockheed and North American SST designs, as well as the expected breakthrough of large subsonic aircraft. It also suggested eliminating the FAA delivery-position policy and making such radical organizational changes as establishing a "new agency" for SST affairs or assigning the Defense Department more responsibility for SST management. Finally the review lashed out at the evaluation's "unsophisticated" economic analysis.[3]

Sensing his weakening position, though his agency strongly defended its handling of the design competition and aggressively criticized Califano's review, Halaby held his fire when he reported to President Johnson in early April. Overruling Bain, Halaby actually was willing to recommend funding both the Boeing-General Electric and the Lockheed-Pratt & Whitney versions for further design work as a "spur" for competition in the design contest.[4]

Halaby's action to compromise failed to stem the tide. By the first PAC meeting it was clear that McNamara was the dominant SST decision maker in the government and that he intended to play a vigorous role. At the meeting he appeared well informed and determined to maintain the PAC's power in the SST area. McNamara expressed strong opinions, insisting especially that the committee evaluate the SST above all on profitability criteria.

Halaby quickly discovered how diminished his position had become. He was supported in the PAC only by NASA administrator James Webb in arguing that the SST was on schedule, defending the FAA's economic analysis, and pushing for a quick comparison of the Boeing and Lockheed designs and rapid prototype construction. Others were less confident and wanted more analyses of alternative designs and economic feasibility. CIA director John McCone called for a broader look, believing that the manufacturers had not been allowed "enough latitude" and consequently had submitted "marginal" designs. Stanley Osborne felt that it would be premature to set up a rigid timetable. There was no hurry, he said, because "most of the airlines wish the whole thing [the SST program] would go away."

McNamara, the most powerful figure and for whom economic considerations were paramount, supported such skepticism. He even declared, "If the Russians could build 707s cheaper I would buy them from Russia." He thought that too much emphasis had been placed on technical issues and not enough on economics and doubted that either the Boeing or Lockheed models could operate profitably without a subsidy. He commented, "I would put myself in the investor's position. I wouldn't touch this deal, to think of putting a billion dollars of my money into this, without knowing more about it. . . . I have invested a lot of money, and I have never invested as loosely as this." McNamara also believed that a wider range of SST design alternatives should be considered; specifically he questioned an FAA limit on SST size since Pan American and Boeing appeared to favor a larger aircraft. Like Califano, he recommended paying more attention to the prospect of large subsonic jets, which were a "hint of things to come."

The FAA then suffered a new severe blow, from which it would never really recover. McNamara and most other PAC members ignored the FAA's recent evaluation findings and recommendations and instead invited the heads of the key manufacturers and airlines to express their views. McCone explained that he did not want to get "the thoughts of the airlines and the manufacturers reflected through the bureaucracy of the FAA." McNamara clearly agreed: "I don't see how we can advise on a matter as complex and as important as this without having direct access to the primary sources of knowledge."[5]

In contrast to the poor impression made by Halaby and Bain, the high-level airline executives—including C. R. Smith of American, Charles Tillinghast, Jr., of TWA, George Keck of United, and Juan Trippe of Pan American—made an enormously positive impact on the PAC members at the next PAC meeting in mid-April. All of them expressed great concern about the economic performance of the proposed SST designs. "I think the present supersonic designs don't look very profitable," Tillinghast declared. Smith was somewhat more confident, though he placed great importance on engine development for improving the SST's economics. Keck was concerned about the implications of a premium SST fare and estimated the SST's seat-mile cost to be 40 percent higher than that of current subsonic DC-8s. Trippe appeared the most optimistic, seeing a potential free-world market for American SSTs of approximately three hundred to four hundred aircraft. He enthusiastically called the American SST the key plane for the 1970s and 1980s and declared Pan American's intention of replacing subsonic aircraft in "the mass-market with supersonic." Both Trippe and Colonel Charles Lindbergh, aviation pioneer and a member of Pan American's board of directors, called for continued design competition.

None of the airline executives appeared to be in a hurry. Tillinghast stated that TWA would try to postpone choosing between the Concorde and the American SST, adding that TWA could wait about three years after introduction of the Concorde for delivery of the American aircraft. Smith and Keck echoed this line of thinking. Trippe also commented that he did not desire a "crash program." Moreover, reflecting a general American suspicion of European aviation technology, Trippe claimed that Pan American had been forced to order the Concorde to remain competitive with European air carriers and implied that Pan American still possessed more "confidence" in the American aerospace industry and in "the tremendously more important" American SST. He believed that the Concorde would not be "a successful heavy-duty ship for long-distance world operation."

The PAC members were now all the more anxious to hear from the manufacturers. Halaby, objecting on the grounds that the manufacturers had already stated their findings, was again overruled, and his cherished hope for quick SST action by the PAC evaporated.[6] The president told Halaby that it was "no longer appropriate for me to hope for a recommendation by May 1," and he directed the PAC to study the SST program thoroughly before submitting its recommendations.[7]

At its next meeting on May 1 the PAC heard from top-level executives in the six manufacturing firms participating in the SST competition. Boeing's representatives were optimistic about the aircraft's economic feasibility but were clearly worried about financial risk, though Boeing president William Allen wisely cautioned that all return-on-investment estimates were "really smoking opium" at this stage. Boeing offered to pay about 14 percent of the forthcoming detailed design-phase costs and for the subsequent prototype construction period suggested a cost-reimbursement-plus-incentive contract. North American Aviation's president, J. L. Atwood, focused on the need to improve the SST engine but was more optimistic about the SST's economics than was Boeing, estimating 1970 operating costs per seat-mile at 1.0 cent for subsonic jets, 1.7 cents for the Concorde, and 1.5 cents for the American SST. Lockheed's outlook was also favorable. The company indicated that it had improved its own SST design and stressed its unique experience with high-altitude, high-Mach aircraft.

The PAC also heard from the three engine manufacturers: General Electric, Pratt & Whitney, and Curtiss-Wright. The real competition was between General Electric's turbojet and Pratt & Whitney's turbofan engine. The turbojet engine, or true jet engine, derives all of its thrust from its exhaust. Air flows into an intake aperture where it is compressed by a compressor. Fuel is then injected into the compressed air and ignited. The combustion products are forced through a turbine that powers the compressor and are expelled through a tail pipe to provide the jet thrust. Generally speaking the turbojet is relatively efficient for high-speed, high-altitude flight but is relatively inefficient under low-altitude, moderate-speed (subsonic) conditions. A turbofan design can raise the efficiency of the basic turbojet by increasing the air flow using a large, front-end fan and then channeling the air around the rest of the engine assembly. By the 1960s the turbofan was found to augment thrust and reduce fuel consumption at subsonic speeds and low altitudes and by the middle of the decade it had come to dominate the commercial jet engine market. At the PAC meeting General Electric claimed that its turbojet was less noisy and had good subsonic economics and was therefore appropriate for the Boeing swing-wing design. But Pratt & Whitney vigorously defended its turbofan design; because it had developed the engine for the recently unveiled supersonic A-11 aircraft, the company also stressed its "unique insight."

The manufacturers' presentation reinforced the PAC's caution. McNamara especially was in no mood to plunge into prototype development and suggested a limited and allegedly risk-free "interim program" for three to six months, at the rate of $1 million per month to each engine manufacturer and $500,000 per month to each airframe company. Halaby then retreated completely, saying that the FAA was "also eager and ready" to implement a continued design competition called "phase II-A," essentially the program McNamara suggested.[8]

Meanwhile the FAA's public image deteriorated, and Halaby even complained to the president about negative press reports.[9] In late April *Aviation Week and Space Technology* mentioned "the growing prospect that management of [the SST] program would be transferred from the Federal Aviation Agency to the Defense Department" due to "White House dissatisfaction" and industry's support of Defense as program manager. *Aviation Daily* echoed the same theme: "It was no secret that McNamara and his Pentagon whiz kids had been running the SST program since early in April." The publication reported that the SST program had "slipped into an atrophied state" and that "SST advocates are gloomy at this point over the future of the program." Similarly *Business Week* told its readers in early May, "President Johnson this week took a step interpreted as shifting development of the supersonic airliner away from the Federal Aviation Agency and into the hands of Defense Secretary Robert S. McNamara. Johnson had been dissatisfied for some time with FAA's handling of the project, and considers the Defense Department better qualified."[10]

Halaby's fall from power appeared practically irreversible at this point. A draft of the PAC's first interim report to the president, which was prepared by Califano and his staff, and subsequent PAC discussion on this draft on May 8, clearly showed that the FAA lost on a number of key fronts. Although Bain and Webb generally favored selecting only the General Electric turbojet engine, most participants energetically disagreed, and McNamara was quite skeptical of General Electric's estimates. He noted Pratt & Whitney's excellent track record and felt strongly that both companies should continue SST engine design work. In the end all PAC members, including Webb and Halaby, voted to award contracts to both firms.

In the increasingly important areas of SST economics and financing, the draft painted a very dim picture, emphasizing the SST's "unusually heavy commercial risk": "If supersonic aircraft cannot operate economically without premium fares over the subsonic aircraft models of the mid-1970's, then consideration must be given to terminating the program." Without effect, Halaby bitterly objected to this opinion of the SST's economics, calling such views "worst case" reasoning. With regard to SST financing, in contrast to Halaby and Webb (who pushed for significant private contribution—in the range of 25 percent—and wanted at the very least a specific level stated in the contract even if it were later modified downward), Osborne, McNamara, McCone, and Black all wanted a more flexible arrangement. McNamara again prevailed. The final report simply recommended that the contractors' contribution be "appropriate to this phase."[11]

The draft undercut the FAA's power by recommending that the Department of Commerce oversee the all-important SST economic studies, a move that was of great concern to Halaby. He had been warned that such a transfer would result in "diffusion of FAA authority" and "fragmentation of responsibility." Halaby had also received a rather disparaging "detailed rundown" on the Commerce Department by Alan Dean, who noted that operations within the department were "chaotically organized" and lacked effectiveness. But McNamara quickly dismissed Halaby's pleas, and the PAC voted in favor of assigning SST economic analytical responsibility to Commerce. McNamara simply suggested that Halaby footnote his objections, which Halaby did in the final report.[12]

The final recommendation in the draft was that current sonic boom studies "be expanded under the direction of" the NAS. Again, Halaby feared a further weakening of FAA power and asked the PAC instead to assign the NAS "an advisory role, rather than management of a program." The PAC allowed the FAA to manage the actual sonic boom tests but gave the NAS the right to make independent judgments about test findings.[13]

McNamara was in total control and at that time had the complete confidence of President Johnson. Late on the same day of the May 8

PAC meeting, McNamara personally discussed the PAC recommendations with the president before leaving for Vietnam.[14] Also the final PAC "Interim Report to the President," which was submitted on May 14, 1964, followed the lines favored by McNamara and approved at the last PAC meeting in recommending contracts for further detailed work to Boeing, Lockheed, General Electric, and Pratt & Whitney for the next six months, authorizing the Commerce Department to conduct systematic SST economic studies, and expanding current sonic boom studies "under the guidance of" the NAS. Halaby was allowed only to footnote a number of his objections, though he also issued his own memorandum to the president expressing his opposing views and recommendations in some detail—all to little effect.[15]

At the White House, the president was quite aware of the basic disagreements and would not back Halaby. Johnson's close aide, Bill Moyers, had accurately summarized the situation for him, emphasizing the debate over financial participation and assignment of economic studies. He informed Johnson of Halaby's opposition to the PAC's views and of Halaby's own confidence in the FAA's abilities.[16] But Johnson sided with McNamara. The president on May 20, 1964 directed Halaby, Secretary of Commerce Luther H. Hodges, and NAS president Frederick Seitz to undertake the actions laid out by the PAC.[17]

Further Fragmentation

The decisions of May 1964 led to increasing fragmentation, bureaucratic proliferation, and delay. Halaby's decline accelerated as he lost more and more control of the effort that he had promoted since the beginning of the Kennedy administration, and he continued to be beleaguered on several fronts. Moreover, in a fatal error, Halaby failed to build a capable and high-powered analytical capability in economics in the SST Office. Such skills were needed desperately by the FAA to counter McNamara's continued objections, which often were supported by relatively sophisticated analysis that overwhelmed the agency. Although a superb promoter, Halaby could not cope with the substantive economic and associated technological issues first raised by the manufacturers and later by McNamara.

Commerce quickly created a unit to conduct SST economic studies and attempted to establish procedures for coordinating its work with other agencies. Assistant Secretary of Commerce J. Herbert Hollomon was designated the Commerce Department's "point of contact" with the FAA. Hollomon had never been a particular favorite of the FAA (the year before he had supervised some rather negative economic studies of the SST), and it was not surprising that friction immediately developed. Halaby objected to dealing directly with Hollomon and refused to provide him with detailed technical data and with the SST contracts. He directed Hollomon instead to work through Bain. Later Abraham Katz was appointed Commerce's project manager for SST economic studies. Katz was a former operations analyst with RCA and, like Hollomon, had also been involved in earlier Commerce Department SST economic studies; his appointment came as little surprise to the FAA.[18]

Bureaucratic bickering between the FAA and Commerce continued throughout the rest of 1964 in such areas as allowing Commerce contact with the SST contractors, funding Commerce's economic studies, and providing Commerce with sonic boom data. But in all these areas the FAA ultimately backed down and gave Commerce what it demanded.[19]

As the FAA had feared, Commerce's SST economic studies grew increasingly complex. Task groups were established, outside research firms and consultants were contracted, and airframe manufacturers were visited by Commerce Department representatives.[20] Confronting this proliferation of activity, Halaby worried about further delays, and he directed Bain to review Commerce's SST work.[21] But sensitive by then to political and bureaucratic realities, Bain noted that Commerce had the support of McNamara, Kermit Gordon, and nearly all PAC members and warned that any attempt to change Commerce's schedule would only be harmful to the SST program.[22] Commerce soon advised the FAA that the SST economic studies would not be available until the end of January 1965, and this effectively precluded any mention of the SST in the 1966 budget request, which Halaby had desired.[23] Halaby's worst anxieties over further delays had proved valid.

Outside of the bureaucracy, emerging public criticism also began weakening the program. In May 1964 Halaby and Bain went to considerable effort in responding to a *New York Times* editorial that suggested an international SST joint venture.[24] In addition, the respected Guggenheim Aviation Safety Center at Cornell University, whose chairman was none other than former FAA administrator Elwood Quesada, adopted a resolution on the SST that cautioned against entering "a race to be first," since prestige benefits were not of primary importance, whereas safety questions and economic issues certainly were. Halaby quickly denied that a race with the Concorde even existed.[25] At about the same time Bo Lundberg, the early SST critic, publicly called attention to SST's potential lack of safety, its small impact on national prestige, its poor economics, and the volatility of the sonic boom issue.[26] In the same vein, Raymond Bisplinghoff, NASA associate administrator, wrote in the *Scientific American* that although the SST was technically feasible, current technology was not sufficiently advanced for developing an economically attractive aircraft, and he urged more research and development.[27] Finally former American Airlines vice-president William Littlewood declared at a public meeting that there was "no economic need" for an SST and that the sonic boom was a critical problem. He quoted one unidentified airline president as saying that he would be "happy if we didn't have a supersonic transport for fifty years."[28]

By late August 1964 Halaby was exasperated and ready to counterattack. He told the FAA director of information services, Charles Warnick, that most of the "intellectuals and liberals" were against the SST, citing the *Saturday Review*, the *New Yorker*, the *New York Times*, and Congressman Henry Reuss (D-Wisconsin), and directed Warnick to draft a strong "social benefit case" for the SST.[29] Warnick's resulting paper—a comprehensive defense of the SST which argued that the aircraft would maintain American leadership in the aviation industry, would produce jobs and money, would alleviate "the balance of payment problem," and thus would support social programs—led him to propose an article for the *Atlantic Monthly*. Warnick's choice responded to Halaby's concern: the *Atlantic Monthly* is "a magazine of the intelligentsia . . . [and] has a reputation for neutral reporting and for not taking dogmatic stands."[30]

Halaby agreed and brought every ounce of his personal conviction to bear in the SST debate. In December he wrote a strange, revealing, and remarkably lyrical letter to the *Atlantic Monthly* in which he defended the SST program. The dreamer and romantic in Halaby came through clearly. "I don't doubt," he observed, "that there are still persons living who feel that the steam engine was a mistake, the automobile a curse, and the airplane an obscenity.'' But ''history teaches us that the stifling of innovation has been the death-knell of great societies,'' he warned, citing the Phoenicians, the Greeks, and the Romans. If the United States did not build the SST, we would ''relinquish our position as the world leader of the aviation industry.'' Halaby's enthusiasm moved him to even greater rhetorical heights: ''Ever since the ancient myth of Daedalus, flight has been both the dream and nightmare of man.'' One Daedalus led to another in his mind, and soon he was comparing the SST to the style of James Joyce, a style that ''has provided us with keys to labyrinths of the human mind which otherwise might have remained hidden from sight.'' In the same way, he believed, the SST ''will add a new and interesting dimension to our world, [and will enable] us to gain some fresh and remarkable insight into the meaning of existence on this earthly globe.''[31]

Even in the area of its primary responsibility during the latter half of 1964, coordinating the six-month extended design competition and in subsequently evaluating the manufacturers' submissions, the FAA's SST Office ran afoul of McNamara and his staff, and the evaluation became a source of continuing frustration to Halaby.[32]

Joseph Califano was particularly unhappy with the contracts negotiated by the FAA, and he still found the FAA specifications too rigid and too limiting for the contractors, as did a Defense Department review. These opinions were also supported by the Bureau of the Budget. Under such tremendous pressure, Halaby was finally pushed to expand the scope of the contracts, and in late June 1964 he agreed with Gordon and Califano to send a ''letter of clarification'' to the four manufacturers. The firms were told that the government wanted ''to give the widest possible latitude to all competitors to exercise ingenuity and explore all promising alternatives'' in order to ''consider and submit alternative designs you consider worthy of analysis by the government.''[33]

Moreover, because the evaluation needed technical expertise and facilities, the Defense Department's already considerable leverage was further enhanced. Using its legal authority, the department classified much of the technical information related to SST engine design, which forced Halaby to route all FAA requests through Califano's office and therefore deal directly with McNamara or his immediate staff even on strictly managerial or technical matters.

Finally McNamara and the Defense Department pressed for detailed economic analysis and models as an integral part of the evaluation, and this too slowed down considerably the pace of the SST program. The FAA, of course, still wanted rapid development. "SST," commented one dismayed FAA official, really stood for "superstudied transport." And Halaby remarked, "Whether or not it ever flies, it will easily be the most analyzed project in the Government's history."[34]

By November 1, 1964 all four manufacturers had submitted new proposals. (The FAA evaluation summary went to PAC members and key Budget Bureau officers, while the Defense Department summary went only to McNamara.)

Boeing proposed a basic swing-wing SST in two versions, intercontinental and domestic, and expressed a preference for the General Electric engine. The international version had a gross weight of 500,000 pounds and a payload capacity of 215 passengers at a range of 4,000 miles. Lockheed provided three basic fixed-wing aircraft designs and indicated its preference for the Pratt & Whitney engine. The estimated gross takeoff weight for its intercontinental aircraft was about 468,000 pounds.

Boeing estimated the development cost of its intercontinental version at about $1.095 billion and a unit price of about $28 million. Estimated return on investment, using an average flight length of 1,980 statute miles, was about 35 percent for the Boeing model (compared with about 18 percent for the subsonic Boeing 707-320B). Estimated total operating costs per seat-mile were 1.96 cents for the Boeing model and 2.22 cents for the Lockheed aircraft.

The Defense Department found the Pratt & Whitney turbofan engine risky because it used advanced technology and probably would not be available until after the aircraft had been certificated for commercial service. The General Electric turbojet engine, on the other hand, called for a level of technology that did "not appear to be any higher than that previously proposed" for the earlier evaluation. The Boeing-General Electric combination clearly held an advantage over its Lockheed-Pratt & Whitney rival in terms of development and operating costs, return on investment, and technological risk.[35]

Submission of new design proposals gave Halaby renewed confidence, and he pressed once more for acceleration of the program. In a long handwritten letter to McNamara, he again urged prototype construction, arguing, "We must get the most out of the competitive forces at work," and he foresaw positive "hard-headed business judgments" due to sound work "back in the shops" of the manufacturers and airlines. He wanted action. He called for a meeting of McNamara, Gordon, himself, and other required officials to arrange for recommending interim funding, which would "prevent delay or dispersion of contractor work," and stated that the new design proposals increased the likelihood that an "early Presidential decision" could be made. Halaby also sent copies of this letter to the president and to Gordon.[36]

But Halaby's momentarily high hopes again were swiftly extinguished by powerful adversaries. Califano advised McNamara that without Commerce's economic studies and the NAS sonic boom evaluations, there was no reason to speed up the program and that no basic SST decision could be made until early 1965. Instead he favored simply a PAC status report to the president at the end of 1964, with PAC meetings in early 1965 to review the various studies now being completed and, in the meantime, a close Defense Department review of the FAA's evaluation.[37] Similarly Black and Osborne cautioned McNamara against quickly beginning prototype construction. If anything, their fear of a crash SST program had increased since issuing their milestone report in December 1963, particularly because of the wide divergence in estimated development costs, the unresolved sonic boom issue, and the alleged development difficulties with the Concorde.[38]

Armed with such cautious advice, McNamara informed the president in November that the PAC was against investing large sums of money until the evaluation findings were studied carefully and until NAS advice on the sonic boom and the Commerce Department studies on SST economics were received.[39] In transmitting this report to the White House, Califano went out of his way to emphasize to presidential aide Bill Moyers the strong inclination for further delay by certain PAC members like Black and Osborne.[40]

Halaby was dismayed, and he told Gordon in late November that the Lockheed-Pratt & Whitney design was "out of the running" and that an extension of these two contracts would be a "waste of time."[41] Indeed the 126-person interagency evaluation of the manufacturer's proposals, under FAA coordination, which took place in November 1964 had found the Boeing-General Electric model "clearly superior" to the Boeing design submitted in January; in contrast, the Lockheed-

Pratt & Whitney version failed "to show any improvement in aerodynamic performance"; the Boeing design also had somewhat better economics than the Lockheed aircraft.

But the airlines and Defense opposed the FAA. Almost all the American-flag carriers that commented on the design submissions—American, Braniff, Continental, Delta, National, Pan American, TWA, and United—urged that all four manufacturers continue their design work, looking for improvements in technical performance and economics and wanting more attention paid to the sonic boom issue. Only National favored a single design.

The Defense Department, which had closely scrutinized the evaluation, also questioned the unfavorable assessment of the Lockheed-Pratt & Whitney design. It argued that the Lockheed model could be improved and that, in any case, the company's previous experience with military supersonic aircraft warranted including Lockheed in the program's next phase. Furthermore Defense suggested that NASA's information on technical performance, which was used extensively in the evaluation, was inaccurate and that NASA was biased in favor of the swing-wing design. In fact, noted Defense, one of the NASA evaluators was a patent holder on the original swing-wing concept, and another was a former Boeing employee.[42]

Bain still claimed that Lockheed "was not technically convincing" and that there was "no valid reason for the Government to continue to finance Lockheed's efforts." He went so far as to suggest that if the company were so optimistic about improving its design, it should do so at its "own expense." He was equally negative about the Pratt & Whitney engine. Instead Bain remained committed solely to the Boeing-General Electric model.[43]

But ultimately Lockheed was allowed in mid-December 1964 to present its own "rebuttal" to the evaluation conclusion.[44] McNamara also flatly informed Halaby that he was "completely and unequivocably opposed to dropping Lockheed and Pratt & Whitney at this time." Somewhat ironically given his own emphasis on detailed analysis, McNamara actually criticized the manufacturers' submissions and the subsequent government evaluations as "just paper studies." He also refused to schedule a PAC meeting until all the government SST studies and evaluations were available.[45]

The FAA had little real power, and Halaby yielded again. The agency contracted with all four manufacturers for further detailed design work—"phase II-B" of the SST program. Lockheed was to work on both its own proposed aircraft and a NASA fixed-wing design, called the SCAT-15F, with the same level of funds allotted to Boeing. Halaby even conceded that there was value in continuing the competition and acknowledged the previous experience of Lockheed and Pratt & Whitney with titanium supersonic aircraft.[46]

But Defense soon accused the FAA of having already "selected the Boeing-General Electric effort as their preferred approach" and of saddling Lockheed with the disadvantage of carrying two designs forward "with the same financial support as Boeing."[47] McNamara then hinted that Lockheed was performing the SCAT-15F work only "as a price for remaining in the competition—the price for not being rejected" and also argued successfully that Lockheed should not operate "under duress."[48] Soon both Lockheed and Boeing were directed to pursue independent studies of the SCAT-15F configuration.

Again McNamara had totally prevailed. On January 21, 1965 the president signed a memorandum, drafted by Defense, which directed that no further action on the SST be taken until he had received the PAC's recommendations.[49]

The postevaluation debate and outcome left the FAA demoralized and bitter. In December an internal FAA review forlornly acknowledged that "there was simply no prospect of getting a firm decision on the magnitude and tempo" of the SST program before early 1965.[50] Bain also angrily complained that the PAC's actions "tended to impose arbitrary and unreasonable" restrictions and to create mere "research and study" efforts that would continue for years "without being productive" or would not "overcome specific design problems."[51] About a month later Bain was even more pessimistic. He pointed out to Halaby that McNamara was "very doubtful" that an SST prototype decision could be made by March 1965, adding that such a decision could take months or even years. By this time Bain had concluded that McNamara actually wanted to delay the program.[52]

The FAA had clearly lost another bureaucratic battle. Its attempt to accelerate the SST program and to enter prototype construction had again failed, and although its November evaluation was completed on schedule, the agency could not overcome the resistance of McNamara and the general belief that new information was needed from other participating organizations. Fragmenting lines of responsibility and shifting power centers made it impossible for Halaby to regain control of the program that he had championed so effectively for so long.

Halaby's Swan Song

Incredibly, Halaby made still one more attempt at "cutting metal" or building the SST prototype. The extended design competition was now expected to end in the summer of 1965, and he wanted prototype construction to begin during the latter half of the year.[53]

But McNamara forced the FAA to wait until the Commerce Department had completed its economic studies before any crucial decisions could be made, and he ignored FAA reviews ostensibly on grounds of competing commitments.[54] Meanwhile, to make matters worse for Halaby, the press reported that McNamara had opposed most of the previous FAA plans on the SST.[55] "In short," Halaby told Bain and other FAA officials, "it looks like a cold, wet spring."[56]

In February Commerce finally began to stir,[57] producing a four-volume set of reports that dealt with air travel demand, aircraft cost-performance relationships, the economic impact of the sonic boom, and a general analysis on financial matters. These were further reduced into a "preliminary executive summary." After months of haggling, the long-awaited Commerce Department economic studies were distributed to the president, McNamara, Halaby, and Gordon during the second week of March.

The actual findings were certainly less dramatic than the previous bureaucratic conflicts and PAC debates, and they were by no means definitive. Commerce projected overall free-world demand for long-range air transport to be three to five times greater in 1980 than it was in 1964 and concluded that the American SST showed "promise" of lower costs per seat-mile than the current Boeing 707. Still the American SST was predicted to have higher costs per seat-mile than a large subsonic plane, while the Concorde had the highest costs of all. The critical problem for the American SST was seen to be fare competition with future jumbo jets. Disregarding this eventuality, however, it seemed likely that the SST could displace current subsonic jets and could be economically successful. The sonic boom might restrict SSTs to overwater flights and thus could shrink the market, but depending on the willingness of the airlines to incur additional costs and on the public to accept higher noise levels, the market could be expanded. Commerce called for more engineering work and for more studies of the "sociopolitical" reactions to sonic booms.[58]

The FAA quickly issued its own optimistic counterrecommendations to the president, McNamara, and other PAC members, though Halaby again demonstrated his flexibility by actually presenting five alternate plans for SST development. He showed that he was still a champion by pushing the three options that were the least expensive and that scheduled a quick selection of a single airframe and engine design by October 1965. As a compromise alternative, the FAA also offered a choice that permitted competition among all four manufacturers until the end of 1966, when a single airframe-engine combination would be chosen.[59]

But fewer than ten days later the Defense Department again hit the FAA with another devastating blow by issuing an overall review of the SST program and sending it to PAC members. This review was extremely critical of the November 1964 evaluation, faulting its methods, its lack of important data on sonic booms and production costs, its technical analysis, its absence of "risk analysis" for manufacturers, its allegedly unorganized consideration of SST economics, and its negative findings on the Lockheed design. "It would be difficult to draw any realistic conclusions about the economic viability of the SST from the evaluation report alone," the Defense review declared.

The Defense Department then presented two alternative SST schedules. One was essentially similar to the FAA accelerated plan that would have selected one engine and one airframe contractor by October 1965. The second option was somewhat similar to the FAA's compromise plan: after one hundred hours of engine testing, a single engine contractor would be selected, while the two airframe manufacturers would continue their research until the end of 1965 when they would prepare prototype design proposals for submission in October 1966, with a single airframe-engine design to be announced at the beginning of 1967.[60]

It appeared to be open season on the FAA. One high-level Treasury Department analyst agreed that the FAA's economic analysis "leaves a great deal to be desired" and felt that Defense could have been even stronger in its criticism: "The risks involved with the SST program are far greater than articulated in any of the submitted documentation."[61]

At the PAC meeting on March 30, totally frustrated, Halaby finally accused the Defense Department of unfairly criticizing the FAA's studies. McNamara blandly responded that any bias had not been intentional and that the recent Defense review had merely been a summary for the PAC's benefit.[62] His statement did not fool anyone. The FAA saw little in the review that came close to being an objective summary. Bain felt that the review was "misleading" and was "so inaccurate in substantive respect that it should not lie in the Committee's [PAC's] record unchallenged." Halaby then sent each PAC member a paper that documented alleged errors in great detail.[63]

At the March 30 meeting McNamara's real hostility to the SST and the reasons behind his negativism came into clearer focus. Indeed McNamara and Halaby personified two starkly divergent views on management practices, decision making, and developing new commercialization technologies.

McNamara valued analysis highly. He placed great importance on performing economic evaluation and producing estimates before launching into hardware development for commercial projects. He voiced concern about the wide range of cost estimates and about overly optimistic rate-of-return forecasts. McNamara viewed the SST program as primarily a business venture, and he even likened himself to the director of the corporation. He refused to decide on an SST design on the basis of the information available and declared, "I don't know any other commercial project that I have ever come across that has as large a risk financing problem attached to it as this does." Also because of the SST's uncertain technology, he favored retaining all four manufacturers for as long as possible.

In direct contrast, Halaby, the former test pilot, placed little faith in the estimates of either manufacturer or government analysts. For Halaby, true knowledge could come only after construction of the aircraft. "At some point you know the only way to find out what the true costs are is to start cutting metal, to build . . . a prototype . . . and to test it. Then you find out . . . what the airplane will really do. . . . It is time now not to conduct more refined studies, but to get somebody on the line doing the job, get a prototype being tested, and use experience as the study, rather than more mathematical analyses. . . . I may be wrong but that is my opinion." To no avail, Halaby reiterated this position to McNamara in April, after taking his congressional allies, Senators Warren Magnuson and Mike Monroney, on a "supersonic grand tour," visiting Boeing, Lockheed, Douglas, and Edwards Air Force Base. Halaby a few weeks later suggested to McNamara that PAC members had seen too much "on paper" and too little "in hardware" and urged that they make a similar journey.

By this point, however, Halaby had clearly given up on "cutting metal" during late 1965 and instead supported the compromise plan that would allow all four manufacturers to continue design work until the end of 1966. Most PAC members were content. Treasury Secretary C. Douglas Dillon termed this compromise "excellent." Even Webb, Halaby's closest ally, supported continued design competition in order to "keep the pressure on" Boeing.[64] Bain continued to protest but could not change the basic consensus that emerged.

McNamara's strong inclination to keep Lockheed and Pratt & Whitney in the SST competition was soon reinforced. Stressing again its "unique background information on a program of similar technical complexity," Pratt & Whitney defended its proposal as being more realistic than that of General Electric and assured McNamara that its cost estimates were neither padded nor overly optimistic. CIA director John McCone also spoke favorably of Pratt & Whitney and Lockheed and noted that General Electric tended to overestimate its capabilities.[65]

The Commerce Department refined its economic studies, and its two supplemental reports, submitted in early May 1965, were even more favorable toward continuing the SST program than the original studies in March, declaring the Boeing SST "economically viable" and three times as productive as current subsonic jets "in its capacity for generating available seat-miles per year."[66]

But just as Commerce began to give the SST program a solid boost in the all-important economic area, Commerce's own ability to have impact on McNamara was undercut. In mid-April 1965 the Defense Department established its own SST economics task force under economist Stephen Enke, and with this action Commerce's ability to influence the SST program practically vanished. The new Defense group had the immediate ear of McNamara and Califano, and its rather skeptical analyses provided crucial data for supporting McNamara's own negative inclinations.

Enke created a team of economic analysts, whose work was impressive in terms of breadth and quality. Soon this group was playing a key role in reinforcing the arguments for delaying the program.

From April to July 1965 it produced some thirty-six detailed and wide-ranging working papers that dealt with such issues as technical uncertainties, recovery of government funds, evaluation of the Commerce studies, comparison of subsonic and supersonic aircraft profitability rates, SST market analyses, and techniques for modeling the SST market.

It was clear from the very beginning that this group would not accept the economic findings of either the FAA or the Commerce Department. In mid-April, as a forerunner of things to come, Enke produced a "predominantly skeptical" view of the SST program that stressed the need for broader economic studies. A basic policy question—whether the government should encourage faster or cheaper air travel—was still undecided. Air travel demand was seen as a function of fare, comfort, speed, frequency, scheduling, and other factors. Enke was skeptical of the SST's alleged balance-of-payments benefits and believed that "the government should attempt to shift more and more of the remaining risks onto the manufacturers and air carriers as technical and other uncertainties are reduced." Finally he seriously questioned Commerce's recent findings that the SST would be profitable and have a high rate of return. He called for "a recalculation of 'profitability' using less optimistic assumptions" and projected a limited SST market if jumbo subsonic jets were introduced. Unlike Commerce, Enke minimized the threat of the Concorde, and he stressed the importance of psychological and political factors in determining public acceptance of the sonic boom. For the first time both the advent of jumbo subsonics and the possibility of sonic-boom-induced SST route restrictions were given high priority in SST economic analyses.[67]

Enke's group expanded and refined this basically skeptical view in a series of internal working papers. One member of the group concluded that subsonic aircraft in the immediate future could easily be much more profitable for airlines than the SST.[68] Another member was particularly disenchanted with the FAA; he urged consideration of a new institutional arrangement for the program, wanted to reduce the government's financial liability, and favored more testing in various areas.[69]

By the end of April Enke had zeroed in on Commerce's economic studies, which had received the group's heaviest and most bitter criticism. According to Enke, a major fault of the Commerce work was that it concentrated solely on supply without assessing SST demand as well.[70]

In early May 1965 Enke's group summarized its conclusions for the PAC. The document read like a supporting brief for McNamara, stressing the SST's "unprecedented" technical and economic uncertainties and the "inadequacies" of Commerce's economic model. With continued subsonic fare reductions, Enke claimed, "investment in SST's would be far less profitable for manufacturers and carriers than investments in subsonics." He reiterated the looming threat of sonic boom route restrictions and the need to slow down the SST "race" with the Concorde. He later called for even more specialized studies.[71]

McNamara was apparently pleased with the new thrust of studies in the economic area, and he wanted similar analytical work on both the financing and technological fronts. Over Halaby's strong objections, which noted "the mountains of data already available," McNamara insisted that Kermit Gordon and presidential science adviser Donald Hornig be designated formal advisers to the PAC.[72]

The PAC second interim report to the president, submitted on May 8, 1964, confirmed McNamara's overwhelming dominance in the SST program. The PAC recommended that all four manufacturers enter another detailed design competition phase, beginning July 1965 and lasting eighteen months. The government-industry cost-sharing issue was left unresolved, with the PAC simply suggesting that the FAA administrator "seek an appropriate degree" of cost sharing but that the level of cost sharing "should not be a factor in the award of the contract or the elimination of a contractor." (At a PAC meeting three days earlier McNamara insisted that cost sharing was not yet a major concern and that it had "received undue attention." In fact, he declared, he would not drop Lockheed even if it refused to contribute 25 percent while Boeing contributed at that level.) The PAC then recommended that the president ask Congress to appropriate the necessary funds—about $220 million, of which $140 million would be required in fiscal 1966.[73]

A fundamental question had thus been answered: the United States would not rush headlong into building an SST prototype. More study and refinement would be done during the next eighteen months. The FAA had lost again, and Halaby could not overcome the PAC's mood of caution and doubt. He had failed in his various attempts to accelerate the program. Important parts of the program had been removed from his control. Under the pressure of successive disappointments, Halaby had retreated and had softened his positions, at times dissociating himself from the more extreme views of his key SST subordinate, Bain.

After a particularly disastrous experience at the March 30, 1965 PAC meeting Halaby met briefly with President Johnson in the Oval Office at the White House. It is not clear whether Halaby specifically resigned at that point, but by early April Halaby had submitted his resignation. He was leaving the government at a time when the FAA's authority over the SST program was at its lowest ebb.[74] The challenge of guiding the SST program was left to a new FAA administrator, but it was an open question whether the extremely fragmented state of the program permitted any hope of ever making it whole once more.

8 Second Chance

New Leadership

On a Saturday morning in late April 1965 at about 9:30 A.M. the telephone rang at Air Force General William F. "Bozo" McKee's home in suburban Washington. McKee had been commander of the air force logistics command and air force vice-chief of staff and was currently NASA assistant administrator for management development. He had just finished remodeling his kitchen, and the newly installed white telephone had never been used before. His wife answered the phone. She quickly gave McKee an amazed look, and said, "The White House is calling. The president wants to talk to you." After a few minutes the president got on the telephone, and using McKee's nickname inquired, "Bozo, you got on your britches?" McKee replied no, and Johnson, ignoring McKee's response, said that he was sending a White House car to fetch McKee.

McKee was driven to the White House and escorted to the president's living quarters. Johnson told McKee that Halaby had tendered his resignation as FAA administrator and that after having considered a number of candidates he wanted McKee to assume this post. Johnson mentioned that he had discussed this offer only with Defense Secretary Robert McNamara and NASA administrator James Webb and that both had enthusiastically recommended McKee. The president, who was very secretive about appointments, directed McKee not to tell anyone except his wife about the proposed nomination (which McKee soon accepted).

Johnson then told McKee that he wanted him, above all, to assume direction of the SST program: "Bozo, get yourself a good deputy administrator to run the FAA."[1] McKee even attended the next PAC meeting on May 5, 1965.

The appointment of William McKee as FAA administrator presented a new opportunity—a second chance—to halt the rampant fragmentation of the SST program. Unlike Halaby, who had been a Kennedy appointee, McKee clearly had the confidence of both McNamara and the president, having been selected personally by Johnson. Also unlike Halaby McKee was experienced at managing complex programs for the air force and NASA. The short-term pressure on the SST program had also been largely relieved by the recommendations in the PAC's second interim report, which urged further detailed design competition for the next eighteen months and the selection of a single airframe-engine design at the end of 1966. Symptomatic of the new, less abrasive atmosphere, McNamara uncharacteristically called for a wide-ranging "bull session" at the next PAC meeting on May 21, 1965.[2]

But McKee still faced the same anti-FAA onslaughts from the Defense Department's SST economic task force, headed by Stephen Enke, and elsewhere that had ultimately helped overwhelm Halaby. Enke's group in mid-May 1965 even urged that an interagency advisory unit oversee Commerce's SST economic analytical work.[3] In addition, staff members in the Treasury Department indicated skepticism about the results of Commerce's studies, about the prestige value of an American SST, and about the effects of an SST surcharge. They also pressed for more private capital and a separate SST development corporation, arguing that the FAA would face a conflict of interest when it came to consider the SST's commercial certification.[4]

Joseph Califano, McNamara's key SST aide, remained as an active anti-FAA force, making certain that McNamara was aware of the negative views that surfaced. Like Enke, Califano was particularly critical of the Commerce study, calling it cumbersome, incomplete, and unimaginative and in a wide-ranging brief urged McNamara to obtain outside, high-level financing advice. McNamara agreed and in May decided that Eugene Black and Stanley Osborne, who had drafted the milestone anti-FAA financing study in late 1963, should help prepare a new PAC financing report. Although looking forward to improvements under McKee's leadership, Califano also continued to worry about FAA managerial ability. He demanded SST status reports from the FAA to the PAC every sixty days. With little hesitation McNamara agreed to this as well.[5] Formalizing his decisions at the May 21 PAC meeting McNamara, who had bluntly declared that the government, which, after all, would contribute $800 million for SST profits, would want "part of the action" with regard to SST profits, recommended a working group, consisting of Black, Osborne, and John McCone, to study and generate alternative financing arrangements. In spite of strong objections by Halaby and Webb, McNamara prevailed—as usual.

McNamara also had his way on other matters. The PAC accepted his suggestion that the Bureau of the Budget report on the whole SST organizational question, particularly on whether the program should remain in the FAA. Although Webb strongly disagreed, the PAC also approved a recommendation by McNamara to ask presidential science adviser Donald Hornig, in consultation with NASA, the FAA, the Defense Department, and the National Academy of Sciences, to prepare a program of further sonic boom studies and tests. The PAC adopted McNamara's proposal that the FAA submit periodic reports on the Concorde and on the Soviet TU-144 program. Finally the advisory committee also decided that the commerce secretary and the FAA administrator should prepare a program of additional SST economic studies by early August 1965 with (at McNamara's insistence) the participation of Charles Hitch, comptroller and assistant secretary of defense.[6]

In summary, by late May 1965, when McKee came on board, McNamara had acquired enormous power and the SST program seemed fatally fragmented. Economic analysis was taking place in at least three agencies; the Bureau of the Budget was examining SST organization; responsibility for evaluating SST financing policy had been assigned to individuals who were outside the government altogether; and still another group had been placed in charge of the sonic boom studies. Moreover the FAA had been directed to provide the PAC with status reports every sixty days.

Even McKee's nomination initially worked to the detriment of the SST program. As Gordon Bain observed to Halaby in mid-June 1965, the SST was "linked" to McKee's confirmation because a delay caused by a confirmation fight would "make it impossible" to enter into contracts for the eighteen-month detailed design competition, which was to begin in July. McKee's confirmation difficulties hinged on the fact that he was an air force general, while the original FAA statute required that a civilian be appointed administrator. Johnson had submitted a bill that would permit McKee to collect about $8,000 annually in military retirement pay plus the regular $30,000 salary as FAA administrator. Senator Vance Hartke (D-Indiana) led the attack against McKee's confirmation, pressing for the appointment of a civilian, and he almost won on Johnson's pension bill for McKee, losing a thirty-five to thirty-three vote.[7]

The SST was prominent at McKee's swearing-in ceremony at the White House on July 1, 1965. McKee's assignment, the president declared, was "to develop a supersonic transport which is, first, safe for the passenger; second, superior to any other commercial aircraft; and third, economically profitable to build and operate." Johnson also announced at this ceremony that he approved the PAC's second interim report and authorized McKee to begin the eighteen-month detailed design phase in August.[8] In spite of Johnson's bullish statements, the program was clearly not going to be accelerated.

Halaby was especially disappointed, and as one of his last acts as FAA administrator wrote to the president urging him to give McKee the support and authority necessary to expedite the SST program. He claimed that further delays would "play into the hands psychologically and commercially of the French, British, and Russian governments and industries." He recommended that SST management be under McKee's sole direction, subject only to the president's own "policy direction," and he urged that the PAC, the bane of Halaby's existence for over a year and a major cause for the delays, be abolished by the end of 1965.[9]

Johnson went out of his way to praise Halaby. "In a difficult job," he wrote, "you brought a good spirit and good heart to the task and rendered a service that will outlive us both."[10] But Halaby's major goal—developing an American SST under FAA control—remained as far away as ever.

With still another design competition about to begin, strong industry complaints about cost sharing surfaced again. Cost sharing was particularly crucial because Johnson would not request a supplemental $140 million fiscal 1966 SST appropriation until all four manufacturers had signed contracts. As before, Bain took a hard line on cost sharing, requiring a 25 percent final contribution by the manufacturers for the work done during the next competition period (called phase "II-C").

But the manufacturers again balked over the cost-sharing issue. Only after a series of tough discussions did General Electric and Pratt & Whitney agree to the requirement. The airframe manufacturers were more stubborn. Bain was especially exasperated with Boeing. "When their money is on the line," he observed bitterly, "they are much more conservative. . . . If we have this kind of difficulty in Phase II-C where the stakes are relatively small, it is interesting to contemplate what the company's position will be in Phase III [the prototype construction period]. . . . If I ever had any doubts about the advisability of making the manufacturers financially participate in the development, they have been completely removed by this negotiation.'' In the end Boeing and Lockheed signed contracts that included a 25 percent cost-sharing requirement, but industry objections continued to be voiced both privately and publicly.[11]

By late summer 1965 Bain was clearly out of tune with most of the key officials who were involved in the SST effort. At PAC meetings he had taken positions that were even more extreme than Halaby's, and these were all rejected. His relationship with McKee was not as close as it had been with Halaby, and McKee soon turned elsewhere for advice on the SST. During the summer of 1965 McKee met with manufacturing executives and others involved with the SST. He was also informed by FAA officers that Bain's SST Office was exclusive and isolated and had failed to use effectively the services of other FAA units.[12] The Bureau of the Budget's assignment to report on the SST program's management structure also greatly offended Bain. It was not surprising therefore that Bain announced his decision to leave the FAA at the end of August 1965.

McKee's choice of a new SST director was Brigadier General Jewell C. Maxwell, commander of the air force west testing range at Vandenberg Air Force Base and former head of the bomber aircraft division of Wright Patterson Air Force Base. He had served as chief military coordinator for the development of the Boeing B-52 bomber and had been chief of staff for the air force systems command. In 1963 he had been chairman of the aircraft committee of Project Forecast, which had included an analysis of future air force transport needs.

McKee and Maxwell seemed almost like clones. They quickly formed a cohesive and compatible team. Both had military backgrounds and extensive managerial experience with large development projects.

Although the program's intense fragmentation continued, a new sense of confidence and vigor slowly emerged. SST management procedures soon became more formal. McKee upgraded the SST Office's status and changed Maxwell's title from deputy administrator to director of supersonic transport development. He also delegated to Maxwell much broader contractual power.[13]

Establishing Control

The battle over SST economic studies, considered crucial to McNamara, continued to rage throughout 1965. The lines were clearly drawn: the FAA (with some support from the Commerce Department) versus Enke's group in the Defense Department. Friction grew steadily. As usual, McNamara had the final word, stating that he wanted to wait for the completion of all SST economic studies before acting on key SST policies.[14]

The key economic study still to be completed was that of Enke's group, which did not appear in final form until July 1965, and it was distributed to the PAC about a month later. Enke was obviously pleased with the finished product, describing it as "bland" and "objective."[15] But if Enke could term this final draft bland, then earlier versions must have been positively scathing. Using its conclusions, an economic basis for developing an SST no longer existed. To Enke's group an economically feasible SST was possible only as long as no severe route restrictions were imposed, but the group termed such an assumption of no route restrictions "undesirable and imprudent." The group also called for a carefully organized sample survey of passenger preferences (to survey attitudes on reduced time versus decreased airfare) and extensive sonic boom tests over several cities, including Washington, D.C., to be followed by a public reaction survey. In addition, the group's report was an explicit attack on Commerce's work. It highlighted the differences between the Enke group "supply-demand model" and Commerce's "cost-benefit model," and, not surprisingly, the Enke group model was found superior in practically all categories.[16]

The report was another milestone. It added weight to the negative economic view of the SST. The document was also highly praised by the Defense Department,[17] while bitterly criticized in other quarters, such as Commerce[18] and the FAA.[19]

The airlines[20] and the manufacturers[21] also reviewed both the Enke group and the Commerce studies. The industry comments were a two-edged sword: they generally expressed great support and faith in the program's ultimate success but dwelled on the great risks involved and the consequent need for significant government financing. The favorable opinion tended to enhance the credibility of the Commerce Department's work, while the emphasis on risk tended to support the Enke group's views.

In spite of the distance between the two sides in the economic debate, however, the overall atmosphere had calmed enormously, among other things because of the new FAA leadership. Johnson and McNamara clearly had more confidence in the new FAA team. In addition, a new under secretary of commerce for transportation, Alan S. Boyd, had replaced J. Herbert Hollomon as the key liaison between the Commerce Department and the SST Office.[22] Califano had moved to the White House to become a presidential aide, and the new acting PAC secretary, John Steadman, was more neutral and less influential. Even Enke had softened. He suggested to Steadman that the Commerce Department, the FAA, or McNamara oversee future SST economic studies (though Enke was not particularly enchanted with any of these options).[23]

McNamara, too, was less hostile. The day before the October 9 PAC meeting, he met with McKee and Maxwell and told them that he would permit FAA direction of further SST economic work and FAA (rather than PAC) appointment of the chairman of a proposed interagency economic advisory group.[24] McNamara made good his promises at the PAC meeting itself.

For the first time since its inception the PAC exhibited a generally conciliatory mood and went as far as engaging in some self-criticism. Even the thorny matter of cost sharing between the government and the manufacturers had become less controversial. Unlike Bain, McKee and Maxwell never demanded 25 percent from the manufacturers. McKee explicitly told the PAC in October that this level of participation was not expected, and he made the same point on Capitol Hill. McNamara was in full agreement.[25]

But a number of sticky issues would not disappear. SST financing had remained especially intractable since the summer of 1965. Black, McCone, and Osborne had been studying SST financing alternatives.[26] Among other things, the question of airline risk participation gradually assumed increasing importance. Of course, the degree of willingness by the airlines to put their own money into SST development was a direct indicator of the airlines' real view of the SST's prospects. Up to this point the airlines with SST delivery positions had contributed the relatively small amount of $100,000 per aircraft to reserve these positions. As of September 1965, with twenty-one airlines and one aircraft leasing firm reserving ninety-six positions, the government had collected $9.6 million in advance airline payments. The FAA then proposed a financial scheme that called for additional airline "earnest money"—$1 million per aircraft—to support prototype development. McKee told the PAC in October 1965 that pressure was building in Congress for significant airline contributions for SST development.[27]

But the airlines and even the manufacturers reacted cautiously to this plan, though the airlines acknowledged the need to provide risk money.[28] Moreover the entire reservation policy came under attack. William Allen, Boeing's president, predicted that the whole system would have to be "realigned" before the SST production phase.[29]

McNamara questioned the wisdom of trying to obtain substantial risk funding from the airlines for SST prototype construction. At a private meeting he even directed McKee and Maxwell not to include this money in the FAA's financial plan for the prototype phase, commenting that requiring the airlines to advance large sums eight years prior to operational use was a significant deviation from normal commercial practice.[30] At the October 9, 1965 PAC meeting McNamara further observed that United Airlines was not among the first ninety-six delivery positions, whereas many little "Toonerville Trolley airlines" had been included in the allocation plan. He thought that the whole system deserved reexamination.

The McCone-Black-Osborne working group was also highly critical. Osborne declared that the present system discouraged airlines from contributing risk money, and Secretary of Treasury Henry Fowler called the entire allocation procedure "an albatross around the neck of the operation from the beginning."

Ultimately the PAC simply recommended further study of the delivery position and airline contribution issues, with the aim of establishing an "equitable" airline risk participation policy. McNamara also directed that the attorney general provide a legal opinion on the current agreements with the airlines in time for the next PAC meeting.

The matter of government recoupment of its SST investment had also started to grow in importance. McNamara, Fowler, and Osborne favored the government's sharing in the SST's ultimate profitability. But McNamara did not press the issue and said he was willing to recommend that the government not collect its investment until private capital was paid back. But Webb disagreed, based on his broader view of technological spinoffs. He argued that the government should evaluate its SST commitment on the basis of total economic development, not on specific return on its investment.

The FAA had not entirely abandoned its hope of accelerating the SST program, a hope that largely depended on whether a design could be selected before the end of 1966. But when McKee and Maxwell privately raised this possibility of early selection with McNamara, McNamara appeared doubtful. Still, at the October 9 PAC meeting the agency pushed for a one-year overlap between the prototype and production phases. However, when the FAA proposed to allocate $64 million for beginning long lead-time prototype construction work in mid-1966, the PAC responded with little enthusiasm. In the end the PAC merely directed the FAA to provide more information by the next meeting.

More important than the substantive decisions made by the PAC in early October was the new mood. With the appointments of McKee and Maxwell, bitter debate and hostility were replaced with a new sense of trust and respect between opposing parties. McKee was even confident enough to call for greater PAC support in dealings with the public, with industry, with the airlines, and with Congress.[31]

Another boost to the FAA's position was its success in late October 1965 in an intense congressional appropriations struggle that McKee had been fighting since his appointment. At the end of October, after many frustrating delays and funding crises, McKee and Maxwell won a major funding victory: a supplemental SST appropriation of $140 million for fiscal 1966.[32]

After a long and frustrating series of defeats the FAA had regained some of its lost authority over the SST program. It had recaptured control over SST economic studies, had been given responsibility to study the delivery allocation system, had obtained power to negotiate with the manufacturers over cost sharing for the prototype phase, and had won a significant funding battle. When the PAC met in early November to discuss its third interim report to the president, the committee responded to McKee's plea for a more positive stance by inserting in the report the baldly optimistic statement that "there is a high degree of probability that, with further work on the basic technological problems, a commercially profitable supersonic transport can be developed."

The debate over SST profit sharing and financing was momentarily rekindled during this review, with a range of viewpoints being offered. The PAC ultimately recommended that the government bear approximately 80 percent of the total prototype development costs and that the government should recover its investment "if the program is successful" and should share in the profits "if they rise to a predetermined level." The PAC also admitted that it did not anticipate the airlines' contributing large risk payments for prototype development, and it also proposed establishing a recoupment formula prior to the awarding of the prototype development contracts. Both the FAA and the PAC decided that it would be best not to tamper with the current delivery position allocation system since international agreements had already been signed, though McNamara reiterated his own objections to the existing system at the November PAC meeting. The PAC merely stated in the report's annex that "modifying the current allocations presents serious problems, moral and political."

The mood therefore had softened even more by the November meeting. The committee was conscious of the need not to appear extremely negative and tried to strike a note of sensible balance. Discussion of the sonic boom, for example, was fairly positive, and the overall prognosis for the program was favorable. The report declared, "It is believed that there is a reasonable expectation at this time, pending the outcome of continued and expanded studies on sonic boom response and the aircraft's economics, that the overall program can earn a return for all participants comparable to that typically earned in domestic industry."

By the time the PAC's third interim report was submitted to the president in mid-November 1965 the SST decision-making atmosphere had changed significantly. At the PAC meetings in October and November there was a new concern that the committee should not appear overly pessimistic. McCone urged, with McKee's strong backing, that the president's state of the union message in January 1966 include a "very positive statement" on the SST program. A new trust in the FAA also emerged. McNamara in particular was much less antagonistic and much more willing to place significant responsibility on the FAA's shoulders.

People therefore do make a difference, and new individuals can mean new possibilities. Five months after McKee's appointment the FAA had regained substantial control in a number of areas. It exhibited new vigor and confidence, and McKee even looked forward to making a "solid recommendation" concerning a possible early airframe contractor selection by the beginning of March 1966.[33]

There was no longer any discussion of transferring responsibility for the SST program out of the FAA. Major SST decision makers by the end of 1965 were more than willing to place most of this mission in the hands of McKee and Maxwell. The challenge for McKee and Maxwell in the coming months would be to maintain the momentum that they had so skillfully established.

9 Bureaucratic Rivalry and Rational Debate

After years of struggles and delays the FAA was once again on the move. It looked forward to selecting a final SST design by the end of 1966 and to beginning prototype development in early 1967. In typical fashion, Jewell Maxwell, the SST program director, systematically prepared to meet this schedule and to complete the final evaluation of the competing SST prototype designs.[1] The design competition, however, presented not only complex technical questions but also matters of economics and financing; and the economics and financing debates quickly became enmeshed with the ongoing struggle for bureaucratic control of the whole SST program. Once more bitter controversy over a blend of both substantive issues—economics and financing—and bureaucratic rivalry reared its ugly head.

Battle over SST Economics

The FAA had acted quickly after regaining control of the SST economic studies in late 1965. In early 1966 the SST Office created a small unit to supervise the various economic studies that were to form the building blocks for a single important comprehensive report on SST economics. This new unit was headed by Allen Skaggs, an economist previously with the Institute for Defense Analyses (IDA). Maxwell also searched for outside economic advice and in mid-April 1966 selected Edmund Learned of the Harvard Business School as a consultant for his SST Office.[2]

As the PAC had directed in late 1965, the FAA established an SST interagency economic advisory group with representatives from the Departments of Commerce, Defense, and Treasury, the Bureau of the Budget, the CAB, and the FAA.[3] But, as usual, achieving interagency cooperation over SST economic studies proved quite difficult. For example, supporting a strong complaint by Skaggs, Maxwell cut in half a $250,000 CAB proposal for an extremely elaborate series of SST economic studies,[4] and he directed the CAB to study only the relationship between air travel volume and air fares.[5] The FAA was beginning to use its new power.

A more significant and protracted struggle arose between Stephen Enke, who headed the Defense Department's SST economic task force, and the FAA. Although Enke appeared cooperative when McKee and Maxwell first took control of the SST program during late

1965 (Maxwell even requested Enke's assistance in recruiting economic experts), this era of good feeling was short-lived. Enke again grew critical and even sarcastic. He also felt left out, accusing the FAA of treating the recently established interagency economic advisory group merely as a peripheral body to be kept informed from time to time. He also believed that the FAA was stalling the scheduling of the various component economic studies that were supposed to be combined later into the crucial overall economic report.[6] He demanded more frequent meetings and a more detailed review system using a wide range of government economists in order to avoid "irrationally structured" reports that might be biased in favor of the SST.[7]

But Maxwell nonchalantly rejected all such suggestions. Emphasizing the FAA's control of SST economic studies, he characterized the interagency group as only "an advisory body" and mentioned that Skaggs, whom he designated the group's chairman, would convene meetings and approve agendas only when "open discussions with all members are required."[8]

Enke complained to McNamara, but this tactic was not as effective as it had been in 1965. Although McNamara still considered the economic studies extremely important, calling previous reports merely "kindergarten" efforts, he did little to intervene with the FAA's plans.[9] (Sensing McNamara's strong interest over SST economics, McKee pointedly directed Maxwell to do "a real job" on the economic studies.)[10]

Enke meanwhile continued to feel ignored, and he again urged more involvement of the interagency group.[11] For example, reminding the FAA of McNamara's keen interest in the SST economic studies, Enke implied that a "vague" April 1966 FAA "progress report" would have been vastly improved had the interagency group reviewed it.[12]

The intensity of this battle and the FAA's new-found power were reflected in the debate over passenger surveys about speed and air fare preferences. In late March Enke and IDA, which had been contracted to study SST demand, pressed for such surveys, but the airlines, Enke claimed, were "for obvious reasons . . . afraid of questionnaires involving trade-offs with fares."[13]

He was quite correct. TWA, for example, argued that such surveys were unrealistic.[14] Maxwell added that the FAA could not adopt an "airlines be damned" attitude.[15] Enke sarcastically rejoined that no one could ever accuse the FAA of an "airlines be damned" position, while, on the other hand, Enke added that he did not want the SST program accused of a "tax-payer be damned" position.[16] Nevertheless Maxwell again prevailed, and no such passenger survey was conducted during 1966. By mid-year the FAA appeared firmly in control of the economic analysis.

The debate over SST economic studies during the latter half of 1966 simply shifted from the question of who would manage the studies to the value, content, and administration of the resulting evaluation.

The FAA claimed that a final comprehensive economic report could not be submitted to the PAC until the end of November 1966 because "validated" technical performance data would not be available until the manufacturers submitted their final designs in September.[17] Enke blasted this argument and just before the PAC met on July 9 repeated in the harshest terms his criticism of the FAA's entire behavior in the economic studies.[18]

This time McNamara backed Enke. At the July 9 PAC meeting McNamara declared that the FAA's proposed schedule was unacceptable; he clearly wanted the economic findings available early enough to influence the final selection of prototype manufacturers, and he needed time to consider the results. He wanted the economic studies by the beginning of November 1966 (thirty days earlier than the FAA proposed), and he minimized the need for validated data, saying that "it will be years before we have data we can certify with any degree of accuracy."[19]

Enke's attack on the SST and the FAA continued throughout the summer, and if anything became even more forceful. To Enke, increasing costs, the sonic boom, and subsonic competition from future jumbo jets were becoming widespread concerns. "Has anyone informed the President how the prospects are worsening and suggested that he instruct the FAA to mute its fanfares?" he asked.[20] He requested a huge amount of economic data: "the entire output of our economic research contractors," Skaggs complained.[21] Maxwell tried to limit access to the FAA's economic contractors, fearing that the agency might be bypassed in such exchanges.[22]

Again the FAA's luck turned for the better. As so often happened in the SST conflict, a key participant suddenly departed. Enke was replaced in early September as the Defense Department representative on the interagency advisory group and left the government altogether in the early autumn of 1966. But even after Enke left, he continued to threaten the FAA. A paper on SST economics, drafted for publication in November by Enke and a colleague, particularly disturbed Maxwell, who saw it as pessimistic, misleading, and detrimental to the whole SST effort.[23] In any event, for a critical year and a half in the mid-1960s McNamara—the most powerful figure in the SST program at this time and who was basically skeptical himself toward the SST—had in his agency a resourceful expert who consistently had supported such negative opinions with perhaps the most sophisticated economic analysis to date.

Even more threatening to the FAA were the conclusions of one of its own contracted analyses: IDA's study on SST demand and balance of payments. This report was crucial since demand was closely related to profitability and commercial worth. The FAA first became worried in midsummer when it received a draft study.

IDA's final report, submitted in December 1966, was indeed an embarrassment to its pro-SST sponsor, the FAA. According to IDA, the size of the SST market would depend upon possible route restrictions due to sonic booms and upon the aircraft's sales price. IDA projected the 1990 market for American SSTs (at a sale price of $40 million) to be 661 aircraft with unrestricted routes but only 279 aircraft with restricted routes. A rising sales price, which would necessitate a fare increase, would attract fewer passengers and would consequently shrink the market. At a sales price of $30 million per aircraft, IDA forecast a total restricted route market of 454 American SSTs; at a sales price of $60 million, this market decreased to 101 aircraft. The report also predicted that SST sales would vary in relation to the value that air passengers placed on their travel time and would be very negatively affected by a surcharge on air fares. Finally, taking into account total expenditures for aircraft sales and services and for passenger fares and ground services, IDA surprisingly found that the SST's balance-of-payments contribution would be small and "under certain conditions" could even be negative.[24]

Naturally the pro-SST forces were greatly disturbed by IDA's findings. There was lively debate among the members of the interagency advisory group, and Skaggs called for additional outside reviews from "recognized transportation economists."[25] The airlines also generally lined up against IDA's views, finding fault with the report's "unrealistic" analysis of passenger preferences.[26]

A major concern for the FAA was how to incorporate in its overall SST economic feasibility study those conclusions that were not particularly favorable to the SST program, the most troublesome of which were in the IDA studies. Maxwell's consultant, Edmund Learned, even suggested that the most controversial aspects—IDA's conclusions about passengers' preferences, SST demand, and balance of payments—be communicated in a transmittal letter rather than in the actual report because the study might fall into the hands of Congress.[27]

The final draft of the FAA's overall report, which was distributed in early November, clearly had the potential to trigger all the acrimonious arguments on the SST economics that had been lying relatively dormant since 1965.[28] Indeed the interagency advisory group could not reach a consensus. Commerce, the CAB, and the FAA generally found the FAA report to be objective and comprehensive, while Treasury and Defense disagreed with certain basic features, and the Bureau of the Budget favored further review.[29]

To head off a full-scale war over SST economics, Skaggs turned to the FAA's consultant, Learned,[30] and this action marked the beginning of a much more sophisticated and effective use of experts and analysis by the agency. Learned brought more than professional expertise to the FAA. He was respected at high levels in industry and government. In addition, Learned was an old friend of McNamara's and had known McNamara since McNamara's days as a student at the Harvard Business School. Skaggs asked Learned to present the FAA's case to the PAC chairman, and Learned was happy to cooperate. He became a subtle and effective advocate as well as consultant. In mid-November he gave McNamara and William McKee quite favorable reviews of the overall study and went out of his way to praise Maxwell and Skaggs. He warned against individuals who wanted to "sabotage" the economic study and accurately characterized much of the SST debate as a governmental bureaucratic struggle. The situation clearly disturbed him: "This is the only major interdepartmental power struggle in which I have been forced by circumstances to participate during my long consulting service to the government."[31]

McNamara, who had always considered economic feasibility crucial, gave his full attention to Learned's briefing.[32] But in spite of Learned's vigorous advocacy of the FAA and the SST and in spite of McNamara's softening on FAA work since McKee had assumed control, negative criticism from the Defense Department,[33] Treasury Department,[34] and Bureau of the Budget continued. The bureau was the most charitable. Although viewing the FAA's market estimates as overly optimistic, it generally saw the report[35] as technically sound and an "excellent beginning."[36] At the November 17 PAC meeting, where the report was the focus of attention, both John McCone and McNamara expressed reservations about the FAA's work. McCone dwelled on the differences between Pan American's estimates of seat-mile costs and of necessary fuel reserves and those presented by the FAA. McNamara told the agency to rewrite its report as if it were going directly to the president and then to submit the new draft to the Bureau of the Budget for "an independent check." He bluntly remarked, "I tell you now he [the president] can't understand it, I can't understand it, and a number of other people can't understand it. So rewrite it with that standard in mind."[37]

A rewritten version was ready by early December.[38] It was still quite optimistic, concluding that "the SST will prove to be a program in which the airlines and manufacturers will make a profit consistent with their risk" and that the government should recover its investment. The estimated price per aircraft was $40 million, the estimated total demand by 1990 in a restricted route market was about 500 aircraft, and the estimated total demand in a nonrestricted market was about 1,200 aircraft. The SST's return on investment was projected at 30 percent (to be surpassed only by the jumbo subsonic 747), and its operating costs would be lower than those of the Concorde and the 707. A key FAA statistical assumption was that passengers would value their travel time at approximately 1.5 times their earning rate, instead of IDA's use of 1.0 times a traveler's earning rate. The higher valuation would make higher SST fares less of an obstacle. (Using the lower 1.0 rate, according to the December report, would reduce the SST restricted route market from 497 to 366 aircraft.) The airlines backed the FAA on this issue.[39]

The new report ignited a new wave of strong attacks. The Defense Department, already having annoyed the FAA by distributing copies of its earlier draft,[40] came out with a highly negative review, accusing the FAA of overstating SST demand by 50 percent and possibly more.[41] The Bureau of the Budget also disagreed with much in the FAA's report, including its "optimistic" estimate of demand and the valuation of passenger travel time. The bureau's director, Charles Schultze, told the PAC that the SST, as an extremely high-risk venture, "would not pass muster" when judged on commercial standards.[42]

But the latest series of reviews had little bureaucratic muscle behind them. Enke was gone; McNamara was less popular in the White House by the end of 1966, and he had concerns—above all the Vietnam war—that seemed of far greater importance than the SST. Moreover no one wanted to stop the program now that a final design was about to be selected. It was as if acknowledging the risk by a credible source was sufficient to assuage everyone's anxiety. McNamara, who had previously placed so much importance on the SST's commercial prospects, calmly observed at the December 10 PAC meeting that high risk and the possibility of a low rate of return were not unusual in the early stages of a complicated program and that he was not overly concerned about the economic figures; it would be more important to choose the appropriate controls and restrictions. "I don't think we can do more than has been done on the profit analyses," he concluded. "After having been dissatisfied with them for three years, I think that FAA and its contractors and Bureau of the Budget have joined together and produced for us about as sophisticated an analysis as we can expect at this stage."[43]

After years of debate and study the PAC members finally accepted the point that Najeeb Halaby had tried to make repeatedly in 1964 and early 1965: the decision to build a prototype should be separate from the issue of ultimate SST economic feasibility. McNamara and the PAC would live with the SST's economic risk, at least through the prototype stage. The prototype would apparently be built even though the question of commercial worth had not been settled.

There was also a brief epilogue to the whole economics battle as the FAA struggled to protect its newly won economic authority and victories. The IDA study had presented unfavorable projections of the SST's balance-of-payments impact, but in December the FAA report[44] and PAC members simply ignored this issue. McNamara himself had little interest in the matter, claiming that one could not specify with any great precision "uncertain conditions" lying far in the future.[45]

The FAA also mobilized the manufacturers and the airlines, which to the FAA's delight harshly criticized the IDA findings. Lockheed in December[46] and Boeing in January 1967[47] argued against IDA's conclusions, especially in the balance-of-payments area. Pan American too believed that IDA's conclusions were "erroneous and misleading"; the valuation of travel time at the earning rate was inappropriate; and foreign SST competition would be significant.[48]

The FAA also sought outside economic expertise to counter IDA's conclusions. By mid-December 1966, with IDA pressing for wider dissemination of its report, the FAA had already received a critique of the IDA work by John Meyer of Harvard University.[49] Early the next month, the FAA asked a number of well-known economists to review the issue, including William Pounds and Charles Kindleberger of MIT, Gerhard Colm of the National Planning Association, and Walter Lederer, chief of the Commerce Department's balance-of-payments division.

Skaggs was extremely pleased with the conclusions of these advisers. In mid-February 1967, after the critiques had been received, Skaggs happily told Learned, "We now have the rebuttals to counter many of the balance of payments attacks from our small circle of critics."[50] In its final SST economic feasibility report, published in April 1967, the FAA flatly declared, "Without exception, other economists could not accept the assumptions and findings of the original [IDA] study."[51] In addition, Stanley Osborne strongly supported the FAA on this issue and told McKee in early February 1967 that he wanted to correct the "rather unfortunate record created by . . . IDA."[52]

The FAA had become quite effective in using analysis for furthering its own ends. The final April report was both credible and favorable toward the SST.[53] The negative impact of IDA's conclusions had been mitigated, and even if difficulties in the economics area remained, they would not hamper the decision to build a prototype. Economic debate would no longer delay the SST program. For the first time the FAA proved superior to its adversaries in mobilizing expertise and bureaucratic support to protect and enhance its position and its goals.

The Financing Controversy

From the beginning of the program, the notion of the government funding an ostensibly commercial venture was very controversial. Debate over financing reached its peak in 1966 when the government and the manufacturers attempted to arrive at a precise cost-sharing formula and collection method and when a specific financial risk participation level by the airlines was first seriously discussed. This entire financing controversy was also related to the underlying bureaucratic and professional struggle between FAA and non-FAA analysts over SST economics, and ultimately it was the financing question, rather than economic feasibility, that threatened to delay prototype construction.

Not surprisingly, the financial debate began with disagreements between the FAA and Enke. While the FAA at the end of 1965 called for specific financial commitments solely for the prototype development phase of the SST program,[54] Enke demanded specific proposals for all of the program's remaining phases. To Enke, the FAA's approach would create a weapon-systems-like program for the SST that would lack commercial incentives. Instead he favored a contractor cost-sharing level that would "hurt" the manufacturers "sooner rather than later," and he was against paying a "fee" to the manufacturers for prototype development.

Instead of using royalties on aircraft sales, Enke pushed for a financial "pool" or partnership of SST contributors, including the government and the prototype contractors. These contributors would ultimately recover their investments and would share in the anticipated SST profits according to a predetermined formula based primarily on their respective shares of the total advances to the pool and an imputed compound interest. This arrangement, he believed, would promote the interests of all parties, would put the manufacturers and the government "into bed together," and would eliminate considerable government intervention. He claimed, "The master-servant relationship so typical of military procurement contracts would be replaced by a partnership, one in which the private companies take the technical initiative and government is not a sleepy financial backer."[55]

But the FAA's financial plan, submitted in mid-January 1966, still focused on only the prototype phase, calling for the manufacturers to contribute 15 percent of development costs and allowing them an additional 5 percent credit for special SST facilities. Any overrun expenses were to be shared on a 75 percent to 25 percent basis. Airline cost sharing was not considered. The plan did not propose profit sharing or a recoupment scheme, which the PAC had favored in its third interim report to the president in November 1965. McKee had simply eliminated all discussion of pooling and royalties, arguing that current technical and economic uncertainties made any recoupment decision premature.[56] He told McNamara, who had consistently supported government profit sharing and a recoupment formula in the prototype contracts, that "it would not be prudent, or necessary, to provide such detailed plans at this time."[57]

Disagreements, bureaucratic rivalry, and professional jealousy once again surfaced. Enke was particularly irked, for various reasons. On a substantive level, he pressed for an industry contribution of at least 20 percent during the prototype development phase and accused McKee of avoiding the thorny recoupment issue. Most of all he resented the FAA's bypassing the non-FAA economists, whom, he claimed, McKee was "trying to exclude."[58] In addition, Stanley Osborne and other PAC members considered a cost-sharing requirement of 15 percent too severe.[59]

Bowing quickly and deftly to these pressures, the FAA in its revised financial plan (issued in mid-February) reduced the manufacturers' financial contribution by permitting a 10 percent development fee to be credited toward a 20 percent development contribution. The FAA also backed a recoupment formula for the prototype phase that would allow full recovery of government funds for a successful program and government profit sharing if such profits rose above a predetermined level.[60]

While Enke and others wanted high industry contributions, the manufacturers continued to demand significant government financial participation.[61] Boeing's president, William Allen, publicly claimed that such participation was essential given the SST's considerable technical and financial risks and government-sponsored foreign competition, though he also feared that excessive government managerial influence would stifle innovation.[62] McCone, who maintained exceptionally close contact with the manufacturers, reported after visiting General Electric in mid-February that the firm also feared an overly severe cost-sharing plan.[63]

McNamara again heard from Enke at about this time. Enke strove to reaffirm the PAC's authority, limit the FAA's expanding influence, and strengthen his own position with McNamara.[64] In early March he submitted his own financial plan, specifying long-range financial arrangements and a pooling recoupment method, but McNamara did not distribute the plan.[65]

At an informal PAC meeting on March 9, 1966 therefore the financing discussion was based solely on copies of the FAA financial plans and the latest FAA status report.[66] Financing was clearly uppermost in McNamara's mind, though McCone also commented on improvements in the Concorde and the rising concern in Pan American and Boeing over the apparent lack of serious government support for the SST. Finally McNamara directed the Bureau of the Budget to prepare a briefing paper on SST financing for the next PAC meeting in May.[67] Although displeased to see its power in the financing area further undermined in this way, the FAA was relieved that at least for the moment Enke was out of the picture in the financing area.[68] Still the FAA and Enke continued their squabbling, with Enke pressing his arguments right up to the May 6 PAC meeting.[69] Again he had bluntly called into question the FAA's analytical competence.[70]

Meanwhile the Budget Bureau's review, which was transmitted to McNamara in late April, turned out to be an important document. For the first time nearly all of the SST financial issues were discussed succinctly by a credible and relatively objective source. Even Enke was pleased. The bureau primarily focused on recoupment. A desirable recoupment technique, it argued, should give manufacturers "an incentive to behave as if their decisions were risking their own capital—i.e., they should be comparing the *total cost* of the decision with *total benefits,* despite the fact that the Government will actually bear most of the developmental cost"; the technique also had to be flexible, relatively simple, and defensible "on equity grounds"; the most important immediate decision would be whether to include the technique in the upcoming prototype development contracts; the royalty method appeared simpler than pooling but it lacked risk incentives. In any case, the bureau believed that program changes would almost inevitably occur no matter which approach was used.[71]

The FAA now held its ground and, if anything, took a stronger position on financing. Unlike the SST Office in the Bain-Halaby era, Maxwell made sure that all FAA positions on issues to be discussed at PAC meetings were thoroughly researched and that McKee had detailed and well-formulated arguments at his disposal. SST financing was the major focus of FAA preparatory work for the May 6 meeting. The FAA still opposed explicit financing provisions for the program's later phases and now wanted to exclude such provisions in the forthcoming prototype development contracts. And to Enke's dismay, McNamara and most PAC members agreed with the FAA on this issue.

On the other hand, recoupment of investment was still a source of bitter debate. The PAC would not accept the FAA's position that the government should not share in SST profits.

With regard to recoupment technique, the FAA continued to advocate a 10 percent royalty on future SST sales. (It had also reported earlier that the manufacturers were "unanimous in rejecting pooling.") The FAA was backed by its influential adviser Edmund Learned, who stressed that a royalty had "simplicity . . . commercial validity . . . and . . . incentive value." Learned wanted an "open-ended" royalty to counteract the monopoly issue: the possibility that the winning airframe-engine contractors would reap unfair profits because of their exclusive position. As for pooling, Learned doubted its managerial effectiveness, claiming that it would lead to significant government managerial interference and would require a substantial modification of normal accounting and business practices. Pooling was backed at the PAC meeting by Treasury Secretary Henry Fowler and by Budget Bureau director Charles Schultze. Schultze believed that pooling was a better financial management tool than royalty.

McNamara had no particular confidence in either method. Formulating a royalty scheme, in his view, was potentially as complex as pooling. In fact, he claimed, "all the pool is, is an infinitely variable royalty." Following up on a suggestion by McCone that the advocates for each method—the Bureau of the Budget for pooling and the FAA for royalty—continue to work on their respective financing plans, McNamara directed that each agency draft contract clauses for their preferred method. But he also asked the bureau to analyze both plans before the PAC's next meeting in July.[72]

The FAA received some additional powerful support in late June 1966 from Osborne, who came out in favor of the royalty approach in a paper on major problems facing the PAC. Although backing government SST support, Osborne ideologically opposed the notion of the government as an equity partner and therefore opposed pooling. Like Learned, he argued that pooling would cause bureaucratic wrangling and increased government intervention and would give no credit to the manufacturer for such "nonfinancial elements" as managerial expertise and facilities.[73]

The FAA could not have been more pleased. Maxwell wanted the paper distributed to PAC members immediately, but, though agreeing, McKee decided on second thought to wait until just before the July 9 PAC meeting so that Osborne's points would be "fresh" in McNamara's mind and so that little time would be available for a "nitpicking review."[74]

Learned too provided further arguments to buttress the FAA's positions. He urged a rate of return of 5 percent rather than the Budget Bureau's goal of 8 percent, and he supported his case by noting the program's considerable length, the absence of significant military subsidies for industry in aviation research and development, and congressional opposition to government involvement in profit-making enterprises.[75]

But the FAA still faced the Budget Bureau, which strongly favored pooling and profit sharing. The bureau argued that changing a royalty rate would be cumbersome and costly to the government; pooling would be able to accommodate a range of circumstances and would ensure that the government and the manufacturers share gains and losses more equally.[76] Also Enke and his staff continued to produce pro-pooling commentaries and to maintain that the royalty method could be the more complicated of the two approaches.[77]

With this interagency bickering as a backdrop, financing emerged as the major issue at the July 9 PAC meeting. With regard to profit sharing, McCone and Osborne opposed it, Webb questioned it, Budget Bureau director Schultze felt it had some benefits, and McNamara remained fairly neutral. Regarding the recoupment approach, Webb, McCone, Osborne, and the FAA all came out strongly in favor of royalty, Schultze supported pooling, and once again McNamara took a neutral stance.

McNamara's disinterest in the whole issue of recoupment method and his inclination to view royalty and pooling as ultimately similar made it difficult for one side to score a decisive victory. Therefore although interagency rivalry continued, without McNamara fanning the flames of debate, the substantive issue of recoupment technique diminished in importance. McCone even suggested a scheme that appeared to combine the advantages of both techniques: royalties would be determined as a percentage of profits. "Maybe these two things [royalty and pooling] will merge together," McCone commented. "I am inclined to think they will," McNamara agreed.[78]

But the softening on the recoupment method issue should not be interpreted as signifying that the financing controversy was unimportant. The debate over financing was part of a more fundamental, almost philosophical disagreement over governmental funding of civilian technological projects. The FAA was defending the SST program on the basis of broad national interests, while the opposition simply viewed the program as a business venture. Such basic differences became quite apparent in late July 1966 in an illuminating series of discussions among Maxwell, Learned, and J. W. Traenkle, a consultant on loan to the Department of Defense. Traenkle, Learned's former student and assistant, discussed the SST with Learned. Learned observed that the FAA was operating on the assumption that the government was developing an SST out of national interest considerations. Maxwell reinforced this viewpoint, expressing his desire to reach the prototype stage quickly and his faith that the SST's technological fallout would more than justify the program's cost. On the other hand, Traenkle discovered that many non-FAA analysts, particularly those from the Defense Department, viewed the SST program as a commercial enterprise. For them, recouping the government's investment was a "key consideration," and, frustrated by the FAA's financing opinions and managerial outlook, they wanted the PAC to assume more responsibility for the program.[79]

Ironically the manufacturers, who would be directly affected by any recoupment decision, were by and large uninformed of the raging debate within the government over pooling and royalty. Boeing and Lockheed simply suggested that any decision regarding the inclusion of a recoupment formula in the prototype development contracts be deferred. Boeing was surprised to learn that pooling was even still being considered; the firm had viewed it as a dead issue because it had not heard the method seriously discussed since the end of 1965. General Electric's director of SST activities, in a review of the entire recoupment issue, never even commented on the choice between pooling and royalties.[80]

By September the atmosphere was much less abrasive than earlier in the year. The Budget Bureau and the FAA ultimately submitted four alternative contracts to the PAC on September 23, 1966: two pooling contracts, with and without profit sharing, and two royalty contracts, also with and without profit sharing. The intensity of the recoupment debate had subsided. Defense Department analysts were quite pleased, noting that the plans (especially those of the FAA) had been greatly improved and that the areas of disagreement had been narrowed. The FAA's submission was much more detailed and technically sophisticated than its former pieces. Even more remarkable, the FAA admitted that it could endorse any of the four recoupment plans.[81]

This conciliatory mood continued. At the October 6 PAC meeting McNamara again mentioned that the differences between the royalty and pooling plans had "narrowed substantially." Government profit sharing, however, remained controversial, with the Defense Department in favor and the FAA vigorously opposed.

The issue of recoupment interest rates was also causing heated argument. McKee and Schultze suggested interest rates ranging from 6 to 8 percent, not wanting to penalize the SST program with high rates, especially in the earlier and riskier stages. As expected, however, the Defense Department thought that this range was much too low and recommended a level of about 10 percent.

But to cool this new round of disagreements, McNamara portrayed profit sharing and the related need to limit monopolistic practices as relatively narrow, less critical matters. In his opinion lengthy discussion would do little to clarify such complex issues, and he therefore wanted to avoid becoming too specific in financing affairs. He recommended that the Treasury Department and the Bureau of the Budget review all of the FAA's financing agreements and suggested that the PAC allow the FAA to take responsibility for solving any conflicts remaining after the prototype development contract negotiations were completed.[82]

In its evolution and eventual outcome the SST financing debate resembled the concurrent economics debate. The basic choices, pooling and royalties, were vigorously contested, but eventually the entire financing decision became part of a larger bureaucratic struggle. In the end an absolute choice was simply not made. McNamara pushed through a series of increasingly detailed studies, which ostensibly merged the strongest aspects of both approaches and highlighted the uncertainty and ambiguity in the financing area.

During 1966 concern over the substantive issues of economic analysis and financing and the intensity of bureaucratic rivalry over control of the SST program reached a peak and then fell drastically at the end of the year. New factors soon rose to prominence and some of the antagonists—at least on one side of the debate, including McNamara and Enke—departed from the SST scene. In addition, the FAA began developing a more sophisticated analytical capability, and the agency thereby strengthened its authority in the economic and financing areas. With a new sense of power and security, the FAA looked forward to shepherding the SST program into the long-awaited stage of prototype construction.

10 Sonic Boom Two: The Widening Debate

The Community Overflights Battle

While such issues as economics and financing were being intensely argued in 1965 and 1966, the subdrama over the sonic boom continued. It was in relation to the sonic boom that the social and environmental consequences of the SST were first seriously discussed, and during the mid-1960s it became increasingly clear that these factors had assumed immense, and perhaps determining, importance. Indeed the debate over sonic boom in the SST conflict was an important indicator that a new and more encompassing view of technology was emerging in American society.

Although the National Academy of Sciences (NAS) Committee on SST-Sonic Boom was directed by the PAC and President Johnson in May 1964 to serve as the major guide for sonic boom research and testing, in 1965 the center of sonic boom activity began to shift as more powerful forces grew concerned. In fact, as early as November 1964, the most influential SST decision maker, PAC chairman Robert McNamara, had been warned by his aide Joseph Califano that "the sonic boom problem may yet prove to be the most troublesome of all" and that the NAS Committee had "increasing doubt" about the FAA's sonic boom assumptions.[1]

McNamara was inclined to agree and began to be more active in discussions on the topic. In May 1965 he pointed out that the NAS Committee had uncovered serious gaps in the knowledge of sonic booms and that further studies were needed; however, he also added that the current organization for the sonic boom deliberations "needed tightening." McNamara then directed that Donald F. Hornig, presidential science adviser and head of the Office of Science and Technology (OST), be responsible for preparing new sonic boom test programs. Consequently a new individual with his own bureaucracy would play a major role in the SST effort.

Hornig, a former chemistry professor at Princeton University, would clearly be much more skeptical than either Halaby or Webb. He was already bothered by the "major uncertainty of public reaction." Even certain cabinet members, he claimed, thought that "the existence of sonic boom at any level with any frequency is an intolerable intrusion on the privacy of the American people."[2]

The NAS Committee's role in the sonic boom deliberations now began to diminish as Hornig acted quickly to establish OST's new authority. He directed that the FAA, NASA, and other agencies submit recommendations for a new national sonic boom test program.[3] By August 1965 he had called for establishing an interagency coordinating committee, consisting of representatives from all agencies involved in sonic boom tests and headed by an OST representative. He urged that the Defense Department assume overall managerial responsibility for any new overflight tests.[4]

OST's new role and the resulting power shift ultimately led to major bureaucratic conflict over the sonic boom issue. The FAA was particularly worried. In July Gordon Bain protested the creation of "yet another committee headed up by a representative of Doctor Hornig's office."[5] The new FAA administrator, William McKee, pressed for total control by the Defense Department over the tests and over the proposed coordinating committee.[6] This FAA opposition greatly disturbed the key OST staff member for the sonic boom, Nicholas Golovin, who was adamantly against McKee's suggestions.[7]

By the October 9, 1965 PAC meeting the design of and responsibility for the next sonic boom tests had emerged as full-fledged bureaucratic controversies. The FAA now wanted control, while Hornig continued to press for an OST-led coordinating committee. New SST program director Jewell Maxwell argued that such a committee would not function effectively or capture much of Hornig's time.[8] Less than two weeks later, however, McKee capitulated to McNamara and Hornig, and came out in support of the proposed OST-sponsored committee.[9]

Hornig formally established the OST Coordinating Committee for Sonic Boom Studies in late October 1965, expecting it both to monitor and guide the sonic boom tests and to serve as a link among the NAS Committee, the government agencies, and contractors working on sonic boom matters.

The committee's chairman was Golovin, a geologist and senior member of the OST staff. He was characterized as a strong-willed, ambitious individual of considerable vision and was already deeply involved with the issues of sonic boom and aircraft noise. As far back as early 1963 he had worked on SST matters as a member of an SST review committee chaired by Vice-President Lyndon Johnson. Other members of the OST Coordinating Committee included an air force representative (who was also an anthropologist), a high-level officer in the SST Office, and the chairmen of the NAS Committee's subcommittees on physical effects and on public response.[10]

Almost as if to bypass the OST Coordinating Committee, the FAA formed still another body at about the same time, an interagency "working group," with representatives from the air force, NASA, the CAB, the FAA, and the Commerce Department, to prepare preliminary recommendations for the next sonic boom tests. This group met for four days in late October and issued a detailed report. Among other things the group backed exploration of the feasibility of a short-range 2,500-mile SST, various alternatives to highly controversial community test overflights, and an independent study of the data on damage claims in the forthcoming tests. The OST Coordinating Committee in November generally accepted the working group's recommendations in these areas.

The working group clearly had a pro-SST bias. It called for writing a "public information brochure," that would be "designed to explain simply the sonic boom and its impact upon the individual as a member of the 1970 society." It even proposed limiting sonic boom damage liability through legislation. The group was confident that technical advances would alleviate sonic boom intensities and criticized the "limited scope" of the 1964 Oklahoma City public opinion survey.[11]

By the PAC meeting on November 6, 1965 key PAC members had become extremely concerned about apparent delays in further sonic boom testing. Hornig warned that the sonic boom problem was so serious that he could not guarantee any overland SST routes. McNamara demanded for a second time that Hornig personally chair the OST Coordinating Committee, and PAC member John McCone accused the sonic boom studies effort of being "bogged down because no individual has taken responsibility on his shoulders."[12] In mid-November, in the PAC's third interim report to the president, the committee formally recommended an expanded series of community reaction tests under the operational control of the air force, with design and planning to be determined by Hornig's OST Coordinating Committee.[13]

This recommendation was well received at the White House where pessimism over the effect of the sonic boom was similarly growing. In late September 1965 both Hornig and Joseph Califano had come out against a suggestion by McKee for convening a White House conference on aircraft noise. Califano commented that the sonic boom problem "is almost sure to get worse before it gets better" and then warned that new "sonic boom tests, worse than those conducted in Oklahoma City [in 1964], will have to be conducted, and these tests will be the most controversial."[14]

The community reaction tests became the major sonic boom area of technical and political controversy. Generally the SST opponents or skeptics, such as McNamara and Hornig, favored elaborate tests along a corridor, which would provide data under a variety of conditions, instead of over a single city, and they wanted to test day and night booms and the use of larger supersonic aircraft like the B-58 and B-70. The SST proponents, such as the FAA, strove to limit or even eliminate the community overflights due to their likely unfavorable political fallout and instead wanted to focus on more structured human response experiments concerning the sonic boom on air force bases.

The importance of the proposed community overflights was underscored by Stephen Enke, head of the SST economic task group at the Defense Department and arch SST skeptic. At the November 22 meeting of the OST Coordinating Committee he dwelled on the detrimental economic consequences of possible route restrictions and characterized the 1964 Oklahoma City tests as actually biased in favor of the SST because of the small number of booms, their precise regularity, and the total absence of night booms. In the name of McNamara, Enke pressed for community overflights, possibly over Washington, D.C., itself.

Not all OST Coordinating Committee members agreed. The NAS Committee representatives in particular favored a slower, more careful approach, questioning the value and reliability of the forthcoming tests and wondering whether public acceptance levels for different sonic boom overpressures could be determined. They expressed doubts about test procedures, scheduling, and the ability to extrapolate SST-scale results from B-58 sonic boom data.

At this time, however, the White House and the PAC were applying intense pressure for quick results. One air force representative commented that the goal of the proposed tests was "to provide information on specific questions in time to be useful to decision makers"; the design and performance of "tidy experiments" was "strictly incidental." It seemed likely that a series of overflight tests would take place during the first six months of 1966.[15]

But disagreement increased in December when the Stanford Research Institute (SRI), which had been hired to design the community reaction tests, submitted a comprehensive test design consisting of two complementary proposals. One proposal called for controlled overflights at Edwards Air Force Base in California to correlate the human response to subsonic aircraft noise and SST sonic booms, using a limited number of test subjects. These overflights would also provide data on the sonic boom's physical effects. Much more controversial was SRI's second test proposal: four cities would be boomed with overpressures ranging from 1.15 psf to 2.30 psf; two of the four cities would have a population of at least 500,000 and the remaining two would have populations of 75,000 to 100,000.[16]

FAA opposition to the community overflights immediately flared, for political considerations as much as questions relating to test methodology. By late October one FAA official had already warned McKee of the political dangers of a community overflight test: "It appears to me entirely possible—and perhaps probable—that we may have an adverse reaction at the outset. If the political and community leaders involved refuse to support the program, I think it would be impossible to carry it out."[17]

Therefore beginning in late 1965 many of those who favored rapid prototype development, including the FAA and some members of the OST Coordinating Committee, mounted a campaign to delay and drastically limit the sonic boom tests, especially the community overflights. Maxwell, for example, claimed that the tests would be of little value for either upcoming congressional hearings or for the design competition and urged greater focus on the sonic boom's structural effects by using the planned tests at Edwards Air Force Base.[18]

The FAA was immediately successful in achieving a delay. Even the SST skeptics in OST, such as Golovin, cited the methodological problems of implementing SRI's ambitious schedule and program. The director of SRI's sonic boom studies project, Karl D. Kryter, had also softened by the time the OST Coordinating Committee met on February 11, 1966. He acknowledged that the Edwards test could be "accelerated" or "enlarged," while the community overflight tests could be delayed at least three months and perhaps longer, but he still made it clear that the Edwards tests were no substitute for the community overflights.[19]

In reality little had been resolved, and the controversy became even more bitter. The FAA now worked to postpone community overflights for as long as possible. In mid-February Maxwell and another FAA officer attacked the SRI plan as incomplete and demanded further work. They argued that the community overflights, besides adding little to the knowledge obtained at Oklahoma City, would produce "a grand, if not 'grander' fiasco."[20] But on the sonic boom issue, as on practically every contested issue related to the SST, the FAA had to fight Enke, who was playing an increasingly active role in sonic boom matters. Throughout February and March Enke advocated beginning community overflights as soon as possible, perhaps as early as June 1, 1966. He was very suspicious of the FAA, fearing (correctly) that Maxwell would eventually try to eliminate the community overflights altogether by arguing that the Edwards tests and the 1964 Oklahoma City results would provide sufficient information. In late February Enke urged McNamara to approve a proposal by Hornig to begin the community overflights three months after the Edwards tests started.[21]

McNamara's inclinations were quite similar to Enke's, even though the extent of Enke's real influence on McNamara was doubtful. (Enke complained in late February that it had been over four months since he had received "guidance" from McNamara on SST matters. Prior to an informal PAC meeting on March 9, McNamara would not permit distribution of Enke's briefing papers on the community overflight tests.)[22] In early March McNamara approved the use of B-58 aircraft for the overflights and authorized the Defense Department rather than the FAA to absorb any additional costs. He also directed that Air Force Secretary Harold Brown, McKee, and Hornig agree on a flight path.[23] These actions were accepted by the PAC at its informal meeting on March 9, 1966.[24]

In mid-March 1966 the prospects for community overflights seemed high. President Johnson had even approved a test corridor between Louisville and Indianapolis.[25] Official endorsement of the tests was expected shortly, and the NAS Committee had accepted the SRI plan.[26] By the OST Coordinating Committee meeting on April 1, the air force had chosen contractors for all phases of the test program and, in coordination with the FAA, had even drafted a public statement announcing the test program. The air force apparently was waiting for a formal go-ahead from McNamara.[27] In the OST Coordinating Committee's status report, dated April 20, the community overflight tests were characterized as an integral part of the whole sonic boom effort.[28]

The conflict over community overflights, significantly intensifying, had left the issue still very much unresolved, however. The key community overflight proponents, including Enke, Hornig, Golovin, and Budget Bureau director Charles Schultze, continued to favor the tests. In early April Enke warned that the FAA was "trying to bend" a draft public release announcing the tests so that the overflights would be associated with "prestige" and "an element of patriotism"; such claims, he felt, would merely "sugar the pill for those who find the sonic booms objectionable."[29] In mid-April Enke flatly declared, "If the Midwest [Louisville-Indianapolis] overflight program is postponed or cancelled so that the economic studies must be completed without preliminary inputs from that program, it should be assumed that supersonic overland flights are not practicable."[30]

Golovin and Hornig fully agreed. In its April 20 status report the OST Coordinating Committee characterized the community overflights as the most critical part of the planned sonic boom tests. Without the overflights, the scope of the sonic boom tests effort would shrink enormously. Of a total estimated cost of about $7.7 million for all the sonic boom studies, the community overflights were estimated to cost at least $5.9 million.[31] Three days later Hornig reported to the president that there were no technical reasons for delaying the community overflights, though he added that there might be "good political reasons for a delay."[32] Golovin argued against delaying the tests. He viewed them as absolutely essential for determining public acceptability of the sonic boom and even proposed postponing selection of a prototype design if the overflights were delayed. Hornig then sent Golovin's report on to Califano, remarking that it "puts the question of the boom program in a very clear light."[33]

Schultze too told the president that the community overflights should start as soon as possible, maintaining that the government needed to know quickly "what kind of restrictions, if any, the sonic boom would place on the SST" and that the longer the results were delayed, the more expensive any design changes would become.[34]

The FAA strongly opposed all such views and launched a full-scale effort to block the community overflights. On technical and economic grounds the agency argued that the sonic boom tests at Edwards Air Force Base needed to be conducted and the resulting data thoroughly analyzed prior to the overflights.[35] But the FAA's greatest fear was neither technical nor economic: it was political. After encountering substantial opposition during the relatively mild 1964 Oklahoma City tests, the FAA grew very worried about the likely public and congressional response to the much more extensive proposed community overflights. Acting to stem the tide toward community overflights, in April 1966 McKee pushed for a meeting of Webb, Hornig, Harold Brown, and himself to discuss what McKee called "the broader implications" of the community overflight program.[36] At the meeting, held in mid-April, Webb backed McKee in advocating a delay, but Hornig argued that no such delay was necessary.

McKee also discussed community overflights with John Dunning, NAS Committee chairman, and with FAA technical consultant Raymond Bisplinghoff, and both seemed to agree with McKee. Bisplinghoff even recommended disbanding the OST Coordinating Committee and reconstituting a coordinating group under an FAA chairman. He also doubted that community overflights would yield a "definitive" answer on community reaction to sonic booms.

Discussions relating to the community overflights were also taking place at very high levels in the government, where the FAA was effectively stressing the political ramifications of these tests. At a meeting with McKee, McNamara recommended that the FAA administrator deal directly with the president because in its third interim report the PAC had already suggested an expanded series of sonic boom tests, with preliminary results available by July 1, 1966. On April 21 McKee sent the president a memorandum urging that the Edwards Air Force Base tests be conducted and analyzed before beginning the community overflights, which he projected as beginning in June 1967 at the earliest. He argued that it would "take a year to get any significant results, . . . far too late to have any impact" on the prototype selection decision. He added that Webb and Dunning "share my view completely."[37]

At the White House in late April Califano summarized the views of McKee, Schultze, and Hornig for the president. "All agree," he reported, "that sonic boom tests should be run over Edwards Air Force Base and this should have no political repercussions." Although Califano and McNamara still believed that the community overflights were essential, they recommended going ahead with the Edwards tests and starting the community overflights immediately after the November 1966 election—"as far away from the 1968 election as possible." In what was one of McKee's most significant achievements, Johnson quickly accepted this suggestion.[38]

McKee's considerable accomplishment—convincing Lyndon Johnson on political grounds to postpone the community overflights—had changed the whole mood of the PAC. By the time the PAC met on May 6, a new formal description of the sonic boom test program had been drafted that reflected the delay decision by the president.[39] At the PAC meeting itself very little of substance was said about the whole sonic boom issue. Moreover, although McNamara still voiced his belief that the sonic boom tests had "lagged" and should be completed as soon as possible because the test results would affect the design of the airplane, the financial analysis, and the choice of contractor, he basically deferred to McKee and the recent decision by President Johnson to delay the community overflights until 1967. He added that the PAC should not become directly involved in the community overflights; this was an issue for McKee and the president to decide.[40]

After this meeting the Edwards tests began to assume greater importance, while the community overflights receded. About three days later Maxwell made a point of assuring Golovin that many of the areas of concern for the NAS Committee would be answered by the Edwards tests.[41]

By mid-June 1966 news of the delay reached the media. *Aviation Daily* correctly reported that the FAA had won a "major victory."[42] At about the same time the *New York Times* also accurately reported that using only the Edwards tests represented a "low-key approach" favored by the NAS Committee; the tests had been chosen "over a plan to boom millions of citizens in the Midwest," which had been urged by McNamara and Hornig, who had argued that "public reaction cannot be determined accurately without involving millions of people."[43]

Even though the immediate threat of community overflights had been eliminated, opposition to sonic booms did not subside. By the early part of 1966 the beginnings of serious public protests over the sonic boom had become evident.

One nagging trouble spot was Oklahoma City, where damage claims related to the 1964 tests were still being filed and suits were beginning to be heard. The last day sonic boom damage claims from the 1964 tests could be filed according to the statute of limitations was July 30, 1966. The FAA closely monitored the filing in Oklahoma City and maintained constant contact with the Department of Justice, which was representing the government in court. Especially dangerous for the FAA was the constant stream of complaints into Senator Mike Monroney's office. As late as May 1966 Maxwell was still attempting to respond to Monroney's inquiries.[44]

Elsewhere, Bo Lundberg, the Swedish aeronautics expert who had been protesting against sonic booms since the early 1960s, declared in mid-May 1966 at an international congress on noise abatement that sufficient evidence already existed to show that overland supersonic flights would not be accepted and that even over-water SST flights might be banned. Building SST prototypes therefore would simply increase the SST's financial losses.[45]

Even more significant, serious criticism concerning the sonic boom began to surface within the administration. In late June 1966 Secretary of the Interior Stewart L. Udall complained to NAS president Frederick Seitz that the NAS Committee did not include a single "environmental preservationist" and instead relied "on engineers and industrialists who have lacked conservation concern." Udall claimed that even though the 1964 Oklahoma City tests had been "stacked" in favor of the SST, the results were still extremely negative. Clearly he felt strongly; on a copy of his letter to Seitz that he sent McNamara Udall scribbled, "Bob, I'm upset—and feel I should speak out on this question!"[46]

Perhaps most startling, aviation pioneer Charles Lindbergh, a Pan American board member, told the PAC at its July 9 meeting that he was "apprehensive" and "very concerned" about the sonic boom. The airlines wanted to reduce SST seat-mile costs by building a larger aircraft, Lindbergh explained, but the sonic boom limited the size that could be built.[47] A little over a week later Lindbergh poignantly elaborated on his views to Udall, McNamara, and other PAC members. He saw the SST as "simply . . . another step upward on the exponential curve of tempo, mechanization, and distraction followed by our western civilization. . . . Without a basic appreciation of nature, I believe an over-emphasis on science will destroy us.''[48]

The FAA attempted to deal with this budding public criticism by beginning its own public relations campaign. By May the agency had developed a new film, ''Sonic Boom and You,'' which it showed to Dunning and others for review. Dunning actually found the film overly negative toward the SST and sonic boom, especially because the film demonstrated ''no advantages [of the SST] to the individual U.S. citizen.'' Similarly NAS public relations consultant Kenneth Youel urged more emphasis on the ''benefits'' of modern transportation. By early June the FAA, taking such comments to heart, had budgeted funds to revise the film.[49]

Still, the sonic boom debate in mid-1966 was within the governmental bodies directly involved with the SST program, and the focus of discussion was now the Edwards tests. These tests, however, immediately suffered a potentially severe setback when one of the two existing B-70 bombers, the only American supersonic aircraft remotely comparable in size to the SST, crashed on June 8 after completing only three of the seventeen scheduled B-70 runs. The FAA had enthusiastically urged the B-70 flights, but after the crash McKee confidently commented that if the remaining B-70, which had been temporarily grounded, could fly by early September 1966, most of the data needed for the final design competition could still be obtained. The first phase of the Edwards tests (minus the B-70 overflights) ended in late June.[50]

The FAA remained suspicious of SRI, which was to write the overall report on the tests. The agency even urged, unsuccessfully, that an additional report be prepared that would contain no ''weighting'' or opinion by SRI.[51]

The FAA's ongoing concern proved to be justified. The agency had learned by mid-June that Boeing's international SST model would exceed FAA sonic boom limits by about 0.3 psf for acceleration and cruise.[52] The FAA for the first time even suggested building two versions of the SST: an international and a domestic model. SRI's preliminary findings, which came out in early July, showed the sonic boom's ''annoyance factor'' to be much greater than previously assumed. Meanwhile Enke continued to advise McNamara in scathing terms that the postponed community overflights were absolutely critical for design planning and economic forecasting.[53] In addition, at the July 9 PAC meeting McNamara stressed the lack of knowledge about the relationship between sonic boom overpressure and aircraft size and about the whole area of public acceptability of sonic booms. Hornig, McNamara, and McCone all called for expanded sonic boom test programs.[54] Later in the month Golovin also demanded community overflights as the "only practicable method for determining maximum acceptable sonic boom levels." He also attacked the FAA's economic studies, calling for more allegedly objective analyses.[55]

Maxwell and McKee themselves were concerned about the SST designs' increasing gross weight, the advent of a more accurate method for estimating shock waves (which produced higher estimated overpressure levels), and the "relatively severe level of annoyance" that was revealed at the current Edwards tests.[56]

The human ear can sense sound pressures covering a ratio of 1 million to 1. Using this linear scale, a quiet conversation would correspond to 1,000, a loud voice would be 30,000, and a sound pressure of over 1 million would cause physical pain. This linear scale is then squared and changed into logarithmic scale. The original scale is transformed into a 1 to 1 trillion scale, and then into a 0 to 12 scale (based on the number of zeros in the logarithmic scale). The unit on this second scale is called a bel in electronics and acoustics. For convenience it is divided into tenths, or decibels, and it is a nonlinear scale. A doubling or significant increase in noise level will appear as only a small increase on the decibel scale.

The measurement of noisiness also takes into consideration how a person judges sound. On the basis of several government and industry experiments in which people were asked to listen to various sounds and then to judge the loudness of these sounds, a unit for measuring the perceived noisiness of sounds, the perceived noise decibel (PNdB), was developed. The following are the approximate PNdB levels for common noises: forty-three for a soft whisper at five feet, eighty for typewriter noise at the typist's ear, one hundred for a power lawnmower at the operator's position, and one hundred twenty for a room where a hard rock band is playing. Later, by correcting for the presence of any discrete (or pure) tones (for example, the whine from a jet aircraft landing) and for the duration of the sound (the longer a sound lasts, the more annoying it becomes), another measuring unit, the effective perceived noise decibel (EPNdB), was developed, which is the international standard measure of aircraft engine noise.[57]

The data on annoyance that the FAA was receiving in mid-1966 was indeed discouraging. This information was released a year later in an overall report on the Edwards tests. There were three groups of test subjects: the first from the Edwards Air Force Base area (who were normally exposed to four to eight sonic booms per day) and the other two from the California towns of Fontana and Redlands (who did not live under the flight track of any supersonic aircraft and were rarely, if ever, exposed to subsonic aircraft noise). Human reactions to sonic boom overpressures were correlated with human reactions to subsonic aircraft noise. When indoors, the Edwards subjects considered sonic booms of 1.69 psf peak overpressure to be equivalent in acceptability to subsonic jet noise of a 109 PNdB intensity. But Fontana and Redlands subjects in the same circumstances judged the same sonic booms to be equivalent to a subsonic jet noise of 110 PNdB intensity. When outdoors, the Edwards, Fontana, and Redlands subjects judged the 1.69 psf sonic booms to be as acceptable as subsonic noise at intensities of 105 PNdB, 111 PNdB, and 108 PNdB, respectively. In addition, when indoors, 27 percent of the Edwards subjects and 40 percent of the Fontana and Redlands subjects judged the 1.69 psf booms in the range of less than "just acceptable" to "unacceptable." When outdoors, 33 percent of the Edwards subjects and 39 percent of the Fontana and Redlands subjects judged the 1.69 psf booms in the same low range. Finally an attitude survey of the residents at Edwards in June 1966, where 15 percent of the population served as test subjects for the sonic boom experiments, found that 26 percent rated the sonic boom environment at that time, which was averaging about ten booms per day with a median peak overpressure of about 1.69 psf, as also ranging less than "just acceptable" to "unacceptable."[58]

In the eyes of experts on psychoacoustic response and community behavior, the results appeared particularly unfavorable for the SST. At one OST Coordinating Committee meeting, after learning that indoor acceptability levels for a 1.69 psf boom and subsonic aircraft noise of 110 PNdB were perceived by two groups as equivalent, an almost instant consensus emerged that the American SST could never fly overland.[59] Kryter later voiced this conclusion publicly and warned that sonic booms from SSTs operating "over the United States will result in extensive social, political, and legal reactions against such flights at the beginning of, during, and after years of exposure to sonic booms from the flights.''[60]

The FAA meanwhile steadfastly avoided any mention of the need for new community overflight tests, and it fought to stem the renewed tide of growing pessimism. For example, the agency attempted to control SRI's review of the Edwards tests. It met with air force representatives to make sure that SRI focused on analyzing the Edwards data and throughout August closely monitored SRI's draft reports. According to one staff member, the FAA now viewed the sonic boom as perhaps the most critical factor in affecting the FAA's ''readiness'' to proceed with the SST prototype.[61]

Responding to Golovin's strong criticisms, Maxwell continued to maintain that there was no need for community overflights and suggested that more reliable ways existed to obtain information. Maxwell strongly objected to what he perceived as attacks on the FAA itself. In exasperation he exclaimed, ''How sneaky can you get?''[62]

Maxwell received important help for fending off new demands for community overflights. FAA consultant Raymond Bisplinghoff and another FAA officer charged that the OST Coordinating Committee had done ''an incomplete job'' because it had concentrated on overpressure, which was ''only one aspect of the total [sonic boom] problem,'' and that Golovin's discussion contained overgeneralizations and ''incomplete statements.'' Moreover NASA and the NAS Committee were at least ambivalent. In mid-September the deputy director of NASA's aeronautics division wanted to wait for the Edwards findings before making any decision about community overflights. The NAS Committee strongly suggested that the United States build both an overseas and an overland version, which would eliminate the need for immediate community overflights tests; instead careful and increasingly comprehensive sonic boom research should be carried out to determine public acceptability. Dunning urged beginning with controlled laboratory experiments and studies and only then reevaluating the need for community overflights.[63]

In September Maxwell attacked undertaking public opinion surveys, a key part of a community overflight test program, since they would suffer from ''guinea pigitis,'' which he defined as public resentment to being tested and the ensuing political protest due to this resentment. At Washington's National Airport, he pointed out, jet aircraft had been introduced without fanfare and the resulting noise was simply tolerated. Had the FAA run public noise-level tests at the time, Maxwell commented, "I would doubt seriously that there would be jets in Washington National Airport today."[64]

The controversy reached such a level that Hornig and McKee, whom McNamara had directed in July to submit a joint report on sonic boom problems, ultimately had to submit separate statements. Hornig bluntly informed McNamara that if available test findings were confirmed, neither the American SST nor the Concorde would be suitable for overland flight routes, and he continued to press for community overflights.[65]

Pessimistic conclusions about the effects of the sonic boom had also reached the White House by mid-September. Califano specifically warned the president against portraying the SST as an overland vehicle.[66]

For McNamara, the sonic boom had grown into a very serious concern during the period between the July 9 and October 6 PAC meetings. He was continually reminded by Enke of the dangerous implications of the sonic boom, especially that the SST might have "no overland market," which, Enke added, would be economically devastating.[67] Secretary of the Interior Udall continued to urge McNamara to protect the environment from the effects of the boom and relayed a similar plea from Charles Lindbergh. McNamara was clearly affected. He directed that the letters from Udall and Lindbergh be sent to all PAC members and acknowledged to Udall that the sonic boom problem was "a major aspect of the SST program." He considered the matter so important and the relevant information so disturbing that he wanted the issue discussed at the next PAC meeting.[68]

By late September McNamara had received Hornig's pessimistic review of the situation, which he also distributed to the PAC, and at the beginning of October, again calling attention to the dangerous economic impact of the sonic boom, Charles Schultze warned McNamara that the prospect of restricting SSTs solely to over-water use would deny the aircraft a large share of the predicted market.[69]

At the October 6 PAC meeting Hornig was even more gloomy. He noted that although total aircraft noise currently affected not more than the 50,000 people who lived near airports, the sonic boom from a single coast-to-coast SST flight would affect about 5 million people, and no one at Edwards had been tested for the startle effect because they had all been warned approximately one minute before each overflight. Hornig declared that little more could be gained from the kind of controlled tests performed at Edwards, and he too again advocated community overflights.

McNamara wanted to "accelerate" the sonic boom test effort and appeared to agree with Hornig on the need for community overflight tests. Osborne, McCone, and Commerce Secretary John Connor also favored this kind of action.

McKee quickly moved to counter this new effort to initiate community overflights. He used the standard FAA tactic of calling for more information and analysis and urged that the OST Coordinating Committee submit an overall report with recommendations; community overflights would be inconclusive and perhaps unnecessary, given other sonic boom data; these overflights could have damaging political repercussions and could seriously threaten the entire SST program. He also invoked presidential support in delaying community overflights. "I took it up with the president," McKee told the PAC, "and the president agreed with me." To drive home his point, McKee told a story about a woman who called the FAA on a Monday to protest jet noise at Washington's National Airport. The FAA official said, "Lady, I am sorry, they [the jet aircraft] are not going to start until Sunday." She said, "Okay, I will call you up on Sunday."

The FAA won another delay, albeit a brief one. McNamara directed the OST Coordinating Committee to submit to the PAC within less than a month a report on the sonic boom's design implications and on the value of community overflights.[70]

The resulting OST Coordinating Committee report became a key policy statement on sonic booms. Even the FAA was forced to accept the document as its own sonic boom position paper.[71] Above all, the document concluded that neither the Boeing nor the Lockheed SST would be publicly acceptable for commercial overland supersonic operations due to their sonic boom characteristics. If this judgment were sustained by subsequent tests at Edwards, there would be no need to reaffirm this finding using community overflight tests. (The Concorde, according to the report, probably would also not be acceptable.) The committee therefore recommended an American SST that was designed purely for over-water operations and a research program on an "economically viable overland" SST.[72]

In one sense, the report seemed a major setback for SST advocates. By restricting the SST to over-water flights, the SST's estimated economics deteriorated drastically. Therefore Maxwell disagreed with the "overall tenor" of the report, believing that its "sweeping conclusions" were not justified by the "limited data" provided by the Edwards tests.[73]

But politically and bureaucratically, SST advocates had scored another very important victory: they had successfully eliminated the threat of community overflight tests and had avoided an opportunity for triggering congressional and more massive public protests. Moreover the relatively poor sonic boom characteristics of the Boeing and Lockheed models were less significant since the aircraft would fly supersonically only over water. In the PAC the document legitimized the notion that the currently designed SST would be an over-water aircraft. The FAA, immediately sensing the report's bureaucratic benefits, informed the PAC that the agency in effect had accepted the prospect of sonic boom–induced route restrictions, but the agency was quick to add that the FAA still projected "a very dynamic" SST market.[74] At the December 7 PAC meeting Maxwell could calmly admit that both the Boeing and Lockheed SST designs exceeded FAA sonic boom objectives. He simply reiterated the FAA's assumption that henceforth SST operations would be limited to over-water and unpopulated areas.[75]

But in early December 1966, just as the debate over the sonic boom issue was apparently leveling off, a crucial related problem had emerged as equally significant: the SST's engine noise.

Both the Defense Department staff and Maxwell reported at the December 7 PAC meeting that an "optimum" over-water SST would be larger than current models and would generate enormously increased engine noise levels. "The bigger the airplane, the bigger the engine, the bigger the noise problem," Maxwell observed. "These are the things that are constraining our choices. It is not the sonic boom," he added. The noise characteristics of the Lockheed model appeared particularly poor, and airport operators seemed to view SST noise as a "severe problem."[76] By the next PAC meeting on December 10 the FAA had reported to the PAC that the SST's growth in weight was "motivated by the single factor of economics''; although the international air carriers, particularly Pan American, wanted better range and payload characteristics, such weight increases and related rises in engine size clearly had an adverse effect on noise and sonic boom characteristics.[77] To make matters worse, according to the FAA, pending congressional legislation would give the FAA authority to regulate noise and sonic boom levels for only new aircraft. The Concorde, as an ''old'' aircraft, therefore would have the advantage over an American SST of being constructed according to the less stringent current noise standards.[78] At the December 10 PAC meeting McKee stressed that the FAA did not have the authority to prevent current airplanes with unacceptable noise levels from operating and that pending legislation would give the agency such authority for only ''new aircraft.''

McNamara, who was becoming as disturbed about aircraft noise as he had been over sonic booms, was astounded that the government should lack such authority. He found the FAA's legal brief on this matter unconvincing; the PAC was already aware that the American SST and the Concorde probably would not be acceptable for overland flight; and this knowledge should be transmitted officially to the British and French governments to avoid any misunderstanding. Others were similarly disturbed. McKee then retreated slightly. He acknowledged that the current law could probably be stretched and asserted the country would ''obviously'' not permit operations by aircraft exceeding current sonic boom and noise levels.[79]

McNamara's pessimism about both sonic booms and noise was reflected in the fourth PAC interim report to the president, transmitted on December 22, 1966. The report found that the ''sonic boom and airport noise appear to be potential limiting factors on operations of all commercial supersonic aircraft'' and stressed the probability of route restrictions. It called for continued studies under OST direction, information dissemination to Britain and France, and ''necessary legislative or administrative steps . . . to insure adequate government authority to control excessive sonic boom and airport noise in the United States.'' The PAC formally put an end to the current controversy that had occupied most of 1965 and 1966: there was no need for community overflights to test the SST's public acceptability, given the negative findings from the Edwards tests.[80] On the other hand, engine noise had emerged as an overriding concern by this time.

These conclusions did not surprise the White House. Throughout 1966 Califano, Hornig, and others were warning President Johnson about the sonic boom and noise problems, and they were receiving continuous reports on the ongoing sonic boom tests.[81] In late December Califano highlighted the PAC report's key elements for Johnson, especially Hornig's belief that the SST's sonic boom would be "intolerable to millions of Americans." The sonic boom factor, Califano continued, would restrict the SST to over-water routes and would raise unit costs by reducing the number of aircraft produced.[82] Fully alerted to the ramifications of both issues, the president continued to avoid any comment or official policy on sonic boom and noise matters.

The FAA appeared victorious at the end of 1966: governmental bodies would continue their deliberations over the sonic boom, and massive community overflight tests had been averted. But the agency paid a certain price. It acknowledged the strong likelihood of SST route restrictions; the sonic boom debate had contributed to the SST program's delay; the sonic boom remained a nagging point of controversy; and engine noise had grown as an issue.

The Controversy Persists

In spite of the FAA's success at the end of 1966 in containing the sonic boom controversy within the government, it still faced the potentially damaging final report on the Edwards tests by SRI that was due in mid-1967. In addition, the agency took little comfort in an SRI proposal in February 1967 for overflights of two or three cities over a two-year period,[83] or in a generally favorable review by Karl Kryter, SRI's sonic boom project director, of a paper by sonic boom critic Bo Lundberg in March.[84]

It was not surprising therefore that the FAA again attempted to intervene in the actual drafting of the Edwards report, which was SRI's prime responsibility. Maxwell said that he had "serious reservations" about an initial draft, questioning the conclusions and data reporting procedures. He wanted the government simply to release sonic boom information without speculating about the implications of the data. He expressed his doubts about the worth of correlating acceptance to noise and sonic booms at various levels (the central methodology of the Edwards tests)—an approach that he said used "rubber yardsticks"—

and recommended that the findings be presented in a less definitive fashion. Ultimately finding the draft unacceptable, he declared that it should be completely rewritten with FAA assistance.[85] The agency continued to object to the report until its release in July 1967.[86] The FAA was supported by NASA's key representative for the sonic boom tests, who recommended in mid-April 1967 a strictly "factual" report, deleting "conclusions and judgments."[87]

Dunning held similar views. Maxwell had already complained to Dunning about the report, and Dunning in turn urged Hornig to limit its judgments and conclusions "only to those statements which are completely supportable by the data included in the report." He even suggested delaying the report in order to incorporate more data.[88]

The FAA met bitter resistance from OST and SRI. Hornig vigorously defended the report: "Every word of it had been gone over meticulously in my office and reviewed carefully with all the participants."[89] Although there was some delay in releasing the document, the original draft was never substantially changed.

The actual report, entitled *Sonic Boom Experiments at Edwards Air Force Base, Interim Report,* covered various topics, including structural response, meteorological factors, and animal response. But the primary controversy concerned the extremely discouraging psychological experiments, which correlated human reactions to sonic boom overpressures with human reactions to subsonic aircraft noise.[90]

Although the 1964 Oklahoma City tests and the 1966 Edwards tests were the most crucial sonic boom field study efforts in terms of SST decision making and debate, at least one other potentially important test occurred. The FAA took advantage of training flights of the air force's high-altitude supersonic reconnaissance aircraft, the SR-71, which were to begin on July 1, 1967 and to last until September 1967. Tracor, a consulting firm that was already engaged in a NASA study of community attitudes toward aircraft noise near commercial airports, was subsequently contracted to investigate community reactions to the SR-71 sonic booms. The study would deal with individual, family, neighborhood, and community attitudes.[91] Both the air force and the FAA carefully downplayed the sonic boom study. Maxwell argued that the principal advantage of the test was that it took place under normal operating conditions. No FAA public relations activity related to the SR-71 flights was permitted in order to avoid any bias in Tracor's sociological survey. Still the FAA and the air force remained in close contact during the test period.[92]

Certain conclusions of the Tracor SR-71 sonic boom study were hardly startling. For example, Tracor found that "respondents have a negative attitude toward the sonic boom, and this attitude increases rapidly in strength as the number of booms per day increases." But like Maxwell, Tracor did question the efficacy of the key Edwards tests methodology, correlating the acceptability level of sonic booms and subsonic noise. Although such a view might have cast new doubt on the Edwards results, the Tracor study actually had little effect during the critical period of sonic boom debate in the mid-1960s since it was not released by NASA until September 1970.[93] By that time the SST conflict had broadened considerably, involving many more participants and issues.

The sonic boom debate within the government in the mid-1960s raised the general level of concern about the SST and contributed to the significant delays in the whole SST program. The key leaders in evaluating and designing the sonic boom tests at this time—Hornig and Golovin of OST and Kryter of SRI—supported the idea of massive community overflight tests and wanted quick release of an Edwards report with broad conclusions. They were especially troublesome to the pro-SST forces.

Ultimately, however, the debate on the sonic boom did not delay the final selection of a prototype contractor at the end of 1966. Moreover, by the spring of 1967, with the FAA now absorbed into the new Department of Transportation, the SST supporters had more bureaucratic power. The larger agency appeared to provide greater management capabilities for undertakings such as directing sonic boom studies.

The SST advocates began attempting to transfer authority concerning the sonic boom study from OST to the Transportation Department. In June 1967 the FAA's own sonic boom evaluation functions were given to the department's noise abatement staff, and the new secretary of transportation, Alan Boyd, sought control of all activities related to sonic boom and aircraft noise. During the latter half of 1967 he implemented a plan that called for gradually assuming responsibility for nearly all sonic boom activities formerly under OST leadership. In late August 1967 Hornig formally agreed to the transfer of such functions. In early October 1967 Boyd consolidated the aircraft noise and sonic boom activities under a single interagency noise abatement program. The Department of Transportation now possessed effective control over nearly the entire sonic boom and noise evaluation system in the government. The department's office of noise abatement served as secretariat for the chairman of the new interagency noise abatement program's coordination committee. The most powerful panel of this program, which focused on the "operational procedures" to reduce subsonic aircraft noise and sonic boom effects, was chaired by an FAA official.[94]

The real internal debate over the sonic boom was now over, and pro-SST forces at the Department of Transportation appeared victorious. They had assumed the key positions of power, and they had the continued support of such allies as Webb at NASA and Dunning on the NAS Committee. The skeptics had either disappeared from the SST scene or had been relegated to roles of lesser authority. After the leadership role in noise abatement and sonic boom studies had shifted from OST to Transportation, Hornig, Golovin, and Kryter no longer had significant influence.

The pro-SST victory was, however, a Pyrrhic one. During the internal sonic boom debate in the mid-1960s, doubt had been planted in government and industry and began to grow. Even the most ardent SST advocates began to acknowledge that the SST could not operate supersonically over populated areas. This factor, in turn, reduced the SST's projected market and led to a further decrease in the SST's already marginal estimated economic performance. Moreover a new related issue, engine noise, was taking on greater prominence. The SST's sonic boom difficulties were beginning to be reported in the press, causing considerable concern at the FAA and creating diplomatic problems in dealing with the French and British.[95]

Most significantly, the sonic boom debate had consumed great amounts of the pro-SST forces' most precious resource: time. Like other controversial areas of the SST conflict, including economics and financing, the internal governmental debate over the sonic boom caused significant delays in the program, and by the time the issue appeared under control, it was too late. New doubts had arisen and a whole new set of opponents had begun to emerge.

11 Apparent Victory

The Last Competition

Although bureaucratic warfare raged over such issues as economics, financing, and the sonic boom throughout 1966, it was the prototype design competition that occupied the attention of most FAA SST-related personnel. Staff members spent the greatest amount of their time managing the evaluations of the engine designs submitted by General Electric and Pratt & Whitney and the airframe designs of Boeing and Lockheed.

The FAA's record in this area was not particularly inspiring. Previous FAA recommendations for a single airframe-engine combination had been rejected in favor of extending the competition. The outcome of the 1966 competition therefore would be a key indicator of William McKee's and Jewell Maxwell's ability to reestablish confidence in the FAA. Selecting a final prototype design had become the most important SST objective of the year for the FAA.[1]

The final competition began at the end of 1965 with an interim assessment of airframe designs. Lockheed, which had previously been a very weak competitor, had improved its design considerably, while Boeing's design had begun to show some fundamental configuration problems. As a result, the competition between Boeing and Lockheed unexpectedly tightened, and the FAA's hopes for an early airframe selection before the end of 1966 vanished.[2]

To bolster its shaken confidence, Boeing organized a thorough reappraisal of its SST design. By late January 1966 it had restructured and strengthened its entire SST effort by elevating its SST group to division status, appointing a new general manager and a new chief engineer, and transferring much of the company's top-level technical personnel to its new SST division. Boeing then became increasingly optimistic when communicating with the FAA.

Boeing's action clearly pleased the agency. Maxwell remarked that he was looking forward to "a real horserace of a competition." On the other hand, the firm had not regained all of its past goodwill. Some FAA members remained convinced that Boeing was more concerned with winning the competition than with designing the ideal aircraft and that its numerous design experiments would lead to further scheduling delays.

Even more significant, Boeing's recent decision to finance the 747 jumbo jet privately raised suspicions in the press and in Congress; if the company were financing the 747, should it not be required to support more, if not all, of the cost of an SST prototype? Boeing's president, William Allen, responded by stressing the SST's economic and technical uncertainties, the comparative reliability of subsonic technology, and the favorable 747 market predictions. He even suggested that 747 development would help train SST managers and would generate more capital to finance SST production.

Still by early May 1966, in spite of ongoing criticism, Boeing's status had again risen as its design improved, though the gross weight of its SST configuration, which incorporated a swing wing and a very large horizontal tail that could be integrated with a fully swept-back wing, had increased substantially from 500,000 pounds to 750,000 pounds.[3]

Lockheed meanwhile was attempting to capitalize on its recent show of strength. During the first third of 1966 the company concentrated on refining its basic model, a fixed-wing, triangular, delta design. This version was referred to as a double-delta configuration because it consisted of a smaller triangular wing in the front with a larger one behind. But Lockheed too encountered problems. By March its design's range-to-payload characteristics had deteriorated, with its payload for a 4,000-mile range decreasing from 40,000 pounds to 30,000 pounds. At the same time, the model's gross weight had increased to 550,000 pounds, 10 percent higher than Lockheed's November 1965 estimate. Lockheed even requested permission to modify its contract with the FAA to allow these worsening estimates. Its design also appeared to be vulnerable in such areas as aircraft noise and the sonic boom. Nevertheless, Maxwell advised the PAC members on May 6, 1966 that the FAA was "satisfied," that Lockheed was making "good progress," and that he thought Lockheed would be able to resolve its remaining design problems.[4]

Both Lockheed and Boeing were having difficulties with their increasing gross weight estimates, changes that seriously affected the SST engine work by General Electric and Pratt & Whitney because a larger airframe called for greater engine power and thrust. But the FAA was obviously more concerned about airframe development than about engine work since the agency was previously satisfied with General Electric's turbojet and now claimed that Pratt & Whitney had also made significant improvements in its turbofan design. The disparity in engine price estimates—$0.9 million for General Electric and $1.4 million for Pratt & Whitney—also did not cause the FAA much concern. Maxwell felt that the first was an underestimate and the second an overestimate and that a final evaluation would bring them closer together.[5]

As the FAA worked during the first five months of 1966 to establish organization and evaluation criteria for the final competition, the competition itself became intermeshed with the ongoing bureaucratic struggle within the SST program, again pitting the FAA against Stephen Enke.[6] Enke pressed for greater PAC involvement, claiming that the committee was the most impartial and competent body to judge the designs on the basis of all key factors, including economics and the sonic boom.[7] But Robert McNamara appeared singularly unenthusiastic about PAC involvement in the competition because he did not want to interfere with the FAA's direct SST management responsibilities. The PAC basically let the FAA continue directing the competition for the rest of 1966.[8]

In contrast, the FAA wanted active airline involvement in the prototype design evaluation. McKee and Maxwell encouraged the formation of an Airline SST Committee in October 1965, consisting of representatives from almost all major American-flag airlines, and the FAA maintained a close relationship with this committee. At Maxwell's urging, the committee established teams of airline specialists for the evaluation.[9] Maxwell proudly told the PAC that this type of cooperation would open up "a real communications channel"; the FAA's work with the airlines was "perhaps one of the best things we have done."[10]

But economic and financial factors were always of great concern for the airlines. With the expected introduction of the 747, they feared being stuck with a new fleet of subsonic aircraft just as the public was turning to the SST. "Were it not for the SST," one airline official stated publicly, "we could see all our way clear to buy the larger [subsonic] jets right now."[11] For example, Pan American, the airline most heavily involved with purchasing 747s, was the last of ten American-flag air carriers to participate actively in the SST competition.[12] The airlines were also worried about excessive government involvement, were solidly opposed to government profit sharing, and favored a dual prototype effort.[13] (Their concern about government intervention evoked a highly sympathetic response from the FAA.)[14]

The July 9 PAC meeting provided an unusual opportunity to hear the unbridled views on the SST of key airline officials. Juan Trippe, chairman of Pan American, and Pan American board member Charles Lindbergh had come at McNamara's invitation to offer their opinions. McNamara had always regarded direct discussion with the airlines as extremely valuable, and he was especially interested in Pan American's suggestions because of its close association with the Concorde and its pioneering role with the subsonic 707 and the 747.

Trippe expressed great optimism about the 747, of which Pan American had already ordered twenty-five. He mentioned a number of the 747's advantages, such as low operating costs and an attractive cabin design; Pan American had even agreed to raise $600 million to help finance 747 production. But surprisingly Trippe was not enthusiastic about the Concorde. He now acknowledged that Pan American maintained its Concorde purchase options as "insurance" to support its market position and to "stay in the ring" with the Europeans. In contrast, he felt that both Boeing and Lockheed had proposed well-designed airframes, and he and Lindbergh strongly favored continuing the competition into the prototype phase. Referring to the "simplicity" of Lockheed's double-delta design, to the "efficiency" of Boeing's swing-wing configuration, and to his general "apprehension" about the whole program, Lindbergh admitted that at the moment he "would not know which [design] to take." (Later that fall, on grounds of technological uncertainty, Trippe publicly favored the construction and flight testing of both prototype designs.)

Regarding financing, Trippe proposed to the PAC that airlines with reserved SST delivery positions contribute $1.5 million per position, adding that this would demonstrate to Congress that the airlines seriously supported the SST. He also thought it essential that foreign air carriers be included.[15]

In September 1966 the various airline specialist teams submitted detailed design reviews, which generally supported the SST Office's technical conclusions. These airline reviews also helped the FAA politically by demonstrating strong airline SST support, and the FAA made it a point to praise effusively the work of the Airline SST Committee and its specialists.[16]

By late November ten U.S. airlines and over twenty foreign air carriers had submitted evaluations of the manufacturers' submissions. The quality of these reviews varied. As McKee told the PAC at its November 17 meeting, some were "not worth the paper they were printed on," while "others have been extraordinarily well done." He particularly praised those of Pan American, TWA, and United. All of the American-flag airlines urged proceeding with prototype construction. American, National, TWA, United, and Eastern recommended a single prototype program and suggested a choice of manufacturers. American chose the Boeing airframe and the General Electric engine, Eastern favored Lockheed and dual engine development, and National backed a Lockheed-General Electric combination. Continental, Northwest, Braniff, and Delta recommended a dual prototype program. Qantas and Air Canada also submitted detailed reports, but no foreign airline made any clear selection. Pan American, the most important of all the airline participants, continued to play cat-and-mouse. Although it had initially urged a dual prototype effort, it eventually admitted that it favored the Boeing design.[17]

Other aspects of the competition were going well from the FAA's perspective. Boeing's design had advanced significantly by the end of June. Its new swing-wing configuration, called the B-2707, had a greatly enlarged horizontal tail and exhibited a single integrated surface when its wings were fully swept back; the engines were mounted directly on the tail. Boeing claimed that the model's direct operating cost would be less than those of current subsonic jets. Still its gross weight had grown considerably, primarily because of airline insistence on greater fuel reserves.[18] Lockheed's basic SST configuration did not undergo any major changes, although its gross weight gradually increased to 590,000 pounds.[19]

Generally the FAA appeared extremely satisfied with the manufacturers' progress. By July it had become evident that the Lockheed airframe, with its lower aerodynamic efficiency at subsonic speeds, was better suited to the Pratt & Whitney turbofan engine, which had greater subsonic thrust. At the same time, the performance characteristics of the Boeing model made it a better match for the General Electric turbojet engine. Lockheed named the Pratt & Whitney engine as its primary choice in early September. Boeing was more cautious, waiting until November to voice its preference for the General Electric model.

Both airframe manufacturers, however, were prepared to use either engine.[20]

The promotional efforts of the two manufacturers contrasted markedly. Boeing was surprisingly subdued, merely trying to relate its SST effort to its obvious successes in developing commercial jet aircraft.[21] Lockheed's promotional behavior, on the other hand, was flamboyant, public, and aggressive. In mid-July William M. Magruder, Lockheed's assistant manager of its SST effort, extolled the technical virtues of the Lockheed design at public and private conferences and in the media, placing particular stress on Lockheed's unique supersonic expertise. In August a two-page Lockheed advertisement appeared in the *New York Times* that claimed that the firm's double-delta configuration would be "virtually stall-proof" and would create a "cushion of air" under a landing plane; this model would incorporate aerodynamic principles that had already been tested in Lockheed's SR-71, a supersonic, high-altitude reconnaissance aircraft, which it termed "the flying prototype of America's SST." Lockheed's president and other company officials similarly lost no opportunity to stress the firm's experience with high-speed, high-altitude aircraft. Lockheed also actively supported the FAA's own ongoing pro-SST public relations campaign.[22]

The structure of the competition evaluation was taking definite shape, and it would be more elaborate and thorough than any of the other previous SST reviews. In the spring SST Office personnel were assigned specific roles, the scoring technique was refined, and the rules for an economic model were drafted. The FAA decided to involve potential SST customers from abroad since they accounted for at least 50 percent of the total estimated SST market; their inclusion would also generate considerable goodwill. By July such government agencies as NASA, the Defense Department, and the CAB had formally agreed to supply experts for the evaluation.[23] At the July 9 PAC meeting Maxwell was able to report that the evaluation was "proceeding according to schedule."[24]

The manufacturers' proposals were submitted in early September, and the technical evaluations began immediately. The overall evaluation structure grew increasingly elaborate. A source selection advisory staff was formed as an internal FAA advisory unit reporting directly to Maxwell. The group helped organize and schedule the complex evaluation activities, prepared briefing documents on the evaluation for high-level FAA officials, and identified future tasks. An interagency unit, the source selection evaluation group, was established to perform technical evaluations, with approximately 235 government experts participating.[25]

The FAA also made use of outside technical evaluations. The Cornell Aeronautical Laboratory had been hired to compare the characteristics of the swing-wing SST design with the fixed delta-wing model, and Raymond Bisplinghoff of MIT had been called in to advise Maxwell on a number of technical matters. In mid-November, at Maxwell's suggestion, Bisplinghoff formed a seven-member SST advisory board, consisting of well-known technical experts from several universities and research institutions, that would report directly to McKee and Maxwell.[26]

In September the FAA created still another group, the source selection council, comprised of several high-level FAA officers. The group prepared the supporting material for McKee's final competition decision in December. Not surprisingly the council "conclusively" found that the technology existed to build a safe and reliable SST and that estimated operating costs indicated that an economically feasible SST could probably be produced.[27]

Management control was clearly important to McKee and Maxwell, experienced and respected project managers who wanted to avoid the kind of charges of loose or ineffective management that had plagued the FAA in 1963 and 1964. Therefore in 1966 the FAA instituted elaborate management and control procedures that enhanced the credibility of both the agency and the evaluation. Maxwell, relying on his air force administrative experience, applied various Defense Department and NASA managerial techniques to organize the evaluation's scheduling. He also sent daily summaries of all evaluation activities to McKee and held daily evaluation staff conferences. At PAC meetings Maxwell dwelled on the evaluation management structure that he had devised, presenting complex flowcharts. At the October 6 PAC meeting he unfolded a "network" chart, which, he said, hung in his "control room," and, using this chart, he discussed the evaluation's organizational structure in characteristic detail. Information flow was also controlled. The four manufacturers were directed not to release publicly any information that "could be construed as being intended to influence anyone's judgment,'' a move that was criticized as censorship but that remained in force throughout the evaluation. In October Maxwell even refused to allow a representative from the CIA, which was monitoring the Concorde and Soviet SST programs, to attend an airline briefing.[28]

The technical evaluations by the government interagency group were submitted to McKee on November 1 and transmitted to the PAC in early December. The match in both size and engineering features between the Boeing airframe and the General Electric engine and between the Lockheed airframe and the Pratt & Whitney engine was readily apparent, and the FAA henceforth focused its comparison on these design combinations. Three areas of particular concern had emerged: the sonic boom, the increase in gross weight and the consequent deterioration of the range-to-payload ratio, and high aircraft noise levels. There seemed to be a fundamental design trade-off between reducing aircraft noise and sonic boom levels and improving the SST's range and payload characteristics by increasing its gross weight.

Even the FAA acknowledged that the most important factor affecting the SST's economics was ''whether the aircraft can be operated over populated land areas because of sonic boom''; route restrictions could reduce the market by as much as 64 percent. To make matters worse, technicians using a new, more accurate form of measurement calculated that sonic boom intensities would be far greater than previously estimated. Although Boeing's design had considerably better sonic boom characteristics than Lockheed's (2.39 psf for climb acceleration, 2.06 psf for cruise, and 1.67 psf for descent for the Boeing-General Electric international SST model versus 3.05 psf for climb acceleration, 2.10 psf for cruise, and 2.17 psf for descent for the comparable Lockheed-Pratt & Whitney model), both SST models generated intensities that exceeded government limits, and public acceptance of such levels was considered unlikely. Consequently all evaluation estimates presented to the PAC assumed that SST operations would be restricted to over-water or unpopulated regions.

The SST's increase in gross weight continued to worry the FAA because the airlines had demanded greater range-to-payload capabilities. According to the FAA, the SST's most important performance characteristic was its ''ability to carry large payloads over long ranges at high speeds.''

The SST's engine noise, too, appeared to have been underestimated, and the FAA and most airport operators considered this problem to be quite severe. The Boeing-General Electric model came reasonably close to airport noise limits, but the Lockheed-Pratt & Whitney design had noise levels that were considerably higher, and it seemed that little could be done to correct this.

With regard to the engine evaluation, the Pratt & Whitney turbofan engine generally operated more efficiently at subsonic speeds than did General Electric's turbojet design, while the opposite was true at supersonic speeds. The FAA also claimed that the total available engine thrust at takeoff and for transonic speeds (crossing the speed of sound) was smaller for the Pratt & Whitney model than for the General Electric engine.

In spite of the remaining problems, prototype construction appeared to be close. The FAA's careful management of the evaluation seemed about to yield results. At the December 7 PAC meeting the basic prototype construction choices discussed were either to select a single airframe-engine design immediately—a less costly and time-consuming but technologically riskier choice—or to continue ahead with two prototypes and eliminating one after one hundred hours of flight testing. But whichever plan was chosen, SST supporters felt that at last a prototype would soon be under construction. Ultimately the PAC decided in favor of constructing a single prototype design.[29]

Final Decisions

The evaluation findings, however, soon became enmeshed in the ongoing bureaucratic battle. The Defense Department's SST economic staff was particularly critical, focusing on the high noise levels associated with delta-wing aircraft and flatly rejecting the dual prototype plan as uneconomical. The group favored extending the design competition as the best means of improving the aircraft and of keeping funding at levels low during the uncertain early stages of prototype construction.[30] But with the exception of Treasury Secretary Henry Fowler, who also favored continuing the design competition, most PAC members at the December 7 PAC meeting agreed with McKee's recommendation to proceed with a single airframe-engine model.

Other obstacles remained. McNamara was still pessimistic about the SST's technical and economic prospects and flatly admitted that he was only interested in the SST "from the point of view of profitability." He asked Budget Bureau director Charles Schultze to "validate" the FAA's claim of SST economic success. McCone envisioned operational problems. He recalled the severe problems that Lockheed had experienced with the Mach 3 SR-71, which, he said, took "almost an Act of God to get . . . off the ground," and he predicted that the Concorde would also undergo unexpected difficulties. He was especially disturbed by the American SST's considerable increase in weight and claimed that the current SST configuration and the initial designs were "entirely different animals," in terms of payload and engine thrust.

It was the airlines, however, that emerged by the December 7 PAC meeting as the most significant delaying factor in the program. They pressed for greater amenities and fuel reserves than specified by the FAA and presented widely varying estimates of operating costs per seat-mile, which according to the Defense Department merely confirmed both the great uncertainty of all SST economic estimates and the FAA's faulty statistical methodology.

Above all, airline financial participation became particularly critical. McNamara insisted on December 7 that a greater manufacturer and airline risk capital involvement was both a prerequisite for prototype development and a political necessity; such risk capital would be needed for congressional approval of the program. These views were shared to greater or lesser degree by most PAC members. The airlines' hesitant attitude toward financing was reflected in statements made by TWA's president, Charles Tillinghast. He envisioned a bright economic future for the SST, which he claimed would be extremely popular with "big business" and "wealthier" travelers who "just won't be seen in an ordinary jet." According to Tillinghast, the SST would evolve into a status symbol because "people like to be associated with the newest, fastest, and the shiniest"; even the Concorde, which Tillinghast said had "lousy seat-mile economics," would fly 75 to 80 percent full with a premium fare. But Tillinghast's enthusiasm diminished considerably when it came to airline financial participation. He claimed that the airlines' current financial burden was significant, especially with Boeing pressuring them to enlarge their subsonic fleets with 747s.

McNamara, however, bluntly informed Tillinghast that if the airlines did not invest risk capital in the SST program, Congress might reject the fiscal 1968 SST budget request of $450 million.[31] At the PAC's next meeting on December 10 McNamara again stressed that financial participation by the airlines and manufacturers was necessary for the coming prototype development phase. He adamantly refused to go ahead unless there was a "reasonable" chance that the aircraft would be commercially profitable. He did not want to be paced by the "fever" of Concorde competition and flatly declared that fear of Concorde competition should not influence the American SST effort, especially in view of such continuing design problems as noise levels, operating inefficiency, and the deteriorating payload-to-weight ratio.

The FAA opposed making airline financing a condition for entering prototype construction as it opposed any other factor that might lead to further delay. McKee did not want the SST contracts with the manufacturers to be contingent on such airline involvement because negotiations with the airlines would probably take months; airline interest was certain to increase anyway after contracts had been signed and the prototype phase had been announced.[32]

But McNamara again, for the last time, had his way. Over McKee's objections,[33] the fourth and last interim report of the PAC, which was transmitted to the president on December 22, strongly reflected McNamara's views. It was not an optimistic document. It stressed "commercial feasibility as the ultimate criteria for success"; it recommended that the prototype contracts require each manufacturer to bear a "significant 'cash' equity risk" in the program and provide "normal" commercial incentives and penalties for the manufacturers; and it strongly backed airline financial participation at an "early stage of prototype development." The report's appendix, a status report on the SST program, was even less enthusiastic than the report itself: the threat of both the Concorde and Soviet TU-144 programs was minimized; it predicted that neither the American SST nor the Concorde would be able to fly overland at supersonic speeds because of the sonic boom; the SST's estimated overall rate of return appeared "far too low for the high risks involved"; and the aircraft's projected balance-of-payments impact seemed unfavorable.[34]

The report's general skepticism was not lost at the White House. Joseph Califano, formerly a close aide to McNamara and now the key presidential assistant for SST affairs, made sure that the president noticed that although the report contained "six unanimous recommendations to move ahead with the development of the SST," it also exhibited "several pessimistic overtones." Califano dwelled on the Senate's possible opposition to SST funding, on the difficulty of obtaining public acceptance of sonic booms, and on the SST's doubtful commercial viability. Like McNamara, he also minimized the threat of the Anglo-French and Soviet SSTs.[35]

Fearing even greater delays, the FAA continued to argue against the PAC's recommendations for manufacturer and airline financial participation. (For manufacturers, the PAC recommended 10 percent cost sharing for development costs and 25 percent for cost overruns).[36]

Airline equity participation, especially, was hotly debated. Striking fear into the FAA, McNamara implied that requests for additional SST appropriations for fiscal 1968 would be temporarily delayed in order to increase the pressure for greater manufacturer and airline financial participation. In response, if the PAC did not recommend additional SST appropriations for fiscal 1968, Maxwell angrily declared, he would urge cancellation of the whole SST program. Similarly McKee told McNamara that such a delaying tactic would be a "serious mistake" and would result in a "loss of momentum."[37] He later told Califano that it would be politically unwise and wasteful to extend the current design contracts of all four manufacturers.

At the White House, as usual, Califano agreed with McNamara. He said that the president should have more than four or five days to make a complex set of decisions that involved potentially $5 billion. He had been informed that a portion of the SST funds already awarded could be used for extended detailed design work in 1967, which eliminated the pressure on the president to act by the end of 1966.[38]

But McKee had not completely given up. On December 29 he sent the president a letter that again outlined his arguments against extending all four contracts.[39]

Johnson ultimately gave McKee only part of what he desired. On the morning of December 30 McKee expected Johnson to make an SST go-ahead announcement at a December 31 press conference. McKee even directed his information officer to hold an FAA press briefing immediately after this conference. But later that day McKee was informed by the White House that the FAA alone would make an SST announcement. When questioned, Johnson merely said that he would have an SST announcement "shortly" and that the administration had no definite date for a decision to begin prototype construction. But a design selection was made: the Boeing-General Electric model was chosen over the Lockheed-Pratt & Whitney design. It was still left to the FAA to inform the four competing firms of this action.[40]

Although there was natural disappointment at Lockheed, the company looked forward to recovering substantial cost-sharing funds that it could apply to other projects, particularly in the subsonic commercial aircraft field. Even at Boeing the reaction was restrained. William Allen declared that he would make no optimistic predictions or hire any new personnel until there were firm government commitments.[41]

The administration's handling of the key SST actions at the end of 1966 followed a typical pattern. Important areas of debate—financial participation, economic feasibility, balance-of-payments effects, and the sonic boom—that had been discussed endlessly disappeared, at least for the moment, without really being resolved. In addition, although other issues, such as the design competition, seemed settled, new controversies, such as airline financing and engine noise, emerged.

The design selection had been made, but the SST conflict was by no means over. At the end of 1966 none other than the FAA's old protagonist, Stephen Enke, who was no longer in the government, declared publicly his opposition to developing the SST because the aircraft would not be profitable.[42] A great many crucial points of controversy lurked just below the surface, ready to appear if given sufficient cause—for technical, bureaucratic, political, or personal reasons.

12 Hiatus

In spite of several unsettled issues at the beginning of 1967, the choice of the Boeing–General Electric model generated a new sense of optimism. The long process of design competition had apparently come to an end, and it seemed that prototype construction would soon begin. *Aviation Week and Space Technology* captured this mood, declaring that the contractor selection gave the SST program "a more clearly defined shape" and would allow the aircraft to become "a firm competitor in the international market": "It should be clear to the airlines of the world, particularly those who participated in the final design selection, that the U.S. program is indeed a sound technical effort that will press forward at the best possible pace to certification and airline service." The journal praised the leadership of William McKee and Jewell Maxwell and acknowledged the need for a "hiatus" between the selection of the prototype contractors and the actual beginning of prototype development, adding that Boeing still had to incorporate certain recent design changes before it could start work on the prototype.[1] Significantly the journal failed to mention two key obstacles: manufacturer cost sharing and airline financial participation.

Encouragement also came from a number of prominent individuals. Elwood Quesada, FAA administrator during the Eisenhower administration, wrote McKee in mid-January in support of SST development and of government assistance to industry. "The dynamic drive that you [McKee] and the agency [the FAA] are providing the SST Program is a source of personal satisfaction to me," added Quesada.[2] Vice-President Hubert Humphrey, chairman of the National Aeronautics and Space Council, declared in mid-February, "I don't think we ought to be second best in anything that we can be first best in. We know from the economics of the SST that it can be designed, that it can be marketable, and it can—over a period of time—repay the government every nickel that has been put into it for design and for prototype and for engineering. . . . I think we ought to build it and get on with the job."[3] Najeeb Halaby, then a high-level executive with Pan American, exclaimed, "Supersonic transportation is on its way, and it's the next logical, normal, attainable step in the evolution of aviation." Senator Jacob Javits (R-New York), after stating that he supported SST construction because of its "revolutionary effect on air commerce," immediately inserted excerpts from Halaby's speech into the *Congressional Record*.[4] Top airline executives also expressed general satisfaction. Marion Sadler, president of American Airlines, praised the FAA's management of the evaluation and pledged his future support. And Charles Tillinghast of TWA publicly declared that the SST was coming "whether we like it or not."[5]

The greatest enthusiasm for the SST, not surprisingly, was within the FAA itself. In early January the deputy director of the SST Office characterized the SST as the ideal aeronautical research and development project for the United States to undertake.[6] Alan Boyd, the new secretary of transportation and McKee's new superior (the FAA had become part of the new Transportation Department at the end of 1966), assured the National Press Club in Washington that "the U.S. will build a supersonic transport that is commercially viable, successful in every sense."[7]

But this optimism was again premature. Prototype development had not yet begun, contracts with Boeing and General Electric had not been signed, and sufficient airline financial participation had not been pledged. The key reason for this delay was political. McNamara, the White House, and others were convinced that congressional support for a fiscal 1968 SST funding request, needed for prototype development, was not totally guaranteed unless there was significant manufacturer cost sharing and airline financial participation during the prototype development phase. Above all, President Lyndon Johnson demanded certainty of the outcome of any SST vote on Capitol Hill before he would act. It was in relation to the airline financial participation question that Johnson as president played his most active and overt role in the SST program.

There was ample reason for concern. A surprising growth of congressional opposition to the SST, especially in the Senate, had emerged as a key bottleneck in the SST program. The SST opponents included Senator William Proxmire (D-Wisconsin) who for years had publicly opposed the SST because of its high cost and its technical and economic uncertainties. He called for total private SST financing and demanded access to government SST economic studies. In addition, as early as 1963 William Fulbright (D-Arkansas) had ridiculed the notion that maintaining national prestige hinged on winning the supersonic race with Great Britain and France. Stuart Symington (D-Missouri) believed that limiting the SST program was a good way to reduce federal expenditures and favored building a Mach 2 aircraft similar to the Concorde. Robert Kennedy (D-New York) argued that more money should be spent on social programs and less on the SST. Ernest Gruening (D-Alaska) expressed similar views, maintaining that the government should not interfere in what was essentially a commercial project. George McGovern (D-South Dakota) agreed, also raising questions of safety, economic feasibility, and the need for the FAA to focus more attention on the airport congestion problem. Senatorial opposition was not limited to Democrats. George Aiken (R-Vermont) considered the expansion of a nuclear-powered naval force more important than the SST, and John Cooper (R-Kentucky) questioned the wisdom of funding the SST at a time when huge sums were being expended on the Vietnam war effort.

The House also contained SST opponents. Congressman Frank T. Bow (R-Ohio) proposed a bill that called for private SST financing and the establishment of a government corporation. Congressman Clark MacGregor (R-Minnesota) supported the Bow bill.[8]

The FAA also did not receive significant public support from the Defense Department. "Except in the most indirect way," McNamara publicly commented, "I anticipate no direct military benefits from the production of the SST. My interest in the project . . . is to see a successful development of a financially profitable enterprise, and I personally don't believe we should proceed with it as a national program unless it can be developed to be financially successful."[9]

The enormous pressure to limit budgeting due to the ever-increasing demands of the Vietnam war plagued the FAA. McKee learned from Budget Bureau director Charles Schultze that his agency probably would not be allocated the entire remaining fiscal 1967 SST budget of $200 million and also discovered that $250 million probably would not be forthcoming in fiscal 1968, which, McKee had claimed, was the absolute minimum needed.[10] Hence appearing sufficiently tough in cost-sharing negotiations, especially to the Bureau of the Budget, was essential for the FAA.[11]

The critical issue at this time centered on whether the prototype contracts with the firms required cash contributions or whether such noncash accounting methods as overhead allocation on existing facilities would be allowed.[12] At two cost-sharing conferences in mid-January Robert McNamara and McKee were the main protagonists. McNamara initially demanded greater hard cash participation, especially in view of the large amount of private funds Boeing had recently committed to its 747 program. McKee wanted to avoid the potentially protracted delays that reopening contract negotiations might involve. In the end, McNamara agreed to accept prototype cost-sharing terms under which manufacturers would assume 25 percent of the cost of class II design changes—changes that the manufacturer was at full liberty to make as long as the terms of the original contract were not affected—and 10 percent of the cost of class I design changes—changes in basic performance objectives, which required prior FAA approval. Both the manufacturers and the FAA could accept these conditions.[13]

In the meantime the debate over airline financing had become more critical to the continued progress of the entire SST effort, for essentially political reasons. The president was adamant in delaying the prototype development phase until the airlines committed additional risk contributions in order to prove to Senate skeptics that the airlines were truly confident about the SST.

The FAA fought this position. McKee declared to Schultze that the decision to enter prototype development "should not be contingent upon significant cash contributions . . . by the airlines at this time."[14] Maxwell defended a view that TWA's Charles Tillinghast had recently expressed at a December 1966 PAC meeting that the airlines could not support additional large investments because of already heavy financial commitments to new subsonic aircraft and because of considerable uncertainty about regulatory and tax policies regarding SST contributions. But even more important for McKee and Maxwell, the demand for airline financing was preventing entry into the prototype development phase. In a draft memorandum to the president, McKee argued that it was the wrong time to force the issue and that he was sure that substantial airline contributions would be secured "in the not too distant future," after the sales agreements between the airlines and the manufacturers had been settled. McKee wanted to proceed directly into prototype development without month-to-month funding.[15]

But throughout January McNamara, Schultze, and the White House pressed the FAA to devise some method for obtaining airline risk capital for the prototype development phase. In mid-January McNamara said categorically that the SST program would not be approved by Congress without airline risk participation and urged Boeing to take the initiative in obtaining such contributions.[16] At the end of January, Schultze provided substantive support for McNamara in a detailed memorandum that he sent the president. Schultze strongly favored airline risk financing, noting that the airlines were in a financial position to contribute and that their share, at $1 million per aircraft reserved, would be small in terms of their overall cash flow. He argued that airline investment would be a reliable indicator of SST prospects and that airline participation would increase the SST's "success potential" by tending to hold down costs. He therefore backed a further delay in requesting fiscal year 1968 SST funding until the financial issues were settled.[17]

This advice, as well as that of McNamara and Califano, fell on receptive ears; at the end of January Lyndon Johnson met with McKee and again emphasized the need to obtain airline risk financing for the prototype.[18] Bowing to White House pressure, in early February 1967 McKee was forced to inform Congressman George H. Mahon, chairman of the House Committee on Appropriations, that he was extending the ongoing design phase contract with Boeing and General Electric on a month-by-month basis and that he would be using part of the $200 million that had originally been earmarked for the prototype development phase.[19]

Although the FAA made negative comments on the likelihood and efficacy of airline risk financing, it actually had received certain positive signals from the most critical airline, Pan American. The airline was becoming increasingly optimistic. As late as October 1966 Pan American still favored a dual prototype program, but in early December it came out in support of the Boeing model, and a few weeks later it substantially lowered its estimated direct operating costs per seat-mile for the Boeing design so that now the aircraft's seat-mile costs were 20 percent less than those of the Concorde for the Paris-to-New-York run.[20] Even McKee and Maxwell were taken aback, believing that these estimates were somewhat out of line with the facts; the airline seemed to have gone from excessive pessimism to excessive optimism. Still McKee was not shy in using Pan American's new data, quickly telling McNamara that the airline's estimates confirmed the FAA's own economic conclusions and indicating that the SST would be competitive on North Atlantic routes.[21]

Even more significant, Pan American offered the FAA in mid-

January an important proposal for airline prototype financing: the airlines would deposit $1 million per airplane position eight years before delivery and in return would receive a $3 million credit against the purchase price of each airplane. The $1 million would be "at risk"—that is, not returnable if the SST program were terminated—and could be used, if lost, as an income tax deduction or claimed for investment credit purposes upon delivery.

The proposal was a breakthrough, and McNamara quickly brushed aside any technical objections to it.[22] The intense pressure for airline risk participation was beginning to have a discernible effect. At a February meeting between top airline executives and the FAA, the airlines agreed to contribute $1 million for each SST delivery position. McKee quickly informed the president.[23] By mid-February ten of the twelve U.S. airlines holding delivery positions had agreed to participate in the FAA airline contribution program.[24]

As the news of airline commitment surfaced in the press, optimism rose among SST supporters. *Aviation Week and Space Technology* predicted that President Johnson would shortly announce the beginning of the prototype development phase and would request from Congress about $250 million for fiscal 1968 SST funding. McKee was further encouraged by McCone's promise to promote the SST with Stuart Symington, a key Senate voice on aviation matters.[25]

But again optimism was premature. American-flag airlines feared that Boeing might work out more generous arrangements for selling SSTs abroad and wanted to be certain, through a so-called favored-nations clause in any contract with Boeing, that they would be treated as favorably as foreign air carriers.[26] At first Boeing hedged, but in response to tremendous pressure by the FAA[27] the company eventually capitulated. The standard SST base sales price for participating American-flag airlines with delivery positions would not exceed the equivalent price for nonparticipating airlines.[28]

The airline financing problem finally appeared settled. By mid-March Boeing had negotiated agreements with ten airlines for a total of $52 million—$1 million for each SST delivery position—and by mid-April a payment schedule for these risk contributions had been established.[29]

The FAA still had to formulate a general policy for foreign air carriers and for delivery allocations. As of early March approximately half of the 115 SST delivery positions were held by foreign airlines. The

FAA had already received $100,000 per delivery position ($11.5 million) and would receive another $100,000 per delivery position no later than six months after the signing of the prototype development contracts. Although the FAA felt it had no option but to let these old agreements stand, it decided in late May, after extensive review by various agencies, that all new SST delivery positions would require a $750,000 risk payment. The agency also invited foreign air carriers to make SST risk contributions under the same terms as those established for American-flag airlines: $1 million for each position held, with a potential return payment of $1.5 million per position out of future government royalties.[30]

In early June the FAA publicly announced its new policy on delivery priorities. By this time 113 SST delivery positions had been allocated to twenty-six airlines, with fifty-seven positions belonging to twelve American-flag airlines and fifty-six belonging to fourteen foreign air carriers. After nearly half a year of negotiation and frustration, a comprehensive new SST reservation policy apparently had been established.[31]

Many of the delays that characterized the first five months of 1967 were political in origin since the president demanded that congressional approval for his fiscal 1968 funding request be certain. Consequently even though the FAA had obtained preliminary airline financing agreements in early February, President Johnson still called for "a better reading on the Senate" before he would enter the prototype development phase and request fiscal 1968 funding. He directed Secretary of Transportation Alan Boyd to take a careful vote count.[32]

The White House and the FAA expended a good deal of effort trying to determine the precise outcome of the forthcoming Senate SST vote. In early March Boyd submitted what he termed a "conservative" vote count that listed fifty-eight senators supporting the SST, twenty-two undecided, uncertain, or unknown, and eleven opposed. Now certain of a favorable outcome, Boyd recommended executing the prototype phase contracts with the manufacturers and requesting the fiscal 1968 SST funding.[33]

But Johnson remained cautious, declining to make the SST announcement and asking Califano to review the entire Senate situation with McNamara and Schultze. Boyd then provided Johnson with an even more favorable poll: sixty-one senators supported the SST program, eleven leaned toward it, seventeen were uncertain, undecided, or unknown, and eleven were opposed. Again Boyd urged a go-ahead. But Califano was not so optimistic, and he drew the president's attention to several "key senators" who had not yet committed themselves, including Russell of Georgia, Symington of Missouri, Aiken of Vermont, Case of New Jersey, Edward Kennedy of Massachusetts, and Robert Kennedy of New York. With Johnson's approval, Califano directed Boyd to "take another run" at those senators in the undecided, leaning for, and opposed categories.[34]

On March 20 Boyd submitted yet another vote count; sixty-three senators now supported the program, twelve leaned toward it, fifteen were uncertain, and ten opposed. A fourth poll taken about a week later was even more favorable.[35]

This gradual increase in Senate support was due in part to the first active pro-SST Capitol Hill lobbying in the program's history. A variety of tactics was used to win over undecided senators. The two Democratic senators from Washington, Warren Magnuson and Henry Jackson, assumed lead roles, contacting fellow senators and providing intelligence to the FAA and to the Department of Transportation. Private firms were also quite active. For example, the president of Ling-Temco Vought, an SST subcontractor, contacted Texas Senators John Tower and Ralph Yarborough, both of whom were uncertain in early March; by mid-March Tower was listed as supporting and Yarborough as leaning for the program. Representatives from General Electric visited or contacted Senators Lausche, Dirksen, Case, Morton, Percy, Spong, and Edward Kennedy. The airlines also did their share of lobbying. Juan Trippe of Pan American saw Senators Symington and Russell at least twice. Symington's vote was particularly crucial. In 1966 he had supported an amendment by Proxmire to strike a $200 million fiscal 1967 funding request for prototype development because of doubts about technical and economic feasibility. Now Russell declared that if Symington voted for the SST, he too would favor the program and its current funding requests. Tillinghast of TWA contacted Senators Clark, Dominick, Brooke, Edward Kennedy, and Symington. George Keck, president of United, contacted Senators Morse, Hickenlooper, Hatfield, Dominick, and Church. Donald W. Nyrop, president of Northwest, spoke to Senators McGovern, Mundt, Nelson, Proxmire, Church, and Gruening. Gruening was firmly against the SST program,

Proxmire refused to discuss it with the SST promoters, and Morse and Nelson had as yet taken no position. Floyd Hall, president of Eastern, contacted Senators Brooke, Ellender, Ervin, Gore, Talmadge, Edward Kennedy, and Dominick. Boyd also asked Trippe and C. R. Smith, president of American Airlines, to contact all senators; Smith later sent a letter to all one hundred senators that strongly supported the SST.

In late March Boyd informed the president of the growth in Senate support for the SST, emphasizing past congressional backing and adding that even Symington was no longer opposed. He stressed that "every effort" was being made to have Symington meet with Jackson, Magnuson, McKee, and leaders from the aerospace and airline industries in order to ensure his support.[36]

By mid-April Johnson still had not acted. Jackson and Magnuson "were very anxious to sit down with the president" to discuss the SST and made it clear that they would "come at any time, day or night, at the president's pleasure," a White House aide reported.[37] Maxwell argued that the current month-to-month contract extensions were adversely affecting the program, particularly in attracting skilled personnel and obtaining subcontractor commitments.[38] Pressure from SST supporters kept building in the face of Johnson's extreme caution. On April 20 Johnson learned that Boyd's latest Senate SST count had sixty-four supporting, fourteen leaning toward, nine uncertain, and eight opposed. McCone, too, strongly pushed for a prompt SST fiscal 1968 funding request, but Johnson simply directed McCone to call on those senators who did not favor the SST.[39]

On April 21 two of Johnson's closest and most skeptical SST advisers, McNamara and Schultze, were urging a declaration of prototype development and congressional funding request.[40] But Johnson waited eight more days, this time for a final report from Boyd. Among other things, Boyd told the president that Senators Byrd, Mansfield, Jackson, and Magnuson would speak out in favor of the program (with Magnuson and Jackson responding particularly to any charges that the SST was taking funds from Great Society social programs), and that Russell Long wanted to trade his support for a commitment to build a portion of the aircraft at a government facility in his home state of Louisiana.[41]

Finally, with Boyd's report in hand, on April 29, 1967 President Johnson announced the beginning of the SST prototype development phase and a fiscal 1968 appropriations request of $198 million. Although he acknowledged that the SST had great promise, he cautioned that its development still entailed "high technical and financial risks," and he stressed private industry's "willingness to share those risks." Two days later the president approved the prototype development phase contracts with Boeing and General Electric, which called for the construction of two SST prototypes and one hundred hours of flight tests at an estimated total cost of $1.444 billion over a period of about four years.[42]

Johnson's decision was a welcome relief to SST supporters. One industry backer saw the move as "energizing" the program. Boyd, who took questions from the press at the White House after Johnson's announcement, appeared much more confident than the president. He emphasized that the decision to build the prototype had the full backing of the PAC, that the Concorde and the Soviet TU-144 offered serious competition, and that the SST would be "a success" even if its routes were restricted because of the sonic boom.

At Boeing William Allen enthusiastically predicted that at its peak, work on the two SST prototypes would generate 9,000 new jobs in the company (currently 1,700 Boeing employees were involved with SST work).[43] TWA's Tillinghast wired the president, applauding the decision, and also congratulated McKee for the effective government-airline partnership he had established.[44] In mid-May McKee personally thanked the heads of the major participating airlines for their support, especially for contributing risk money for prototype development.[45]

SST supporters now even claimed to discern certain benefits resulting from delay. *Aviation Week and Space Technology*, which had favored a "hiatus" back in January, argued that the interval between the design selection decision and the announcement of prototype construction had been useful because it had provided an opportunity to obtain airline risk capital and to minimize congressional opposition. Actual prototype development, the journal claimed, had not been delayed all that much.[46] Even Maxwell, who had earlier warned of the "adverse effects" of a long delay, now pointed out that the manufacturers had been kept working during this interval under contract extensions and had been "able to use this time to good advantage." He admitted that airline risk contributions made the SST program sounder financially than it had been four months earlier.[47]

The key events that finally broke the impasse during the first half of 1967 were the achievement of airline risk participation for the prototype development phase and, to a lesser extent, the resolution of the manufacturer cost-sharing question. The airline participation issue in particular had been a major challenge for the FAA. Without continuous pressure from McNamara, Califano, Schultze, and especially Lyndon Johnson, the FAA would not have acted so vigorously to obtain airline funds. The president's instincts regarding the mood of Congress had been sound; obtaining airline contributions had made a critical difference. "In my opinion," Symington told McKee in mid-April, "you have been wise in your insistence upon the increased financial participation by the various airlines of the world."[48] Of course, the real insistence had come from McNamara and the White House, not from the FAA.

The FAA, however, had met the challenge, and it took the credit. McKee commented to Edmund Learned that perhaps the FAA's "biggest single accomplishment" had been to request airline risk money for further SST development: "While this decision doesn't change the problem of financing, or has little or no bearing on the economics of the program, it is clear that the fact that the airlines are willing to put up risk money will have a major impact on the attitude of the Congress."[49]

By early May 1967 contractor selection had finally been completed, and prototype construction was underway. The SST program had apparently seen its last major delay. Even the crucial political question of congressional support, which had been the underlying cause for the entire delay and which had surfaced as a major consideration for the first time in the life of the SST program, seemed settled. In addition, problems relating to SST economics, to sonic boom impacts, and to continuous PAC interference had been, if not eliminated or fully resolved, at least removed from center stage. The only discernible factor that could delay the SST program further would be technical problems in developing the prototype. But purely technological considerations seemed much less imposing when compared to the difficult hurdles that had already been overcome.

13 The Limits of Technology

In early 1967, with the selection of the Boeing swing-wing design, the prospects for completing the SST program never seemed more favorable. But once again the program was stalled; this time the key negative factors were managerial and, especially, technological.

The Boeing SST management structure had disturbed the FAA for some time. In early 1967, for example, the company's public relations pieces seemed ineffective, and the firm was having trouble in certain production areas.[1] Under FAA pressure, in February 1967 Boeing's SST unit, formerly a separate division, became the responsibility of two groups. One, headed by H. W. Withington, was in charge of development, construction, and testing; the other, under W. T. Hamilton, was concerned with new ideas and design concepts. Boeing also established a high-level company-wide SST technical advisory council, with E. C. Wells as chairman.[2] Then in March 1967 Boeing placed all of the SST units except Hamilton's into its Commercial Airplane Division.[3] Although Boeing's president William Allen argued that integrating SST activity into the larger division would provide the program with better cost control and scheduling, the FAA wondered if Boeing's SST activity was not being seriously deprived of experienced technical personnel.[4] To make matters worse, by late summer Boeing was also having difficulties with its subcontractors, a significant problem because the company had already indicated that it would subcontract nearly 70 percent of the SST work.[5] One of these firms, Martin-Marietta, in fact, eventually left the program because of "internal problems."[6]

But it was the SST design itself, not Boeing's management practices, that presented the gravest hazard to the program. Several technical problems with the firm's B2707-100 model had already emerged in the final SST design competition evaluation during the autumn of 1966. Technical criticism centered on the range-to-payload ratio of the aircraft. There was some question as to whether the fuel reserves would be adequate. If they were not, either the estimated range would have to be reduced or more fuel reserves would have to be added, with a consequent reduction in payload or an increase in total gross takeoff weight.[7]

After the B2707-100 was selected, Boeing was given until June 30, 1967 to submit a modified design. Meeting this deadline was crucial; if the firm could not, other stages of the SST program would be delayed.

Doubts became more intense by spring. At a general program review in April attended by airline representatives, Boeing even proposed to submit a new model, an offer that appeared to be a sign of weakness. Participants left this meeting feeling that the final SST design was still unsettled, and concern grew.[8] Although certain FAA staff members argued that technical problems were a normal part of the preconstruction stages of a complex aerospace project, others remained skeptical.[9] The SST Office received new and disturbing reports in May that the Boeing configuration's weight was continuing to increase.[10] Moreover recent wind tunnel and computer tests had identified certain stability and control problems.

Withington attempted to eliminate the design problems by lengthening the fuselage and by fastening canards (small wing-shaped control devices on the forward part of an aircraft's body, just behind the cockpit) to either side of the aircraft. These changes added weight to the aircraft, which led to greater fuel requirements. In effect, a new design, the B2707-200, had evolved.

The June 30 deadline for a final SST design was not met because much of the new configuration's technical information would not be available until November or possibly early 1968. Therefore Boeing requested that the deadline be shifted to September 30, 1967.[11]

The weight of the aircraft, the paramount design problem at this time, continued to grow. In August the FAA granted a Boeing request to allow an increase in the prototype's gross takeoff weight from 650,000 pounds to 675,000 pounds. Although Boeing tried to reduce the aircraft's weight, the newly added canards had made the plane substantially heavier.[12] SST engineers agreed that the weight was too high, but they also felt that the B2707-200 design was superior to its predecessor.[13] By September, however, the weight estimate had increased to about 750,000 pounds.

The FAA at this point became extremely dissatisfied with Boeing's current "weight reduction program," calling for more comprehensive design approaches.[14] According to one FAA official, the current model's range and payload characteristics probably would never be ideal, and such pessimistic findings were transmitted to McKee at the end of September in preparation for a series of important discussions between McKee and Allen.[15]

At Boeing too there was a gradual realization that the prototype design needed fundamental reexamination. In September Withington submitted a gloomy evaluation on the current model; the required drastic modifications on the B2707-200 would be unacceptable to both the FAA and the airlines. Boeing's high-level technical advisory council then concluded that a higher allowable weight or a radical SST configuration change was needed.[16]

In November FAA consultant Raymond Bisplinghoff, Maxwell's key outside technical adviser, submitted a damaging evaluation of, and ominous warning about, the B2707-200 design. Bisplinghoff doubted the advantages of adding canards, questioned Boeing's approach toward weight reduction, and claimed that the firm placed "an inordinate emphasis" on projected improvements for the production aircraft; Boeing was simply not being "tough enough" with the prototype's range and payload characteristics, and if these continued to remain poor, the SST would lose the airlines' financial support, and the whole program would "flounder."[17]

Both Boeing[18] and the FAA[19] were careful in public to stress the normal aspects of Boeing's proposed design changes, with Boeing publicly maintaining that the B2707-200 was essentially the same aircraft that had won the 1966 SST design competition. Missing was any mention of Boeing's continuing frantic effort to reduce the weight of the aircraft. But by December 1967 the FAA had made detailed charts showing the Boeing design's dramatic deterioration in operating efficiency and range and payload characteristics since September 1966.[20] In fact, by this time both Boeing and the FAA had in private become singularly unenthusiastic about the B2707-200 model.[21] In what was to be a profoundly significant step in the SST program, both organizations had already accepted the fact that a new SST model was required. By late January Boeing was considering a number of other configurations, including a fixed-wing design.

In a "general review" to McKee in January 1968 William Allen defended Boeing's new policy of "deliberate speed," a euphemism for seeking a better SST design. He based his defense on the notion that a sound technical foundation had to be established before large-scale spending for hardware could be approved; also without further design work, there could well be cost increases later in the program. Allen then assured McKee that all of the required resources at Boeing were being dedicated to the SST and that vigorous efforts were underway to solve the aircraft's technical problems.[22] The agency was already well aware of Boeing's problems. Allen's review was clearly as much for public relations purposes as for substantive communication. (The agency transmitted Allen's views to the chairman of the congressional subcommittees that handled SST appropriations; McKee and Transportation Secretary Alan Boyd endorsed Boeing's general delay decisions.)[23]

"Deliberate speed" was soon spelled out. After considerable debate within the company, Allen finally decided to ask the FAA for a one-year delay for a comprehensive and thorough design review. Both the airlines and the FAA heartily agreed, and the agency decided without hesitation to grant Boeing's request for a slowdown, instructing the company to concentrate on solving its technical difficulties and to defer major prototype expenditures until these problems were answered.[24]

A February meeting of Maxwell, Bisplinghoff, and Boeing officials in Seattle sealed the fate of the B2707-200 design and resulted in approval of the year-long design review. As Bisplinghoff later observed, Boeing had finally recognized that the model was simply "not good enough and that a substantial redirection will be required." (Like Edmund Learned earlier, Bisplinghoff was more than just a technical consultant for the FAA. He offered the FAA some sound political wisdom as well, advising Maxwell not to request further appropriations until the program was on a more solid technical foundation; funding could be arranged on a month-to-month basis so that the program could be terminated at any time that progress was judged unsatisfactory. Turning to political tactics, he also suggested that the knowledge that the FAA might withdraw its fiscal 1969 SST appropriations request and would continue design modification be made available to only a few key officials in the administration, on Capitol Hill, and in industry; it would not be desirable to inform the general public of this new delay.)[25]

During this period the airlines were, if anything, more dissatisfied with Boeing's SST work than was the FAA, generally stressing the firm's mismanagement and its less-than-candid dealings with the SST Airlines Committee. Many airline representatives were willing to give Boeing only one last chance. If the company failed to come up with a satisfactory design, the airlines would not object to cancelling the program.[26]

At the White House, frustration with Boeing and the whole SST program also grew. President Johnson was kept informed of SST progress through brief reports from the Department of Transportation; he also received advice from Califano, Schultze, and McNamara, who with others in the White House were becoming increasingly skeptical about the SST's development. Obviously the program had severe problems. Califano reminded the president in February 1968 that "some of this trouble was predicted by the McNamara committee [the PAC] and this will clearly create some problems on the Hill."[27] When the FAA formally recommended that Boeing concentrate on fundamental design problems for a year and defer major expenditures for prototype hardware (which would reduce the fiscal 1969 SST budget request from $223 million to between $50 million and $75 million, according to the FAA), Johnson responded by directing that no immediate public announcement be made.[28]

Unfortunately for Johnson and the FAA, the press learned almost immediately about the one-year extension. The *New York Times* reported it on the front page, stressing that the delay was motivated by a political desire to reduce the fiscal 1969 funding request to a level that would be acceptable to congressional SST critics. On the same day, February 21, 1968, the *Seattle Post-Intelligencer* also reported an imminent SST slowdown.[29]

The *New York Times* report caused considerable consternation among SST proponents. Both Senator Jackson and Magnuson's administrative aide called the White House that day. White House aide Harry C. McPherson reported to the LBJ Ranch, "They said substantially the same thing: The *New York Times* story is political. They urged early separate press conferences by FAA and Boeing to explain that the slowdown is caused by technological problems." By evening Johnson had spoken with Jackson and McPherson, and Boyd and Maxwell agreed that a public announcement was urgent. Boyd told McPherson that all the "industry people" with whom he had talked were discussing the news report and that he would have the FAA announce the slowdown, referring technical questions to Boeing. A Transportation Department memorandum to the president on the same day recommended an immediate announcement. It also called attention to the "intense" public interest stimulated by the *New York Times* report and mentioned that many people outside of Boeing—in the airlines, Congress, and the Transportation Department—were already aware of the SST's range and payload difficulties.[30]

On the next day, February 22, Boeing announced that it had recommended to the FAA further SST design and development work before beginning prototype construction. And the FAA publicly stated that it had accepted Boeing's proposal.[31]

Both Boeing and the FAA assumed a brave and confident public stance. Allen defended Boeing's performance by maintaining that the SST program was not on a "rigid schedule"; the current B2707-200 was airworthy but not economical. Allen was optimistic that the required improvements would emerge during the new review period.[32] The FAA continued to maintain that design changes were to be expected in such a technically complex undertaking and even commented that Boeing's decision would be advantageous in the long run.[33] Maxwell publicly reiterated that the SST was not on a strict timetable and professed not to be discouraged by the delay.[34]

The SST contract amendment that authorized Boeing's year-long redesign effort was finally signed in April. The due date for the new Boeing submission, which would be judged solely by the FAA administrator, was January 15, 1969.[35]

But both Boeing and the FAA realized the true precariousness of the SST program at this time. Allen informed McKee in early April 1968 that the prototype Boeing would develop might not meet the FAA's production and performance objectives and suggested modifying these objectives.[36] McKee, however, vigorously rejected this appeal, stating that the FAA had no intention of lowering SST standards simply because the company could not meet performance requirements.[37] The FAA was clearly worried about Boeing's ability to produce an acceptable configuration. In April 1968 Maxwell stressed the need to limit the government's liability in the event Boeing failed to perform as stipulated in its contract and warned that the possibility of such a default was "a real consideration."[38]

Boeing's new redesign effort marked a crucial turning point in the company's SST work. The firm, with a new, expanded, and upgraded centralized SST management structure under the control of Withington, began a comprehensive review of several alternative SST configurations in April 1968; these would be evaluated and eliminated one by one until December 1968, when a single design would be selected. The company had previously concentrated on developing a swing-wing aircraft, but as of April 1968 other designs were explored; of the six or seven alternatives to be considered, two were fixed-wing models. In addition, Boeing had further reorganized to facilitate the greater involvement of top-level officials.[39]

By June the company was concentrating on three configurations: a nonintegrated variable sweep wing, a fixed wing with a broad horizontal tail, and the NASA SCAT-15F fixed-wing configuration, an arrow-shaped fixed-wing design with flipper-like extensions at the tip of each wing that appeared extremely promising.[40] Each of these three designs had unresolved problems, but all of them appeared clearly superior to the B2707-200.

The competition soon narrowed. The SCAT-15F derivative developed certain technical problems. By late June Boeing was focusing on the other fixed-wing and the remaining swing-wing design.[41] By August the fixed-wing model had emerged as dominant, though Boeing continued to make detailed technical and economic comparisons of all three alternatives. After visiting Boeing one FAA official reported that only the broad-tailed fixed-wing configuration was essentially "uncomplicated and problem free." By late August Boeing had acknowledged that it was focusing primarily on this fixed-wing model. The design's estimated prototype gross weight was 640,000 pounds, its production gross weight was 710,000 pounds, and its estimated payload was two hundred passengers. In October 1968 Boeing formally selected this design.[42] The company abandoned the design concept that it had advocated for so long, the swing-wing SST. Boeing was now publicly committed to a fixed-wing aircraft.

The actual announcement of Boeing's shift caused little surprise because the Congress and the public at large had already been forewarned by the press. As usual with the SST program, the redesign effort had been plagued with news leaks. As far back as April 1968 *Aviation Daily* and the *New York Times* reported that Boeing was seriously evaluating the fixed-wing SCAT-15F. The *Tacoma News Tribune* told its readers in early May that Boeing was reviewing "SST basics," and later that month *Aerospace Technology* speculated that the SCAT-15F might be a leading contender in the year-long redesign.[43] Throughout the summer of 1968 the fixed-wing approach gained prominence in the press. In early July the *Washington Post* reported that Boeing might scrap the swing-wing concept altogether. About two weeks later *Aviation Week and Space Technology,* in a remarkably accurate story, wrote that the B2707-200 design was no longer being considered, that the SCAT-15F was a "long shot at best," and that a fixed-wing design with a separate horizontal tail seemed to be favored over a swing-wing design. In September the *Washington Post* and the *New York Times* had both reported that a fixed-wing plane was definitely planned.[44]

The evaluation of the new Boeing submission, termed the B2707-300, in view of Boeing's radical shift in SST design philosophy was particularly critical in substantive and political terms. The FAA clearly wanted to preclude any move to reopen the whole SST design competition. After all, the design that won at the end of 1966, the Boeing swing-wing model, had proved unacceptable. In 1968 anti-SST members of Congress, such as Sidney Yates (D-Illinois), were inquiring whether the current SST contract permitted the kind of radical redesign effort then underway. The agency was already arguing that it was "inconsequential" whether Boeing selected a fixed-wing or swing-wing configuration so long as the performance criteria were met and so long as a costly extension of the design competition was avoided.[45] But in order to justify its decision to allow only Boeing to redesign the SST, the agency needed a thorough and credible evaluation.

There were at least three parallel centers of evaluation activity: an interagency government group under the direction of top officers from the SST Office and an official from the FAA General Counsel Office; a committee of outside technical advisers, established by Maxwell and chaired by Bisplinghoff; and various assessments from the nine major American-flag airlines that were contributing risk capital to the SST program.

The interagency evaluation, which took place in December 1968 and January 1969, was performed by a group of approximately one hundred scientists, engineers and other specialists from the FAA, NASA, and the air force. The results of this review were generally quite positive. Its technical review concluded that the B2707-300 was further advanced than other American SST designs or comparable military and commercial aircraft projects at the same developmental stage; although certain technical problems were noted, the model seemed to present only "normal" development risks. In the system integration area the findings were somewhat less optimistic, especially regarding engine noise and the sonic boom. Certain noise characteristics compared unfavorably with those of subsonic jets, and sonic boom overpressures were estimated at extremely high levels. But the interagency group was more hopeful about the SST's economic performance. The reduced size and simpler design of the B2707-300 seemed to indicate a lower production cost, a lower sales price, and a higher demand (450 to 500 aircraft).[46]

Not surprisingly, the overall coordinating body of the interagency evaluation, consisting of FAA officials, came out with an equally optimistic report, concluding that the plane was well designed and adequately supported by test data and engineering analyses; it required no major changes before being moved into the prototype construction phase. Although the officials cautioned Maxwell to pay special attention to such problems as engine noise, they characterized the overall current status of the SST program as one of "medium risk, normal for a program of this magnitude."[47]

The evaluation of the Bisplinghoff committee was crucial to the FAA and to Boeing. A favorable assessment obviously would give the new design increased legitimacy. Members of the group, in addition to Bisplinghoff, were Arthur Raymond of the Rand Corporation, formerly senior vice-president of engineering at the Douglas Aircraft Company, and Ernest Sechler, professor of aeronautics at the California Institute of Technology. The FAA and Boeing supplied these committee members with numerous government evaluation reports and considerable technical information.[48]

The Bisplinghoff group in practice was more like a loosely knit collection of three individuals than a real committee (before February 1969 they had met together only twice). In the end, each member wrote his own report, all of them presented to the FAA on February 7, 1969. All three concluded that the government evaluation had been competently carried out and that the B2707-300 design was promising enough to start prototype development, though certain technical problems were noted. In a highly expansive vein, Bisplinghoff argued that the SST would act as a focus or pacemaker for American aeronautical research, much like the Apollo moon program did for space research, and that the SST would lead eventually to the development of hypersonic aircraft, which would fly at six to eight times the speed of sound. Both Bisplinghoff and Raymond stressed that the sonic boom made the SST a strictly over-water vehicle and that noise was the greatest problem to face. In determining the SST's future, Raymond predicted, "the public will prove more influential than any government agency." The FAA was delighted with the reviews from Bisplinghoff's group. One FAA official told Bisplinghoff in mid-February 1969, "Certainly we are very pleased with your report and feel that it will be extremely helpful to the SST program."[49]

The airlines, which Maxwell also considered critical for the SST's ultimate success, were given preliminary summaries of the government's findings.[50] In February 1969 the FAA held an important meeting with representatives of the nine airlines that had contributed risk capital to the SST program. The conclusions of the interagency evaluation were reported in considerable detail, and the Bisplinghoff group presented its findings.[51] At this time the airlines generally were assuming an increasingly serious view of sonic boom and engine noise issues. The head of the Airline SST Committee acknowledged that the SST's sonic boom characteristics almost certainly made it a solely over-water vehicle,[52] and individual airlines also informed the FAA that successful SST operations would depend on limiting airport and community noise, operating cost, and airport congestion.[53] In addition, a significant split in the attitude of American-flag air carriers toward the SST had emerged: the most enthusiastic airlines, such as TWA and Pan American, were those with major over-water routes and with potential competition from either the Concorde or the Soviet TU-144; the more doubtful airlines were those with primarily overland, domestic routes.[54]

The FAA evaluation strategy was successful. After the highly favorable findings on the B2707-300 by both the interagency group and the Bisplinghoff committee, the demands for reopening the entire SST competition practically vanished. Maxwell formally recommended approval of the B2707-300 model in February 1969. On a "technical basis," he felt, the program was ready to proceed at a "normal pace."[55]

Maxwell had accomplished his greatest achievement: he had steered the SST program away from potential disaster in 1967, when it had become clear that the original Boeing design was not acceptable, and by early 1969 the government had approved a radically different configuration. Maxwell had contained the process. The redesign phase had not expanded into a full-scale renewed prototype competition and thus had not become an effective target for SST opponents. The FAA at last was on the verge of building an SST prototype, something it had wanted to do since 1964. On the other hand, Maxwell had lost something even more precious than the new SST design he had gained: time. Two years had passed since the Boeing configuration had been selected, and during that period technical problems had retarded SST development seriously. At the beginning of 1969 SST managers became part of a new administration, and it became clear that technical difficulties were being solved just as problems in other areas would create new delays.

14 The Foreign Threat

The SST program encountered an array of obstacles during the 1960s: technical difficulties, uncertain or questionable economics, bureaucratic conflict, fragmentation of programmatic control, growing concerns over noise and the sonic boom, doubts over management capability, and general skepticism at the very highest levels of government and parts of industry. Any one of these factors might well have killed the program. Given the great diversity of vulnerabilities, it is remarkable in many ways that the program survived the decade. Yet when all other pro-SST arguments failed, supporters could always point to the foreign threat to American hegemony in commercial aviation. Across the Atlantic with their Concorde program, the British and French appeared to be making a determined effort to wrest from the United States its post–World War II dominance of the aviation industry. The Soviet Union, with its TU-144 SST program, also seemed eager for part of the spoils. A natural concern for American officials, even for those skeptical about the SST, was whether the United States could afford ignoring this challenge. The existence of a foreign threat was crucial for the survival of the American SST program in the middle and late 1960s.

American perception of foreign SST programs was ambivalent and complex. Extreme SST opponents disparaged all SST efforts as wasteful, harmful, and futile. SST supporters, on the other hand, emphasized the soundness and potential success of the Concorde and, to a much lesser extent, TU-144 programs, but they were also careful to stress that American aviation technology was superior to that of state-controlled European industry. Therefore, although the Concorde program was seen as serious competition, it was characterized as a threat only if the United States failed to act.

Views on the Concorde

When the President's Advisory Committee (PAC) was established in the spring of 1964 the Concorde project was not the center of policy-making discussions. As in the past, almost all significant American SST activity with regard to European SST efforts was limited to purely technical SST information exchanges with the Europeans. For example, in 1964 tripartite SST meetings involving the United States, Britain, and France began, which dealt with such technical areas as flight performance, cockpit vision, airframe design, propulsion, fuel reserves, air traffic control, airfields, the sonic boom, engine noise, atmospheric effects, and medical issues.[1]

To the Americans at this time the Concorde effort soon seemed to be in deep political and technical trouble. In late 1964 the newly elected Labour government called for a thorough review of British participation in the program. At the end of October the CIA reported that the Labour party's stance would have serious repercussions for the Concorde, and PAC chairman Robert McNamara was even warned in November 1964 that Britain might withdraw completely. McNamara's key SST aide, Joseph Califano, however, believed that the British would probably continue to participate, though keeping a withdrawal option as insurance against any change in the "political and/or economic climate." Califano also discounted press reports of West German involvement in the Concorde effort. The CIA reported that the French would be unable to develop the Concorde alone because of the heavy financial and managerial burden involved. In sum, the British reappraisal had clearly weakened the Concorde. "Whatever the outcome," Califano told McNamara, "the introduction of so much strain and uncertainty into the Concorde program because of the political factors makes it doubtful whether the degree of cooperation that has thus far prevailed between the British and French can be maintained."

Meanwhile design revisions had also set the Concorde program back as much as two years, development costs were spiraling (estimated at $400 million for the British share in November 1964), and apprehension was growing over effects of the sonic boom. A NASA analysis of the Concorde "optimistically" estimated that Concorde direct operating costs would be 1.4¢ per seat-mile (compared with 1.0¢ to 1.1¢ per seat-mile for the subsonic Boeing 707), and Califano indicated that the Concorde's performance would probably deteriorate even further.[2]

At the end of 1964 only the strongest SST proponents, including FAA administrator Najeeb Halaby and potential SST contractors like Lockheed, bothered stressing the danger of Concorde success.[3] Although in January 1965 the new British aviation minister, Roy Jenkins, formally announced the British decision to proceed with the Concorde, British doubts about the project and fear over the prospect of a competitive American aircraft remained strong. In early 1965 Jenkins proposed a high-level tripartite meeting on SST policy as well as technical issues. Jenkins and French officials particularly sought to avoid an international SST race of "crash programs."[4] On February 12 the French openly proposed "dividing the world [SST] market" between the United States and the Europeans.

The tripartite meeting took place in mid-February. Although the Concorde officials exuded confidence, emphasizing that the British reexamination had caused no real delay and that both nations were determined to proceed and predicting a Concorde prototype by late 1967, Jenkins (on the grounds of eliminating waste) again hinted at the mutual benefit of keeping the American and Concorde SST programs "in step."

The Americans, however, had no inclination to abandon their long-standing aversion to cooperating with the Europeans on the SST. Halaby, the head U.S. delegate, quickly brushed aside such ideas. He stated bluntly that the recent British and French decision to proceed had effectively precluded direct American involvement in a joint SST development venture; besides the respective SST designs differed greatly, and the American SST program had not yet advanced beyond the review stage. In contrast to Jenkins, Halaby appeared almost to welcome foreign competition; a certain amount of waste and duplication was inevitable and perhaps even desirable because the airlines would want some product choice. He went out of his way to paint a very bright future for the American SST: the American vehicle possessed a larger capacity, faster speed, and better economics than the Concorde; the first prototype flight was projected for 1968; and a 25 percent return on investment was expected from the American SST (compared with less than 16 percent for the Concorde and with 45 percent for the 707).

Halaby, while deemphasizing the issue at home, used the apparent differences in policies concerning the sonic boom to maintain distance between the two SST efforts. He expressed great disappointment that Anglo-French sonic boom tests had been tabled until 1966. He dwelled on the sonic boom's impact on SST economics and on public acceptance of the aircraft. Just when more sonic boom tests on public acceptability were needed, Halaby declared, the British and French seemed to have done nothing of consequence.[5]

The generally disdainful American view of the Concorde effort was reflected at the March 30, 1965 PAC meeting. CIA director John McCone, in presenting the current intelligence on the Concorde (carefully noting that the CIA had not used "clandestine sources" because of the "risk of offending one of the host countries"), appeared quite uncertain and skeptical of Concorde accomplishments. He reported that little work seemed to have been done on the sonic boom problem and that extensive design modifications and economic uncertainties would surely cause further delays. He reminded the PAC members that in moving gradually upward from the Mach 1.5 to the Mach 2 range, unexpected technical problems were bound to arise, and these would take time to correct. He was not worried about the Concorde's alleged two- to three-year lead and suggested that Anglo-French forecasts be taken "with a grain of salt . . . quite a large one." McCone was also supported by Stanley Osborne on this matter.

But Halaby, trying to accelerate the American program, portrayed the Concorde project as a more significant effort than he had at the recent tripartite meeting. He emphasized the high caliber of Concorde personnel and officials, especially Roy Jenkins, who was "quite an able and different man from his predecessor." Halaby warned that without an American SST the airlines would purchase the Concorde; Concorde development appeared to be on schedule; and it had won significant airline commitment (in addition to BOAC and Air France, forty-eight delivery positions had been reserved by other airlines). Halaby added that the Concorde managers "think they will muddle through" and that in any case the Concorde was not "another Comet," referring to the ill-fated British commercial jet introduced in the 1950s. The mere existence of the Concorde was a key weapon in the SST advocate's arsenal.

McNamara, however, had strong doubts concerning the Concorde's effectiveness. McCone's skeptical presentation, in his view, was "a very interesting report, the best we have had so far." McNamara held to his consistent theme that the American SST should be a profitable commercial venture and that the pace of the Concorde work should not dictate American SST development. He felt that the United States would ultimately develop a better SST so there was no need to worry about the Concorde's lead.[6]

Stephen Enke's SST economic task force in the Defense Department meanwhile produced analytical evaluations that strongly supported McNamara's views. The group's estimates gave the Concorde only a minor share of the market, and Enke was convinced that the

Concorde would have difficulty keeping up with American competition. The task force termed "unrealistic" the Anglo-French dates for commercial Concorde operation. The British and French were inexperienced at sustaining Mach 2.0 speeds, claimed the group, especially since the British had recently cancelled their TSR-2 supersonic fighter program, on which the Concorde managers had counted to help develop the Olympus engine for the Concorde. One member of Enke's group noted, "The American SST has great growth potential, the Concorde almost none." According to Enke's group, there was clearly no need for a crash American program.[7]

By their May 5 meeting the PAC members, including Halaby (who would shortly leave the government), appeared even less troubled than before by the Concorde and more confident in the ultimate success of the American SST. McNamara predicted that the American SST would be "far more successful commercially than the Concorde." He believed that the United States could demonstrate the American aircraft's superior characteristics prior to the Concorde's first commercial flight and therefore "need not feel the pressure" of a swiftly progressing Concorde effort. Osborne, Webb, and others agreed.[8] The FAA had failed in its attempt to use the specter of Concorde competition to accelerate the American program.

Although in mid-1965 there was little real anxiety over the Concorde (in fact, such views were reinforced by continuing reports of technical and economic problems and by Concorde officials' pressing again for some sort of tripartite cooperation), American concern increased during the latter half of the year.[9] This new mood arose from a genuine worry about the Concorde as a threat to American aviation interests and to a reinvigorated political effort to influence PAC members by the FAA under the new leadership of William McKee and Jewell Maxwell.

In this respect, McKee and Maxwell surprisingly received some analytical support from Enke's group. The task force produced an important study that concluded that the Concorde would displace approximately 25 percent of the one hundred or so American SSTs expected to be sold by 1985 under a restricted route condition. (Under unrestricted routes, the study found the Concorde offering only nominal competition.) The study also concluded that the Concorde's lower plane-mile costs (in contrast to the Concorde's higher seat-mile costs) would make it more suitable for low-density routes and hours; cheaper subsonic air fares would hurt the larger-capacity American SST more than the Concorde, as would route restrictions because of the sonic boom, given the resulting limited demand for SST air travel; British and French production techniques tended to be less capital intensive than American ones. "The U.S. SST needs a relatively large supersonic market," the study found, "which probably means only moderately restricted routes for Concorde competition to be unimportant."[10]

The FAA was once again lobbying by emphasizing the Concorde threat, and this campaign was well planned and effective. The agency told the PAC in early October 1965 that the gap between the time of announced Concorde commercial introduction (1971–1972) and estimated American SST commercial availability (mid-1975) was sufficient to assure "an adequate market for the Concorde," given a "reasonably economic" Concorde design. The United States "must assume that the Concorde will be a successful program," declared the FAA.[11] Moreover the notion that the Concorde might be a real competitive danger assumed new credence at the October 9 PAC meeting where Osborne and Boeing reported that the Concorde would meet its announced schedule.[12] Media reports, for once, helped the FAA at this time. Although mentioning increases in the cost of the Concorde, the press also reported that the technical feasibility of every Concorde system had been determined and that the plane would fly in March 1968, allegedly giving it a two-year edge over any competitor.[13] Similarly an FAA intelligence summary for PAC members found Concorde management "operating smoothly," the aircraft's advertised performance data "reasonably valid," and development on schedule; the Concorde's technical systems and design generally possessed "no problems"; the aircraft's range had been extended to 4,150 statute miles with "adequate" fuel reserves; and, with fifty total orders, airline confidence was "rapidly increasing."[14] Finally McKee and Maxwell asked the key FAA official for Concorde matters, Raymond B. Maloy, assistant administrator for Europe, Africa, and the Middle East, to prepare an "authoritative view" on the Concorde's competitive position for use at the November 6, 1965 PAC meeting. Although Maloy saw major problems with the Concorde in several technical areas, he stressed the Concorde's "high political significance as representative of the new commitment of Europe to collaborate and cooperate in order to meet the U.S. challenge to the European aircraft industry." According to Maloy, neither the British nor the French, especially Charles de Gaulle, would abandon the Concorde, at least through the construction of two prototypes.[15] At the November 6 PAC meeting McKee read verbatim from Maloy's evaluation, and Eugene Black then spoke in favor of an earlier delivery date for the American SST as a means of slowing the British Concorde effort.[16]

The FAA counterattack in late 1965, based largely on the alleged successes of the Concorde, achieved some success. The PAC concluded in the annex to the PAC's third interim report to the president, transmitted on November 15, that the Concorde could "prove to be a serious competitive threat," especially on low-density routes. Still, certain technical difficulties and high operating costs per seat-mile (compared to the American SST) were also mentioned; weight increases in the Concorde's configuration, from 326,000 pounds to about 360,000 pounds, indicated that the aircraft was approaching or had reached "its limit of growth without requiring major redesign."[17]

Enke immediately attempted to block the FAA resurgence. In early January 1966 Enke hurriedly flew to Paris and London to meet with high-level French and British officials, ostensibly to deal with various economic and sonic boom research problems but really to discuss how to slow down proportionately the Anglo-French and American programs. Enke reported that a number of influential British and French officials seemed to favor this approach. Unlike FAA officials and others recently, he also sent back less than favorable assessments of the Concorde: economic prospects were pessimistic and the airlines were not enthusiastic about the aircraft. Most significant were Enke's reports that the British and French had different performance and political goals for the Concorde. Enke remarked that the French originally wanted "a somewhat larger and supersonic Caravelle" (that is, an SST without the range for nonstop North Atlantic operations), while the British insisted from the beginning that the Concorde should have nonstop transatlantic capability; the French yielded to British urgings; in terms of political goals, the Concorde was a matter of pride and national prestige to the French, and they resented the idea of leaving SST development to the Americans; on the other hand, the British tended to view the Concorde as a price they had to pay to avoid a French veto of British membership in the Common Market. Britain, according to

Enke, was a "reluctant partner," continually stressing the Concorde's poor economics. Enke characterized the mood of one high-level British official as a "fatalistic hopelessness that combined an awareness of financial losses ahead with a belief that little could be done about it." Enke therefore concluded that the time was ripe to explore "time phasing and design differentiation" with the British and French and recommended informal contact using private PAC members Black, Osborne, and possibly McCone, thereby avoiding bureaucratic government channels.[18]

The FAA naturally was quite upset over Enke's European trip, especially when a January 1966 issue of the *Economist,* a British journal that had consistently taken a strong anti-Concorde stance, reported on "feelers . . . to see if some sort of agreement can be reached about the timing of the new supersonics to avoid the otherwise inevitable race—Europe against the United States—to deliver the most aircraft the soonest."[19] Enke's ideas had obviously been leaked.[20]

The FAA quickly mobilized. It received additional favorable Concorde reviews from TWA, Lockheed, and Boeing,[21] and began systematically to organize and assess Concorde information.[22] In a report to the PAC in mid-February 1966 the FAA dwelled on Concorde improvements in range, seat-mile costs, and return on investment. The Concorde program "has not encountered any serious problems which have not been resolved," the FAA commented.[23]

When the PAC held an informal meeting on March 9, 1966 therefore it was receiving somewhat contradictory signals on the Concorde from Enke, the FAA, the manufacturers, and the airlines. Indeed as a result of the FAA's skillful assembly of favorable information, respect for the Concorde had grown. Of course, neither the SST advocates nor skeptics had any real intention of joining forces with the Europeans. The proponents simply wanted to accelerate the American effort, and the skeptics wanted to slow the program, doubting the economic wisdom of either the American or Concorde program.

In spite of the vigorous and increasingly effective attempt to spur the SST program using the Concorde, the PAC again went along with McNamara. The FAA still had not convinced the PAC that the Concorde was a real competitor. In early March 1966 the PAC agreed that the American SST program "should be scheduled at an optimal rate to provide a safe aircraft as soon as possible that will be profitable to its developers, manufacturers, and operators. The pace of the U.S. program should not depend on an uneconomic desire to race the Concorde into commercial service."[24]

Impact of the Concorde

After this brief flurry of optimism the Concorde's stock again declined in the eyes of key American decision makers (though the FAA continued to portray the Concorde as a great rival and a "serious threat").[25] The CIA in late March 1966 reported on the Concorde's engine difficulties and in late April enumerated a number of other problems.[26] At the May 6, 1966 PAC meeting McNamara stated that this negative information demonstrated that although "not a failure," the Concorde "did have a few problems." He added that lack of supersonic experience had led the British and French to underestimate the Concorde's technical difficulties. Both McCone and McNamara explicitly warned against the Concorde's influencing American SST development. McCone did not want the Concorde to force an unhealthy "telescoping" of the American effort. Similarly fearing undue pressure (especially on Boeing to meet a submission date), McNamara instructed the FAA to report any instances where the Americans were "doing something differently than they would do it if there were no Concorde."[27]

As usual, the information on the Concorde that the Americans received, from public and private sources, was contradictory. Some of the data indicated that the Concorde was proceeding smoothly and on schedule. Britain and France seemed to be cooperating well, and the FAA particularly was more than willing to believe Concorde claims. The FAA was assisted in early July 1966 by Juan Trippe of Pan American, who told the PAC that the Concorde's timetable was realistic and that its performance characteristics had improved. He pointedly added, "Any place that we don't have such a ship [the Concorde] covered, as more or less a loss leader for advertising purposes and so forth, we think we would be in trouble during the period after Concorde delivery."

But again a basic lack of American confidence in the Concorde program remained. The most credible U.S. sources throughout the summer and autumn of 1966 stressed the Concorde's long-term problems and the likelihood of a substantial delay. The Americans were particularly bothered that Concorde officials were unable to comprehend why the American government was worried about sonic booms and overland supersonic operations. In addition, published reports indicated ever-increasing Concorde development costs (up to $800 million from the originally estimated $470 million) and a growing sales price ($16 million in contrast to the originally announced $10 million). Perhaps more fundamental, the Americans generally disparaged the European aviation industry. Even Juan Trippe spoke of the "miserable performance in Europe compared to what we have done in this country" and admitted that Pan American's Concorde orders were really "a sort of an insurance program" to cover Pan American in the event that an American SST was delayed.[28] Supporting American poor opinion of European commercial aviation, CIA reports emphasized the Concorde's major technical and nontechnical problems, and, though acknowledging that the Concorde was currently on schedule, it warned of "serious" future delays in the Concorde's production phase. The search for solutions to technical problems, according to the CIA, could delay the program for up to two years. Moreover in the nontechnical area the CIA dwelled on potentially fatal disagreements between the British and the French; the French, worried about the proposed American SST and unsure of their British partners, wanted to enter production quickly and rejected a British proposal to increase Concorde passenger capacity to 167; the British, on the other hand, already doubting the Concorde's economic strength, thought that a larger vehicle was needed to compete with the Americans on transatlantic routes, which would require more development time. According to the CIA, the British had "the uneasy feeling that they are being led into a venture that could prove disastrous."[29] The powerful Bureau of the Budget also was unconcerned about the Concorde's alleged lead.[30] Enke and his group naturally minimized the seriousness of the Concorde challenge, arguing that even a year's slippage in the American effort would have little impact on American SST sales. (Reflecting the intensity of Enke's views, he even accused McKee of substituting "a few pages of trivia" in an FAA report on the Concorde for detailed calculations by FAA staff).[31]

For the rest of 1966, the period of the final design competition evaluation, general American skepticism toward the Concorde remained more or less unchanged or, if anything, increased somewhat. Even the

FAA grew more cautious. The agency received confirmation that the Concorde's gross takeoff weight had increased to about 350,000 pounds.[32] In addition, FAA consultant Raymond Bisplinghoff, after a visit to Sud-Aviation in September, criticized many Concorde development procedures, though he did find the Concorde to be on schedule and a source of national pride to both Britain and France.[33] At the October 6 PAC meeting Maxwell admitted that no solutions had been found for the Concorde's problems; development costs and sales price were increasing, but so were airline orders. Presidential science adviser Donald Hornig added that the Concorde developers had "sort of shut their eyes to the [sonic boom] problem and resigned themselves to at least having the overseas market."[34] Similarly the CIA, in a report at the end of October, emphasized the Concorde's long-term difficulties, its lack of good "growth prospects," and its increasing development costs.

Still, as the CIA also acknowledged, the Concorde did exist and was apparently on schedule. Moreover Concorde orders had increased from fifty-four to a tentative sixty-four since August, and the project had assumed a high order of diplomatic and political importance. "General de Gaulle," the CIA observed, "continues to view the Concorde as an important step in demonstrating the technical competence required of a major power. He sees the project as a means, also, to enhance French prestige, particularly vis-à-vis the U.S., and has taken a personal interest in it. The [French] government's determination that the project be completed, despite growing British disenchantment [because of mounting costs] also stems from Gaullist assertions that France's 'independent' foreign policy has not harmed its friendship with its allies." As another high French official succinctly stated, "For technological, commercial, and also political reasons, our European countries cannot allow themselves to sink to the level of mere subcontractors."[35]

Even more positively TWA president Charles Tillinghast told the PAC on December 7, 1966 that the Concorde could indeed be a real threat. Although noting that the Concorde had "lousy seat-mile economics" and that TWA "would love to skip the Concorde," he maintained that if the American SST fell further behind, TWA would have no choice but to buy the Concorde. Tillinghast estimated that TWA could afford an eighteen-month lag but not much longer. He warned,

"The British and French are in. They may have been silly to have done it. They are in. They are going ahead. I think anyone who has a tendency to write off the Concorde as a lot of flop is being very unrealistic. Its economics are considerably less than sensational but it will fly, it will fly well."

Finally both the CIA and Raymond Maloy felt that the desire to join the Common Market would force Britain to continue with Concorde production, which the French desperately wanted for political, technological, and economic reasons. The CIA concluded, "If Britain withdrew from the project while EEC [Common Market] negotiations were in progress, de Gaulle would almost certainly cite it as proof that the U.K. was not sufficiently European-minded. On the other hand, should de Gaulle veto the British bid for EEC membership, [Britain] . . . would be under strong pressure to withdraw from the [Concorde] project." Both McNamara and McKee agreed at the December 10 PAC meeting that it was a "fair assumption" that the Concorde would be produced, either through a joint Anglo-French effort or, if that should fail, by the French alone.

But even the prospect of a Concorde ultimately flying never really frightened most PAC members, at least not enough to accelerate the SST program. McNamara declared on December 10 that the United States was "unduly concerned" and that both the Concorde and the American SST were sure to face serious technical problems and subsequent delays. McCone and others agreed. Moreover, after having been told by Eugene Black that the British officials "just laugh off" the sonic boom problem and by the FAA that the agency did not currently possess the right to prohibit commercial flights on grounds of intolerable sonic booms or excessive aircraft noise, McNamara and other PAC members strongly supported the stricter legislation in this area that was pending (although McKee said that the new bill would apply only to "new aircraft, not the Concorde").[36]

The PAC therefore ended its life with the generally skeptical view of the Concorde that it had held at the beginning. In its fourth and final interim report to the president, submitted on December 22, 1966, the committee emphasized the technical and economic superiority of the American SST and observed that many aircraft development problems typically do not become apparent until the prototype stage. The PAC expected significant delays with the Concorde and predicted little performance improvement given the aircraft's small engine thrust and resulting limited range. The committee also claimed that the Concorde's estimated direct and total operating costs were, respectively, 25 percent and 15 percent higher than those of the American SST and were increasing rapidly.[37]

Beginning in 1967 the U.S. decision-making structure for the SST began to change significantly. With fundamental policies related to the basic design, contractors, sonic boom tests, and financing apparently decided, the PAC did not meet again after December 1966. Generally SST decisions now became more programmatic, centering on technical problems and on relations with contractors. In the same vein, American officials began to view the Concorde more passively. The FAA continued to monitor its development, but the intelligence effort became less focused and more irregular, and the CIA's role diminished. In addition, the usefulness of the raw intelligence on foreign SST programs from the CIA and the State Department was questionable, adding little to what was already known.[38]

In any event, the past American skepticism about the Concorde was continually confirmed in 1967. By mid-1967 the FAA was receiving news of Concorde delays. British and French Concorde officials appeared to regret their once-confident predictions regarding deadlines and costs.[39] A group of American aviation experts who visited Concorde facilities in the summer of 1967 reported increases in gross weight, limitations in basic engine size, and diminished fuel capacity. Maloy too called attention to these problems and the increased noise of the Concorde. By November 1967 even TWA, whose chief executive had stressed the Concorde's strength to the PAC the previous December, was, in Maxwell's words, "not at all enchanted with the Concorde." TWA decided not to reserve additional Concorde delivery positions and told Maxwell that it might even drop the six positions that it held if the aircraft's performance levels slipped further.[40]

Although the first Concorde prototype was unveiled on December 11, 1967 at Toulouse, France, there were new delays in 1968 for a variety of reasons, including technical problems, management changes and resignations, strikes, and a crash of a jet fighter that was simulating Concorde flight characteristics. Concorde officials had predicted that the Concorde would fly before the end of 1968, but the year came and went without the aircraft's taking off.[41] (The FAA and its SST contractors worked strenuously to counter the resulting publicity from the unveiling. Boeing contacted twenty "media people"—including representatives from the Washington, D.C., dailies, the three major television networks, *Time, Newsweek,* and the *Wall Street Journal*—to supply them with background information and a picture of the newly designed American SST.)[42] Air India cancelled its option for two Concordes while retaining two delivery positions for the American SST and claimed that the later delivery date of the American SST was actually beneficial because it would allow time for the airline to assess its needs more accurately.[43]

Still throughout the post-Boeing selection period—in spite of the predominantly skeptical view of the Americans—Maxwell continued to use the Concorde threat for promoting the SST program. For example, in late June 1968 he told Congressman Philip Philbin (D-Massachusetts) that the Concorde was making progress, that a first prototype was "being readied" for flight in France, and that a second was being built in Britain and was "nearly complete." Maxwell reminded the congressman that the British and French governments had committed over $2 billion to the Concorde in the form of subsidies, loans, and loan guarantees and that the Concorde could possibly enter commercial service in the early 1970s, "three to four years ahead of our U.S. SST." Two months later Maxwell declared that "too much emphasis has been placed on Concorde problems and not enough on Concorde progress."[44]

Moreover during the latter half of 1968 progress was apparently being made. With great fanfare the French rolled out their prototype for its first taxiing trials on September 20. (Unfortunately for the French, Soviet troops in Czechoslovakia stole the headlines in the next day's papers.)[45] In contrast, the British prototype was unveiled in Britain at about the same time with a minimum of publicity.[46] Finally, after a frustrating series of further delays in early 1969, on March 2 a Concorde prototype flew twenty-seven miles over France.[47] (One reason for a Concorde rescheduling at this time was so that the aircraft would not have to compete for publicity with an American Apollo mission.)[48]

American views about the Concorde grew more complex during the Nixon administration, which began in January 1969. On the one hand, doubts continued, despite the success of the flight; reports on payload, fuel consumption, and aircraft sales were pessimistic. Similarly an interagency Ad Hoc Review Committee, which President Richard Nixon established to examine the American SST program, like the now-defunct PAC, did not take the Concorde very seriously. One member, Undersecretary of the Treasury Paul Volcker, believed that the Concorde posed no serious threat to American leadership in aviation, that it would not create a burden on American balance of payments, and that it should not be "an overriding factor in the consideration of our SST project."[49]

But respect for the Concorde as a potential rival was increasingly voiced, and the Nixon administration generally viewed the Concorde much more favorably than did its predecessors. By early April 1969 the French prototype had completed eight flights and had flown a total of ten hours, and on April 9 the British Concorde prototype made a successful first flight. One experienced observer characterized the British prototype's performance as a "good, standard, easy takeoff."[50] In mid-August Transportation Secretary John A. Volpe reported to President Nixon that the Concorde test flight phase was "progressing satisfactorily," with British confidence holding firm. On October 1 the French prototype flew at supersonic speeds for the first time. It was the aircraft's forty-fifth flight.[51] One SST Office official remarked that in spite of skeptical views from the Ad Hoc Review Committee, both the Concorde and the Russian TU-144 had been successful. The SST Office believed that the Concorde would be "a viable commercial aircraft" and would be operational in 1973.[52] Similarly in early May 1970 Undersecretary of State Alexis Johnson declared that the Concorde program was progressing favorably, with twelve aircraft authorized for construction; consideration was being given to a second-generation Concorde, which would be comparable in size and economic performance to the American SST.[53]

Meanwhile there was never any real pressure for collaboration between the two SST efforts. The totally different design philosophies of the Americans and the Europeans—and the Americans' goal of being the preeminent sole technological and market leader in aviation—precluded any meaningful technical cooperation.

Fundamental differences about the issue of the sonic boom also kept the programs apart. Although both the British and French appeared interested in discussing the sonic boom and aircraft noise with American officials, they consistently discounted the significance of these problems.[54] In December 1968 one Boeing officer reported to Maxwell that BOAC officials expected to fly the Concorde supersonically over Britain and actually ridiculed Boeing for its repeated warnings of overland flight restrictions. Maxwell tended to treat such statements lightly until he met with three European air-traffic-control representatives, who also affirmed that there were official plans to permit supersonic flights over Britain and France. Maxwell was surprised at the way in which Concorde officials ignored the implications of the sonic boom: "It is interesting how they have come to such a remarkably different conclusion on the feasibility of supersonic overland flight from that in this country, even though they are basing their conclusions on the same data and test results."[55]

American noise standards were another increasingly important concern for Concorde officials. In November 1969 the FAA issued new regulations, FAR 36, that established American standards for noise certification applicable to all new subsonic aircraft. The Concorde was considered an "old" aircraft; it applied for American certification in the summer of 1965. In mid-April 1970 the FAA issued a notice of proposed rule making that barred civil aircraft from generating a sonic boom detectable on the ground.[56] Concorde planners feared that the noise criteria for new subsonic aircraft would be extended to SSTs. As one high-level FAA official, with uncanny foresight, warned at that time, once standards are established and the public becomes accustomed to them, higher noise levels, even if permitted, in all likelihood would attract increasing opposition; public protest and political pressure could in fact cause subsonic noise standards to be imposed on supersonic aircraft, and local regulators, such as the New York Port Authority, could even establish their own noise limits and thus bar the Concorde from their airports.[57] (Such local regulations and protests were the focus of a battle over Concorde landing rights that lasted far into the 1970s.)

In summary, the mere existence of the Concorde project was an important political and bureaucratic weapon for the American SST advocates in keeping the U.S. effort alive, especially during periods when the Concorde was winning a certain amount of grudging American respect, as was true in the Nixon administration. Still the United States, as reflected especially by the PAC, was consistently skeptical about the Concorde's strength as a competitor, and the successive delays in the Concorde program appeared to confirm these doubts. Competition from the Concorde generally was not viewed seriously, and most American officials believed that a two- or three-year gap was tolerable. The idea of a foreign threat was never compelling enough to accelerate the American SST program, but this argument was crucial at times in helping the SST proponents successfully withstand attacks from the critics.

The Soviet TU-144

If overall skepticism—tinged with latent fear and occasional respect—marked the general American view of the Concorde, clear-cut disdain and irrelevance comprised general American impressions of the Russian SST. The Soviet aircraft, the TU-144, was never put in the Concorde's league as a potential competitor.

The Soviet Union ceremoniously unveiled its TU-144 design at the Paris Air Show in June 1965. The aircraft's slim delta configuration strongly resembled the Concorde, but unlike the Concorde, which had engines mounted in pods under the wings, the TU-144's four turbojet engines were concentrated at the center under the aircraft. The Russians reported that the cruise speed of the TU-144 was Mach 2.4 with a payload of 121 passengers, a range of 4,000 miles, and a gross takeoff weight of about 260,000 pounds. A prototype was scheduled to fly in 1968.[58]

From the beginning the Americans were not particularly impressed. Gordon Bain, director of the SST program, reported to Najeeb Halaby in 1965 that "most of the airlines of the Free World would be quite reluctant to accept a Russian transport if only because of the uncertainty of the manner in which they would live up to their warranties and provide for an effective system of parts supply." Bain concluded that the TU-144 would compete with neither the American SST nor the Concorde for the first-generation SST market, but he warned that the Russians were competent technically and that their aircraft could fly commercially within the Soviet sphere; such a situation could have "prestige implications," and the Soviet Union eventually could become a competitor in the SST field.[59] McNamara's aide, Joseph Califano, however, disagreed with this forecast, noting to McNamara that the TU-144 was slightly smaller than the Concorde and that the "Soviets have had virtually no success with their commercial airline ventures." He was also skeptical about the aircraft's projected flight date.[60] In any case, throughout the rest of 1965 and all of 1966, as evidenced from the considerable lack of attention paid to the Soviet effort in PAC meetings and major PAC reports, American officials viewed the TU-144 as relatively insignificant.

Of course, the TU-144 was still monitored. During 1966 the CIA periodically reported on the Soviet program to the FAA, which in turn relayed CIA data and estimates to the PAC as part of FAA status reports on foreign SSTs. The basic American view of the TU-144 as noncompetitive remained. For example, in August 1966, although indicating that the Soviets were contemplating a Mach 3 SST and predicting that the TU-144 would enter commercial service one to two years ahead of the Concorde, the CIA emphasized that the TU-144 probably would not initially achieve its design objectives. The agency termed the TU-144 display at the 1966 Paris Air Show the beginning of a "propaganda campaign" and suggested that the Soviets were working toward a first TU-144 flight in 1967 in honor of the fiftieth anniversary of the Bolshevik Revolution.[61]

At the end of 1966 overall Soviet prospects still appeared bleak. The TU-144 was behind schedule, and the Russians were trying to buy electrical systems and other technology from Britain. The British and French believed that the TU-144 would not undergo flight testing until late 1969, and the CIA now predicted that the aircraft would fly shortly after the Concorde.[62] In its report to the president on December 22, 1966 the PAC declared that the TU-144 was not "regarded as a competitor in the Free World Market" and that its limited production would be primarily aimed for the Soviet airline, Aeroflot. Four days later Califano echoed this view to President Johnson: the Russian SST possessed "a limited range and a very small market potential."[63]

Still American SST proponents could not resist using the TU-144 for their own propaganda purposes. In January 1967, when President Johnson was delaying the SST prototype contracts to Boeing and General Electric until the airlines contributed additional funds for development, industry representatives pointed to the TU-144 as a definite competitive threat. Union officials, speaking for thousands of workers in the aerospace industry, also predicted that the TU-144's first flight could come as early as 1967 and that the president was "making a big mistake" in delaying the American SST program.[64]

But in April 1968 the TU-144 was still on the ground. At that time, however, Russian aircraft designer A. N. Tupolev stated in a mysterious interview in Moscow (with an obscure correspondent of a Hungarian trade union newspaper) that the TU-144 was ready and "awaiting the order" to take off and that the Soviets were already preparing a "TU-154," which would be faster than the TU-144 and made from titanium and steel. The French were surprised at Tupolev's statement since their engineers, who had recently viewed TU-144 facilities, estimated that the TU-144 was about three months behind the Concorde and would not fly at least until June 1968.[65]

As with the Concorde, further delays then plagued the Russian vehicle. In mid-May 1968 French officials reported that the first of the three TU-144 prototypes would not be ready for four or five months; the Soviets were making a "radical change" in the wing configuration.[66] In October a high-level Soviet aviation official stated in an interview that the TU-144 was "under construction" and that the Russians would "begin to master" the aircraft "at the end of the present five-year plan." This indicated to the U.S. State Department that "the first flight of the TU-144 may not be planned for this year," as had been rumored.[67]

For once the TU-144 estimates were too pessimistic. On December 31, 1968 the TU-144 flew, beating the British, French, and Americans. Aeroflot announced TU-144 service over the vast and sparsely populated regions of the Soviet Union by 1971.[68] The Russians pressed their advantage and in 1969 permitted foreign correspondents to view the aircraft. On July 30, 1969, a group of representatives from TWA, Pan American, and Boeing visited the Soviet Union to make technological, operational, and economic evaluations. This team reported that the TU-144 would enter commercial service in 1973 or 1974. Detailed performance, weight, and cost data were not provided, however.

The American group generally was not impressed with the plane itself. The chairman of an airline advisory committee for the SST program declared that the aircraft "appeared to contain no novel features or extreme advancement in the state-of-the-art." Group members were skeptical about range and payload claims and observed that the TU-144 designs reminded them of the Concorde three years earlier.

But the American team was impressed by the strong Russian commitment to bring the TU-144 to commercial status; the Soviet SST group was dedicated and bound to improve with experience. One American visitor later told the FAA that the Russians were seeking to beat the Concorde into North Atlantic service, the Russians were reported to have flown the TU-144 at supersonic speeds, and the prototype appeared to be of high quality.[69] An FAA analyst at the end of 1969, studying the Tokyo-London route via the Soviet Union, concluded that the Russian aircraft had a substantial advantage in terms of flight time and distance.[70]

Russian hopes, however, were dashed: the Concorde first flew in March 1969, only a few months after its Soviet rival. The TU-144 never became a competitive threat in the world market.

Moreover both foreign SST programs became even less important in influencing the American SST program during the late 1960s. The real threat to the project was at home where other forces were transforming that effort into a mass political issue.

III EXPLOSION

15 The Emergence of a Protest Movement

Throughout the first half of the 1960s, important SST activity was more or less contained within the confines of government agencies, the White House, aviation-oriented congressional committees, the aerospace and airline industries, and certain specialized research and technical organizations like the National Academy of Sciences and various consulting firms. But toward the middle of the decade a significant change began to take place, one that was not immediately apparent: the SST slowly emerged as a matter of public concern. At first only a few newspaper and magazine articles signaled a growing general awareness by society at large of the SST's social and environmental effects. But in the last years of the 1960s the focus of significant involvement in the SST program gradually shifted to a new arena outside the government and the other established centers of interest. This new public perception of the SST evolved according to the following general pattern: initial criticism by the media, then formation of a few organized anti-SST public interest groups, and finally emergence of a widespread protest movement against the SST. By the end of 1969 the SST conflict was no longer just a bureaucratic, political, and technological struggle; it had become clearly visible to the public. It had turned into a major media event and had become a symbol of new concerns that were emerging across the nation.

Although it is difficult to pinpoint the exact beginning of this wave of public concern, an early important indicator was an extremely influential article in the July 1966 issue of *Harper's*, "The Case Against the SST," by John Gibson, dean of engineering at Oakland University in Michigan. Gibson attacked the SST program on a number of fronts, marshaling technological, economic, managerial, and environmental arguments. The FAA clearly viewed the article as a serious threat and organized an extensive in-house effort to refute each of its negative points.[1]

Gibson's critique spurred other anti-SST press pieces. An editorial in the *Wall Street Journal* that same month discussed technical and economic problems associated with the SST. The paper's attack was based primarily on its opposition to the government's SST financial involvement. "The SST Program," the editorial said, "is all too typical of a Government that seems to think that it can fight a war, go to the moon and do practically everything else all at once."[2]

The *Wall Street Journal* carried its anti-SST campaign into 1967. In

February columnist Fred Zimmerman wrote a piece, "Supersonic Snow Job," that argued that the program was running largely on FAA propaganda and that pointed out that many government economists disagreed with the FAA's conclusions. Zimmerman sarcastically mentioned that in the sonic boom the "FAA press agency faces its greatest challenge" and implied that an FAA claim that an exclusively over-water SST would still be profitable was more a matter of image building than sound economics.[3]

Other critical evaluations were published. Newspapers—the *Detroit News*, for example—reported negative findings to their readers.[4] In June 1966 Kurt H. Hohenemser, a professor of aerospace engineering at Washington University in St. Louis, wrote a negative article for the *Los Angeles Times* that received considerable local attention. (This article also caused concern among FAA officials, one of whom said it "contained a smattering of truths about the SST with highly derogatory interpretations.") Six months later in the *Bulletin of the Atomic Scientists*, Hohenemser again highlighted problems associated with SSTs and attracted more public interest. Congressman Ogden Reid (D-New York) asked the FAA to comment on the article, which had been brought to his attention by one of his constituents. Maxwell responded with several counterarguments.[5]

As anti-SST commentary in the press mounted during late 1966, and 1967, the FAA began to monitor all relevant press statements, many forwarded by Boeing or other SST contractors. SST media coverage and related public opinion in the United States and in Europe soon became major concerns of the agency. Important press reports and commentaries were often reviewed personally by Maxwell and McKee, who made a concerted effort to discuss the SST with key media representatives and to write letters to the editor on particular SST issues.[6] After the *Washington Evening Star* had published three separate anti-SST editorials, for example, McKee requested a meeting with the paper's editor, but this action had little effect on the paper's anti-SST stance.[7]

In late March 1967 the *Washington Daily News* carried an article that focused on the "current gloom" at the FAA. The report termed the sonic boom the FAA's "biggest problem" and stated that over-water route restrictions would make the SST economically infeasible; in fact, former FAA administrator Najeeb Halaby had made precisely this statement two years earlier. The article also cited engine noise as a critical problem.[8]

The FAA was working particularly hard to counter negative publicity in the sonic boom area. By March 1967 the National Educational Television network had completed a film, "Noise: The New Pollutant," which discussed the sonic boom and which included, by permission of presidential science adviser Donald Hornig, actual footage of the sonic boom tests at Edwards Air Force Base in California.[9]

The increasing media criticism of the SST, especially over the sonic boom, resulted in a number of congressional inquiries during 1967. Early in the year Senator Frank Moss (D-Utah) informed McKee of reports of damage at Bryce Canyon and Canyon De Chelly National Parks caused by sonic boom. McKee promised an investigation of these allegations. Moss then issued a public statement to the effect that a thorough investigation had been launched and that he had asked the FAA to see if supersonic flight patterns could avoid areas of national significance or of dense population. Local Utah newspapers, however, soon began to complain of sonic booms in other regions of the state. In Richfield, Utah, the *Reaper* carried an angry and sarcastic editorial that spoke of rattling windows and frightened children. This kind of commentary generated still another concerned letter from Moss to the FAA.[10]

In early May 1967 Senator Stuart Symington sent the FAA a *New York Times* editorial that attacked the SST program on various financial and economic grounds, and he asked for comment. In his reply McKee defended the program, saying that the government's involvement would be limited to the prototype phase and that the contracts with Boeing and General Electric would protect the government's investment.[11]

Later in the year Maxwell responded to other SST inquiries from Senators Clifford Case (R-New Jersey), Edward Kennedy (D-Massachusetts), and William Proxmire (D-Wisconsin). "We are continuing with our battles," Maxwell wrote at one point to the FAA consultant Edmund Learned, "and, while I believe we will get the program through the Congress, we still have problems defending it within the Administration against the budget cutters."[12]

Not all press comment on the SST program was unfavorable, and the FAA naturally tried to encourage positive media coverage. Both

Aerospace News, a publication of the Aerospace Industries Association, and the *Air Force/Space Digest* ran pro-SST articles and earned the approval of SST supporters.[13] The FAA was extremely pleased with an article in the February 1967 issue of *Fortune*, "The $4-Billion Machine That Reshapes Geography," which downplayed the sonic boom issue yet presented a comprehensive review of the entire SST program. Maxwell declared that the article was the "best-written story on the SST that has been published." FAA economist Allen Skaggs also offered compliments; the authors had "left no stones unturned in seeking out and understanding the many biases which have confronted us in our diligent efforts to do a good job for all interested parties—the Government, the airlines and the manufacturing industries."[14]

In April 1967 the Air Transport Association drafted a report that defended government participation in SST development.[15] The journal *Airline Marketing and Management* and the *Miami Herald* both strongly supported the program, and the White House was informed of this support.[16] In June the SST Office sent material to the journal *Space/Aeronautics* for a new piece on the SST.[17] Maxwell also complimented *Aerospace International* for its pro-SST article, "Global Air Transport: The Next Decade." The journal, he declared, was "performing a real service."[18] Continuing his active public relations work, Maxwell at the end of 1967 held a luncheon meeting with the *New York Times* editorial staff to discuss the SST.[19]

But unfavorable media reports continued to appear throughout 1967. The July–August issue of *Challenge* carried an article, "SST: Not All Smooth Flying," which argued that financing and managing the SST program should be moved from the government to the private sector. The SST Office responded that the airlines alone could not possibly assume the large financial burden required.[20] In mid-August syndicated newspaper columnists Rowland Evans and Robert Novak accused the administration of covering up a 1965 Treasury Department report that allegedly called for higher cash contributions from the manufacturers and also concluded that the government would not recover its SST investment. Another Evans-Novak column referred to a second secret report that had found that "practically *all* financing of the SST could come from private sources." In response the SST Office informed key senators and representatives that the first of these reports had really been a working paper and that the second had never been classified as confidential; the second report had not proved that the private sector was "willing or able" to finance the SST (although the FAA acknowledged that private financing was being considered for future phases of the program). In December the *Washington Post* reported on the conflict over sonic boom policy within the administration.[21]

Other negative articles appeared in 1967 in such diverse journals as *Business Horizon, Harper's,* and *Commonweal.* An anti-SST remark by an official of National Airlines was mentioned in a mid-October issue of *Aviation Daily,* and the *Washington Post* reported in mid-December that Secretary of the Interior Stewart Udall was establishing his own panel "to evaluate how widespread sonic booms might destroy the natural resource of tranquillity." McKee responded with a point-by-point rebuttal of Udall's views.[22]

In countering these media attacks the FAA received some assistance from certain airlines and manufacturers. In early February 1967 Juan Trippe of Pan American sent McKee a strongly pro-SST editorial from the *Los Angeles Times* and mentioned that the paper's chairman was on the board of Pan American. McKee expressed his gratitude for the support.[23] Four months later Charles Tillinghast of TWA, after a particularly stinging anti-SST speech by Senator Proxmire, assured McKee that TWA's officials were doing everything possible to "place the facts before Congress."[24]

Even Boeing became more active in public affairs. By the end of May it had organized a "public relations plan" for its SST subcontractors that was designed to sell the program to the general public, to government decision makers, and to the financial and technical communities. The plan would use press coverage, advertising, brochures and giveaways, speeches, technical papers, and motion pictures. After a *Wall Street Journal* editor had spoken highly of a recent visit to Boeing, the company took credit for contributing to a "better attitude by the *Wall Street Journal.*"[25]

By the autumn of 1967 the SST Office also had become much better organized. It established a special projects staff under Colonel J. H. Voyles to coordinate and organize the office's public relations efforts. By the end of the year this group was screening all invitations for public appearances, assisting in the preparation of speeches and papers, and distributing copies of such material. Voyles's assistant, Paul Frankfurt, coordinated and encouraged public relations activities with

Boeing and arranged appearances for Maxwell with such groups as the New York Bar Association and the *New York Times* editorial staff.[26]

The FAA conceived some fairly grandiose publicity plans. At one point it was suggested that members of the Hollywood Press Club, along with the actor Jimmy Stewart, be flown to Seattle to discuss the SST with Maxwell. Another bright idea was to promote the SST on the television show "My Three Sons," where the star, Fred MacMurray, was cast as an aeronautical engineer.[27]

Voyles's group actively worked with public relations officers at Boeing and General Electric. Boeing officials emphasized the need for a "total information plan," kept Voyles informed of public relations activities at Boeing, and assisted in preparing the SST segment of the FAA's annual report. Voyles in turn suggested that an industrial publisher be hired to design a pro-SST booklet for the general public. General Electric also supplied Voyles with photographs and assisted in the preparation of the FAA's annual report.[28]

During late 1966 and throughout 1967 critical views by prominent individuals were expressed with increasing frequency. These were the seeds of the first significant and organized public opposition to the SST. In a letter to the *New York Times* in August 1966, John Edsall, a respected Harvard biology professor, expressed worry over the "biological hazard" of the SST, mentioning such problems as cosmic radiation and the sonic boom. Edsall also complained to Congressman Thomas P. O'Neill, Jr. (D-Massachusetts), referring to the anti-SST technical and economic analysis by Hohenemser in his *Bulletin of the Atomic Scientists* article and to the financial criticisms made by Zimmerman in the *Wall Street Journal*. Maxwell responded to these criticisms as best he could, but he was facing ominously growing resistance; the opposition press and the individual opponents were beginning to reinforce one another.[29] Notable figures, including the photographer Ansel Adams, wrote directly to the FAA to complain about the sonic boom.[30] In April 1967 consumer activist Ralph Nader demanded information on the 1964 Oklahoma City sonic boom tests. Fortunately for the FAA, Nader did not take on the SST as a major issue for advocacy at this time.[31]

One highly influential long-term opponent of the SST program was Bo Lundberg, director-general of the Aeronautical Research Institute of Sweden. Lundberg was one of the few technically credible individuals to criticize the SST since the early 1960s, and he provided later critics with a foundation of data and arguments to support their attack. In December 1966 Lundberg wrote to Senator Warren Magnuson, a strong SST supporter, about the hazards of the sonic boom to people at sea and about the adverse economic impact of the boom. In response, McKee told Magnuson that there was no evidence that the SST's sonic booms would cause physical damage to structures or people on land, let alone on water.

Some in the government tried to ignore Lundberg. The FAA was cool to the idea of meeting with him, considering his arguments too extreme. One high-level Transportation Department official advised Secretary of Transportation Alan Boyd in mid-1967, after referring to Lundberg's "exaggerated and subjective efforts to publicize his view [that] borders on hysteria": "I feel you cannot afford to meet every 'crackpot' who has strong opinions on supersonic civil aircraft development. I recommend [that] you not invite Dr. Lundberg to meet with you." But the OST Coordinating Committee for Sonic Boom Studies reviewed Lundberg's paper, "The Menace of the Sonic Boom to Society and Civil Aviation," at an April 1967 meeting. Maxwell eventually became concerned enough to send Boeing executives another piece by Lundberg, which Proxmire had inserted in the *Congressional Record*; Maxwell mentioned that the FAA was preparing a counterstatement. He then said that Boeing should "know what the opposition is saying" and suggested that Boeing too should take some "follow-up action."[32]

In terms of public opposition, the most significant development in 1967 was the birth of the Citizens League Against the Sonic Boom under the vigorous and innovative leadership of William A. Shurcliff, a fifty-eight-year-old physicist at Harvard University's electron accelerator. Shurcliff's specialty was radiation and optics. He held patents to twenty-one inventions, mostly in optics. Among other things, he had invented a formula for military camouflage paint at American Cyanamid and an automatic focusing slide projector at Polaroid, where he had worked for thirteen years. After Polaroid Shurcliff joined Harvard to direct radiation security at the accelerator. As this position implied, Shurcliff was careful, thorough, and highly organized.

Shurcliff had been impressed by Lundberg's pieces and by the letters to the *New York Times* by Edsall and others on the dangers of the sonic boom. As early as October 1966 he had requested information on the sonic boom tests from the FAA, and by January 1967 he had begun his anti-SST campaign. He wrote a number of letters to members of Congress and to administration officials complaining that a NASA report on the sonic boom was biased in favor of the SST, and the FAA then received inquiries on this issue from numerous congressmen who had been motivated by Shurcliff's letters. The debate between Shurcliff and the FAA over the findings of the NASA report continued throughout the first three months of 1967.[33]

On March 9 Shurcliff formed the Citizens League Against the Sonic Boom, with himself as director and Edsall as deputy director. This group was headquartered in Shurcliff's home in Cambridge, Massachusetts, and its staff consisted of Shurcliff, his sister, Mrs. F. J. Ingelfinger, and his son, Charles H. Shurcliff. The Citizens League prepared anti-sonic-boom material, published "fact sheets" and newsletters, distributed news items to newspapers, columnists, and television and radio stations, and reprinted anti-sonic-boom pieces by Lundberg and others. Although small, it had few financial problems. Soon after it was established it began placing advertisements in the *New York Times*, the *Wall Street Journal*, the *Washington Post*, and the *New Republic*. Funds and membership grew quickly, and by 1970 the Citizens League had about 5,000 members.

Shurcliff's Citizens League was the antithesis of the giant governmental and corporate organizations that were the league's adversaries. There was a minimum of bureaucracy, and the turnaround time for a reply or action was virtually instantaneous. Fund raising was never a problem for Shurcliff. Once or twice annually he would mention in the Citizens League newsletter that the group needed money. The response rate of such an appeal was typically about $10,000 during the period between 1967 and 1971. In addition, Shurcliff never neglected members who did not contribute financially since he believed that such people could very well be helping the cause in other crucial ways. Shurcliff's bookkeeping and records were immaculate. An index card was kept on every member, which listed the date and amount of each contribution, and the type of acknowledgment returned. Most of the funds were used for postage, duplicating, and printing. The Citizens League produced a prolific amount of material, all of which was free of charge. Five years later Shurcliff described his highly effective operating style: "Except on two occasions each year, I never (almost never) mention money to members. If they write me, I write them back at once, with news, etc., but no mention of money. When I do receive a check, I try to mail off an acknowledgement the same day, always with at least a minuscule personal touch. Always done by me; never by a secretary."[34]

The organization was extremely effective. Fewer than ten days after it was formed, the *Christian Science Monitor* reported on Shurcliff's activities and mentioned the anti-sonic-boom arguments of Lundberg and others. By late April the Citizens League had published eleven fact sheets, and in the following months the FAA was deluged with related inquiries. Shurcliff warned the FAA on numerous occasions that the Citizens League intended to halt the SST program. He wrote a stream of letters to Senator Edward Kennedy, Senator Edward Brooke, Secretary of State Dean Rusk, the president, and a host of other public officials. These letters invariably led to requests for an FAA response.

Shurcliff pressed to make the FAA define precisely what it meant by "sparsely populated land areas," and he rhetorically asked Edward Kennedy if such regions included Cape Cod. In June he told Maxwell to ban supersonic flight "over all regions where there are people."[35]

Throughout the rest of 1967 the Citizens League continued to run newspaper advertisements, to publish fact sheets, and to write anti-SST letters to influential groups and individuals. It soon achieved considerable notoriety. By the end of August the Citizens League had about 620 members and had received more than $14,000 in voluntary contributions; by December it had 2,000 members. It reported that during July and August alone, approximately 250 editorials and news articles had mentioned the threat of the sonic boom and the possibility that the SST would fail financially. It monitored all foreign anti-SST and anti-sonic-boom activities and regulations, including the formation of the British Anti-Concorde Project. Shurcliff continued to demand that the FAA categorically ban all domestic supersonic flight and continued to claim (using data from a recently released official economic study) that the SST would be a financial disaster. The Citizens League even approached SST subcontractors and suggested that they withdraw from the SST program because of "human and financial factors." In September *Harper's* ran a profile of Shurcliff, entitled "The Case of the Angry Physicist," which generated more letters of protest to the

FAA and to Congress.[36] As a result of the league's activities, more people both inside and outside the organization were encouraged to express their anti-SST opinions publicly, and the number of anti-SST protest letters to the government swelled during 1967.[37]

At this time the sonic boom was the issue that most frequently motivated SST opposition. Senator Mike Monroney pressed the FAA to settle sonic boom claims quickly and fairly.[38] The *Chicago American* editorialized that many people would not "get used to sudden bangs that are irregular, unpredictable, and ear-splitting."[39] Dr. Henry Allen, an eye surgeon at Harvard Medical School, warned that during eye surgery a "startled reaction to a sonic boom" could result in permanent blindness for the patient.[40] The city council of Santa Barbara, California, unanimously voted to ban the sonic boom over that community. (Maxwell thought that this action was important enough to write Santa Barbara's mayor and ask for more information.)[41] By the end of 1967 the anti-SST forces not only possessed an effective, growing, and adequately funded protest organization, but they had found in the sonic boom an issue well suited to capture the public's concern.

At first, the FAA offered a more or less standard response to Shurcliff and other SST opponents: the Concorde and the Russian TU-144, if left unchallenged, would monopolize a lucrative commercial supersonic market. The agency began assembling a package of materials to respond to congressional inquiries activated by the Citizens League. "Even if the United States abandoned the SST Program tomorrow," Maxwell declared, "the future of commercial aviation would not be drastically altered. The airlines would simply buy more foreign-made supersonic airplanes." The FAA also declared on numerous occasions that sonic boom overpressures could be somewhat modified because their intensities would be affected by such factors as altitude, size, and the shape and weight of the aircraft. The American SST, in Maxwell's words, had "excellent sonic boom characteristics for an aircraft of its size and performance."

In July the FAA faced more protest letters as a result of advertisements run by the Citizens League. McKee complained that a particular Citizens League advertisement "grossly exaggerates the effects of the sonic boom and makes it out to be a menace to life, limb, and property."[42] Eventually, almost in exasperation, the number of substantive FAA replies in response to Shurcliff's letters diminished;

henceforth the agency simply acknowledged receipt. One FAA official wrote Shurcliff, "It is clearly evident from our previous correspondence that it serves no useful purpose for us to attempt to respond to your charges."[43] Maxwell himself admitted that the great amount of correspondence between the Citizens League and the FAA had been "generally fruitless."[44]

SST proponents had acquired a new awareness of the importance of public opinion. During 1968 and 1969 there were a series of coordinated SST public relations programs by contractors and subcontractors, more frequent formal contact between top SST managers and representatives of the media, and symposiums on SST matters organized by the aerospace industry and others.

The FAA's public relations efforts during 1968 and 1969 centered around Voyles's special projects staff at the SST Office. Voyles's unit and the public relations executives at Boeing and General Electric coordinated their activities quite closely. The FAA carefully reviewed Boeing's SST brochures, advertisements, and releases for newspaper and television news stories, and it often provided detailed criticisms and modifications.[45] Boeing and the FAA exchanged information on SST media coverage and on public opinion regarding the program. In September 1968, for example, the FAA listed as public relations activities three speeches on the SST by FAA personnel, a *Fortune* magazine interview with Maxwell, and six articles on the SST in the *New York Times*, the *Washington Post*, the *Washington Star*, and *Time* magazine. During June and July 1969 Boeing gave six interviews, submitted five technical papers, prepared SST background information for its SST subcontractors, revised twelve film catalogs, and arranged various showings of a Boeing film, "You and Me—and the SST," which Boeing had produced in early 1969.[46]

In July 1968 Boeing's SST public relations manager, Gregg Reynolds, sent the FAA copies of articles on the SST from *Boeing* and *Sunset* magazines.[47] He suggested that they could get "grassroots exposure" by reprinting these articles in FAA employee publications.[48] Boeing also organized an extensive advertising program. The firm's SST public relations manager boasted to Voyles that Boeing's SST advertising would result in more than one advertising impression per individual in the United States; this, he felt, would assure "solid grassroots support."[49]

Boeing and the FAA exchanged negative press comments as well. For example, at the end of January 1969 Boeing sent the FAA a copy of a "horribly inaccurate SST story," and its SST public relations manager mentioned that he was trying to get more information on the piece, which he would pass on to the FAA.[50] Boeing also kept the FAA informed of its subcontractors' public relations work and of the distribution of its SST literature.[51] In May 1968 Boeing had shipped to the FAA 2,000 copies of an SST program brochure, "The U.S. Supersonic Transport"; a month later the FAA ordered an additional 10,000 copies. Thousands of Boeing brochures went to the National Aerospace Education Council and to other professional associations.[52]

FAA involvement in Boeing's SST public relations effort was active and pervasive. The agency monitored press briefings and reviewed articles, brochures, and press releases before publication. It was generally quite satisfied with Boeing's efforts, and Maxwell particularly praised one brochure, "The SST and Its National Benefits," as an "outstanding job." The FAA also approved a public relations scheme that involved Fred MacMurray and other show business personalities, but it warned against big "Hollywood productions."[53]

The public relations units at the FAA and at General Electric developed a similarly cooperative relationship. General Electric produced and distributed various materials on the SST, such as a brochure, "Financing and Investment Recovery," a pamphlet, "Washington to London," an engine fact sheet, and various progress reports. Because the FAA had to approve the release of such literature, it became involved in nearly all General Electric SST publications; the FAA even reviewed the SST references in General Electric's 1968 annual report. The company was proud of its SST public relations work, especially as it related to the financial community and government. General Electric placed advertisements on the SST in such publications as *Time, U.S. News and World Report, Newsweek,* the *Reporter, Forbes, Business Week, Saturday Review,* and the *New Yorker.*[54]

The FAA too was particularly pleased in April 1968 with General Electric, especially with a one-minute commercial that was scheduled to run three times on network television. Finding the commercial especially convincing Voyles happily told Maxwell that General Electric had hit "bingo" with this piece.[55]

There was also a three-way exchange of information on SST public opinion among the FAA, Boeing, and General Electric. General Electric took many of its cues from Boeing's presentations, and the two companies often worked together to counteract adverse press coverage. Referring to a rather negative article on the SST in the *PR Reporter*, the Boeing SST public relations manager told his General Electric counterpart that the article "smacks of Dr. Shurcliff and company." He suggested that they compose a letter for the FAA to send that would outline the SST program and its "national benefits." His General Electric colleague agreed and told Voyles that Boeing and General Electric would be accused of having their own "axes to grind" if they sent the letter themselves. But the *PR Reporter*'s criticisms continued into early summer. General Electric's SST public information manager complained to Boeing and the FAA that "the anti-SST forces still come through. . . . Anything we can do with these people? Seems sort of a shame that we're being castigated within our own profession." In the autumn of 1968 representatives from Boeing, General Electric, and the FAA met to discuss future SST communications and public relations programs.[56] The SST program also received public relations assistance from such industry groups as the Air Transport Association.[57]

The FAA continued to seek favorable media coverage. In February 1968 McKee vigorously defended the SST program in a published interview with noted economist Eliot Janeway, and later that spring Maxwell taped a fifteen-minute interview for widespread radio use.[58]

But the FAA was not always fully successful. For example, Maxwell agreed to be interviewed for the network television news program the "Huntley-Brinkley Report" (after being assured that he would not be asked any "loaded" questions). With a possible audience of 40 million to 50 million viewers, this would have provided an excellent opportunity to promote the SST. But on June 28, 1968, when the interview was scheduled for broadcast, it was drowned out by other news items and fed only to local NBC-affiliated television and radio stations.[59]

On the other hand, in January 1969 the CBS television show, "The 21st Century," narrated by prominent newscaster Walter Cronkite, featured the Boeing SST simulator. Just before the show was aired, a Boeing news release exclaimed, "What will it be like to fly the U.S. SST? Walter Cronkite found out in the Boeing Visual Space Flight Simulator."[60]

The FAA continued to monitor media coverage closely. In February 1969 the agency praised a Seattle television station for a documentary on the SST.[61] And in early April Maxwell personally thanked a U.S. House member for the "tremendous boost" he had given the SST program with his supportive comments on the "Today" television show.[62]

The FAA's attention to the written press was even more intense than that given to the electronic media. In February 1968 the agency drafted a detailed rebuttal to an Evans-Novak column that accused the forthcoming Johnson administration budget of providing a 60 percent increase in SST appropriations and a "windfall" for Boeing in the midst of wartime.[63] The FAA provided SST information for articles in *Air Travel*[64] and in *Aerospace Technology*, which devoted its May 20, 1968 issue to SST technology. In fact, many of the contributors to *Aerospace Technology* were from the SST Office, and the FAA was delighted with the journal's coverage; "technical reporting at its best," Maxwell called it, and predicted that it would be a real help to the program.[65] The agency was also pleased with a piece in *Holiday*, "Travel Jets into the Future" by Ernest K. Gann,[66] and it provided extensive background material for an article in the July 1968 *Fortune*.[67] Throughout 1968 and 1969 the agency continued to disseminate SST information to writers and to review advance articles on the SST.[68]

The FAA spent as much time countering negative reports as promoting favorable ones. In May 1968 Maxwell wrote a letter to the editor of *Prevention*, objecting to an unfavorable article in that journal, "SST: Soul Shattering Thunder"; after some haggling, Maxwell's entire letter was published.[69] At about the same time the General Welfare Committee of the New York City Council reported on a resolution to prohibit SST landings or takeoffs within one hundred miles of the city. McKee told Mayor John Lindsay that the resolution would not protect New York City against sonic booms generated by SSTs flying supersonically over the city and argued that the prohibition, if enforced, would simply divert the SST travelers to other cities. McKee then urged Lindsay to seek the advice of John Dunning, chairman of the NAS Committee on SST-Sonic Boom, as well as New York City's Science and Technical Advisory Council.[70]

Two weeks later Maxwell wrote *Newsweek* an angry letter to correct a piece reporting "more bad news" for the SST.[71] He also wrote to the *New Republic*, claiming that certain information used in an article had been quoted out of context.[72] And when the *Atlantic Monthly* carried a rather unfavorable article on the sonic boom (which also publicized Shurcliff's Citizens League) and then sent Maxwell copies of the article after it was already in press, Maxwell revealed his exasperation: "My observation is that if your editors had wanted to use my comments I would have received the article before it was printed."[73]

The FAA's efforts to mobilize favorable public opinion became increasingly aggressive in 1968. Maxwell and other key SST officers began to hold meetings and to give speeches on the SST in various parts of the country. In January 1968 Maxwell met in Cleveland with top media representatives, corporate executives, civic leaders, and educators. On the same day he also held a press conference and appeared at a Rotary Club luncheon.[74] In February and March Maxwell gave speeches in Dayton, New York, and Philadelphia.[75]

The FAA placed particular importance on a speech Maxwell was to give at the Wings Club in New York City on March 20, 1968. Because New York was the hub of the communications and financial community, this talk presented an excellent opportunity to receive wide coverage and to meet influential news people. Boeing sent Maxwell some background material. The FAA successfully attracted approximately twenty members of the press to the Wings Club talk, including syndicated columnist Bob Considine, whose subsequent reviews pleased Maxwell immensely. He called Considine's accounts of the SST "one of the finest I have ever read," and he also suggested that McKee send Considine a note of appreciation.[76]

Maxwell spoke at the University of Portland in April 1968.[77] He gave an important speech to the National Association of Manufacturers in May. In coordination with this talk, Boeing and General Electric prepared news releases and offered a newsclip for network and local television stations.[78] In June Maxwell spoke at a dinner in Pittsburgh, supported by Boeing and General Electric pamphlets, a three-minute newsclip, and a television interview.[79] In September he participated in a symposium at the Air Force Association's annual meeting and addressed the Scientific Research Society of America.[80] In 1969 he spoke at the University of Illinois, the Channel City Club of Santa Barbara, California, the Armed Forces Management Association, the

Boeing Management Association in Florida, the Advisory Committee of the Council on Transportation Law in Washington, D.C., and at other meetings of professional associations across the country.[81]

Other members of the SST Office were also mobilized and encouraged (as "an excellent way of getting our SST message across on the local level") to address local chapters of various technical and scientific societies.[82] During 1968 one SST Office administrator took part in a seminar at Dartmouth College; members of the office's flight operations and safety branch participated in a number of technical and nontechnical conferences; another official gave several talks to local chapters of the American Society of Mechanical Engineers and the Management Club in North Hollywood, California; and still another officer addressed the Society of American Military Engineers in Tullahoma, Tennessee.[83]

A key attempt by the FAA to influence a certain sector of public opinion on the SST issue was Maxwell's effort to organize the SST portions of the annual meeting of the American Institute of Aeronautics and Astronautics, which was held in Philadelphia in late October 1968. Maxwell began work on this project in April and was heavily supported by the SST Office staff in his planning. Maxwell had high-level airline executives and John Dunning chair the various SST sessions. Maxwell wanted to invite the Soviets to present a paper on the TU-144 program, but ultimately, after much negotiating, this idea fell through. General Jean Forestier, the French Concorde program manager, and J. A. Hamilton, director of the Concorde program in Britain, presented a jointly authored paper, and another French official gave a talk on the Concorde flight test program.

Speeches were given by at least two persons connected with the NAS Committee on SST-Sonic Boom: John Dunning and Raymond Bauer, chairman of the NAS subcommittee on human response. Maxwell also made an address. Many of the other papers presented were written by individuals in the FAA, Boeing, General Electric, or subcontracted firms. The meeting's proceedings were published in a hardbound book, and each session was televised and videotaped.[84] After the meeting Maxwell was extremely pleased. The media coverage had been quite favorable, and the FAA had been complimented on its technical reports and its management of the conference.[85]

The FAA had focused much of its public relations efforts on its natural allies in technical, professional, and industrial associations. There was a close relationship among several individuals in various organizations that were involved in or supported the SST. During the latter half of 1968 Maxwell served on the national dinner committee of the Explorers Club for organizing an annual event honoring "noted pioneers of the aviation industry."[86] He also attended a Wings Club dinner in October 1968 honoring the four Fathers of Aviation: Juan Trippe of Pan American, Eddie Rickenbacker of Eastern, C. R. Smith of American, and W. A. Patterson of United.[87] In March 1969 he was invited to an "off-the-record" dinner of the American Institute of Aeronautics and Astronautics' aerospace technical committee,[88] and in April he was nominated for the National Aviation Club's award of achievement.[89] Such efforts by the FAA had some success. The presidents of the Air Force Association, the Air Transport Association, and the International Association of Machinists and Aerospace Workers all wrote to the White House in 1969 expressing their vigorous support for the SST program.[90]

The SST Office increasingly grew aware of the need to influence the public at large during the 1968–1969 period. In 1968 it established an SST slide library, contracted for a film on the SST, and developed a number of informational pamphlets. By July 1969 it had organized an internal group of public speakers on the SST program; one officer even suggested establishing an "SST Club" for students, for which the FAA would furnish certificates and pins.[91]

In spite of all these efforts, the SST Office throughout 1968 and 1969 continued to face adverse media coverage, and it was inundated with anti-SST letters. Most of the negative SST comments were related to the sonic boom. In August 1968 Congressman Bob Mathias (R-California) sent the FAA a series of news items on sonic booms that indicated growing concern. One discussed the unusual Santa Barbara City Council ordinance that had outlawed sonic booms, and another reported that a sonic boom had caused a fatal convulsion in a six-year-old child.[92] Other critical articles appeared in such diverse publications as *Reader's Digest*, *Good Housekeeping*, and the *Sacramento Bee*.[93] In June 1969 the California State Senate passed a resolution that called for Congress and the president to recognize the possible harmful effects of sonic booms and to seek ways to alleviate further damage.[94] Many congressional SST opponents sent their protests directly to the FAA.[95]

Prominent individuals and ordinary citizens sent protest letters to the FAA, the Department of Transportation, and the president; constituents wrote to their members of Congress, who in turn referred the letters to the FAA. For example, in a letter to Senator Margaret Chase Smith (R-Maine) in May 1968, the president of the Natural Resources Council of Maine observed that the critical hazard of the sonic boom was its psychological effect on humans rather than its physical effects on property. He feared that the "lure" of profits from an overland SST market would lead to SST flights over wilderness areas, which he said should be "the last place where booms should be tolerated." Maxwell assured Senator Smith that the FAA was "not banking" even on circuitous overland routing to avoid populated areas; the alleged SST over-water market of five hundred aircraft alone was "a sufficiently attractive market to warrant development."

An angry telegram from an audiologist in March 1969 reflected the intensity of much of the growing anti-SST feeling: "An airplane that can travel only over uninhabited regions must still get to and from these regions. If it drags its sonic boom over inhabited areas, it will leave in its wake screaming children, frightened dogs, old people falling downstairs from the shock, and quite possibly serious emotional upsets. The problem is no laughing matter, as any audiologist or musician could explain, for it is we who are the specialists in this field."[96]

In fact, during 1968 and 1969, despite a few favorable letters, most of the SST-related correspondence received by the FAA and the administration was negative.[97] The vast majority of letters addressed to President Nixon and to Transportation Secretary John A. Volpe urged outright termination of the SST program. (One Department of Transportation official, however, believed that these negative letters had simply been "instigated" by Shurcliff's Citizens League.)[98]

The FAA's typical defense was to argue that a successful SST program would directly and indirectly provide jobs and would stimulate the national economy with massive amounts of capital, thereby increasing future revenues. "Reliable economists," it said repeatedly, predicted an over-water, intercontinental market of five hundred SSTs, with sales of only three hundred aircraft required to break even. There were continuous assurances that the government would restrict commercial supersonic operations to over-water, intercontinental routes if sonic boom noise was "excessive"; in any case, recent NAS evaluations had shown that even extremely intense booms would not cause hearing or direct physiological damage.[99]

During 1968 and 1969 Shurcliff's Citizens League maintained and nurtured the SST opposition. While SST proponents met with one another in government agencies, in corporate headquarters, or at annual association meetings, Shurcliff constantly hounded the FAA and diligently and vigorously carried on his attack from his home in Cambridge, Massachusetts.

At the beginning of 1968 Shurcliff expressed doubt about an estimate by Maxwell of the upper overpressure limits of a normal sonic boom and noted that the SST's sonic boom could be twice as great as the 2 psf Maxwell claimed. Shurcliff also asked for lawsuit information, raised doubts about the FAA market projections, and questioned Maxwell about the stability of the SST design. His barrage continued throughout the year, and his arguments were generally well documented with technical reports and articles.[100]

Shurcliff's tactics were reflected in his handling of an edition of his *SST and Sonic Boom Handbook.* In November 1968 he sent the SST Office a copy and asked to be notified of any mistakes. Maxwell angrily refused: "In light of your many similar requests in the past for my comments, which you chose not to heed, and in view of the numerous occasions in which you distorted quotations of mine, I decline to point out the mistakes in the handbook. It is replete with errors and inaccuracies; my staff does not have time to do your work for you." Other top officials in the FAA and in the Department of Transportation also refused to provide comments.[101] But the SST Office actually took Shurcliff's *Handbook* more seriously than it would admit; at least two divisions in the SST Office went over the manuscript in considerable detail.[102]

Shurcliff continued to press his point. He raised questions about the SST's technical risk, requested a set of news releases on the SST from 1967 and 1968, and claimed that Maxwell's speeches provided inadequate information.[103]

Throughout the first four months of 1969 Shurcliff and the director of the FAA's Office of Noise Abatement argued over the extent of the

FAA's authority over noise and the need for noise regulation. The FAA's noise regulations, according to the FAA official, had to be "technologically practical, economically reasonable, and appropriate to the particular type": "If aviation is to survive and serve the public, the public must be protected by restricting where certain types of aircraft fly, rather than by establishing unrealistic noise levels that merely stop aeronautical development."[104]

Shurcliff's attack became broader, more strident, and more vigorous in 1969. In June the Citizens League issued a release on economics that forecasted an SST market closer to one hundred aircraft instead of the five hundred claimed by the SST Office.[105] Shurcliff also battled the notion of allowing solely over-water supersonic flights. He claimed that at any given time there would be 200,000 to 1 million people on the oceans under main SST routes. He estimated that one SST flying from New York to Europe would boom about 4,000 people shipboard alone. Many over-water SST routes would also be above populated islands in the Caribbean, the South Pacific, the Arctic, and along the African coast. In July 1969 Shurcliff transmitted to the FAA a petition by more than one hundred yachtsmen, all of them members of the eminent Cruising Club of America, which requested that the SST be prevented from inflicting sonic booms on American coastal waters. In September Shurcliff protested that projected SST paths for circumventing the tip of Florida failed to take into account the residents of such nearby islands as Grand Bahama and Cuba; such a consideration would effectively eliminate this routing.[106]

Shurcliff's attack was by no means limited to a direct onslaught on the government. He supplied members of Congress with ammunition for their own anti-SST campaigns[107] and wrote countless letters to influential individuals outside the government. For example, he asked Pan American's president in June 1969 if that airline had planned the routing of its SSTs. Pan American answered that its SST routes had not yet been established.[108]

Shurcliff's impact went far beyond the specific responses to his own correspondence. Citizens League members and other individuals, with Shurcliff's encouragement, also contributed articles to newspapers and journals and wrote letters to the president, the Department of Transportation, the FAA, and members of Congress. Often copies of government replies were forwarded to Shurcliff, who then began a new attack. He generously distributed (free of charge) newsletters and information on relevant topics and directed pressure where it would be the most effective. When a crucial interagency committee Nixon established to review the SST issue was deliberating in March 1969, Shurcliff issued circulars and urged Citizens League members to write to each committee member.[109] By June 1969 the Citizens League had attracted about 3,300 members.[110] The next month Maxwell, as he approached the end of his tenure as SST program director, after many months of delegating or merely acknowledging Shurcliff's inquiries, felt compelled to write a five-page detailed rebuttal to his persistent adversary.[111]

Even more significant were the efforts of the Citizens League to nurture other organizational and professional protests against the SST. During the late 1960s such protests proliferated across the United States, and Shurcliff established contact with a number of such anti-SST groups.

Among these groups were the Town-Village Aircraft Safety and Noise Abatement Committee, which represented eleven municipalities in the vicinity of JFK Airport. Beginning in June 1967 Shurcliff and the Town-Village Committee established a creative working relationship that lasted for several years. Shurcliff supplied this group with anti-sonic-boom information, and the Town-Village Committee reciprocated by sending Shurcliff numerous related articles and copies of its newsletter, which Shurcliff praised profusely. The Town-Village Committee took a very strong anti-SST stance. At the beginning of 1969 it distributed 6,000 leaflets declaring that the SST was "being built to serve not the general public but the favored few," that the sonic boom posed a hazard to both people and structures, and that the airlines could not be trusted when they claimed that the SST would not fly supersonically over the United States. In its monthly newsletter the committee emphasized the SST's dangers and substantiated its arguments with references to technical reports, which were reproduced or paraphrased.

New groups were formed later that year. The town of Hampstead, New York and the city of Inglewood, California sponsored a seminar on organizing opposition to airport noise. Out of this meeting was born the protest group NOISE (National Organization to Insure a Sound-controlled Environment). Shurcliff became a charter member and supplied it with the names of other interested individuals.[112]

The Citizens Committee for the Hudson Valley, a conservation group, joined the anti-SST movement in 1967. The director of the Citizens Committee, William Hoppen, in 1967 and 1968 sent Shurcliff small monetary contributions and the names of other environmental groups, such as the Wilderness Society, the National Parks Association, and various local conservation organizations. Hoppen reproduced and distributed Shurcliff's reports and mailing lists, gave the Citizens League widespread publicity, and praised Shurcliff's *SST and Sonic Boom Handbook*. He also maintained contact with the Town-Village Committee.[113]

One person who became involved was a prominent hearing specialist, Zhivko D. Angeluscheff of New York's City Hospital, who was a member of the New York Academy of Sciences and the International Association Against Noise. Angeluscheff had observed the 1964 sonic boom tests in Oklahoma City, and he soon became an ardent SST critic. In May 1968, at the Fifth International Congress for Noise Abatement, he recalled the 1964 tests: "For the first time in history, from the sky over Oklahoma, the power of the supersonic jets created by man descended on a city. . . . I was witness to the fact that men were executing their brethren during six long months. . . . with their thunder, the sonic boom, they were punishing all living creatures on earth." Angeluscheff called the sonic boom the "Achilles' heel" of noise abatement, and he found the SST "a harbinger of gloom for human life and health" that could cause progressive deafness, mental disturbances, and miscarriages. In November 1967 Angeluscheff sent Shurcliff a list of internationally known medical and acoustical specialists who were concerned with noise, and Shurcliff immediately forwarded to each a package of anti-sonic-boom literature from the Citizens League.[114]

Shurcliff corresponded with the Sierra Club, one of the largest and most active environmental groups, with about 55,000 members. The January 1967 issue of the *Sierra Club Bulletin* referred to structural damage that sonic booms had done to the ancient cliff dwellings at Canyon De Chelly and at Bryce Canyon in Utah. Throughout the summer and fall of 1967 officers and members of the Citizens League and the Sierra Club exchanged information on sonic boom hazards. At a

Sierra Club board of directors meeting on September 9, the organization formally declared itself "opposed to the operation of civil aircraft under conditions that produce sonic booms audible on the surface of the earth."

Shurcliff was extremely pleased; the Sierra Club was the largest and most important group that had yet taken such a strong stand. He urged that the formal declaration be given greater publicity and suggested contacting individual reporters, members of Congress, and publications. The Sierra Club encouraged Shurcliff to take on the task of this distribution himself and continued to rely on him to provide valuable sonic-boom material.[115]

The connection between Shurcliff and the Sierra Club became even closer during the summer of 1968 when Shurcliff's son, Charles, worked at the club's headquarters in San Francisco. In June Shurcliff sent his son a three-page letter proposing a number of anti-SST actions for the Sierra Club to take. Shurcliff's suggestions included initiating a focused anti-SST campaign, paying $4,000 for a half-page *New York Times* advertisement, giving financial assistance to sonic-boom lawsuit plaintiffs in Oklahoma City, supporting the Citizens League in a current battle with the NAS over a controversial report, and writing a book on the harmful effects of the SST and sonic boom.

The Sierra Club, however, did not play an active role in the anti-SST campaign during 1967 and 1968. Although its members and officers were sincerely sympathetic with Shurcliff's work, the club focused instead on protecting wilderness areas. Moreover Senator Henry Jackson, a strong SST supporter, was chairman of the Senate Interior Committee, and the club needed his backing in other conservation causes. At that time the organization was devoting most of its effort to defeating the proposed construction of two dams on the Colorado River. As the summer of 1968 progressed, Charles Shurcliff became less enthusiastic about the prospect of a prominent role for the Sierra Club in an anti-SST campaign and felt that the Sierra Club could not be counted on to launch a major protest effort. Consequently his father decided not to encourage Sierra Club members to join the Citizens League since he wanted only dedicated members. The *Sierra Club Bulletin*, however, continued to report on anti-SST activities.[116]

Shurcliff also worked to gain support among the legal profession. In early 1967 he began a long correspondence with William F. Baxter of the Stanford Law School. Baxter was staff director of an FAA study on the legal aspects of aircraft noise and sonic booms and was later an FAA consultant. The fact that Baxter was an FAA consultant was curious given his strong expressions of doubts about the SST as early as January 1967. He distrusted the FAA's public statements and doubted that the FAA really expected that the SST would be permitted to fly over populated areas. Shurcliff supplied Baxter with anti-SST literature throughout 1967, and in April 1968, about a month after Baxter had finished his report for the FAA, Baxter offered his assistance in Shurcliff's anti-sonic-boom campaign. Throughout 1968 Baxter gave Shurcliff his opinion on such matters as an anti-SST initiative in California, prohibition of SSTs at the Los Angeles and San Francisco airports, and lawsuits involving physical damage caused by sonic booms.

In the November 1968 issue of the *Stanford Law Review* Baxter wrote an effective and thorough article, "The SST: From Watts to Harlem in Two Hours," on the legal problems created by overland, commercial SST flights. Shurcliff praised its ironic title, its technical coverage, and its discussion of legal issues. He took great pleasure in sending President Nixon a copy of the article, which Nixon forwarded to the FAA.[117]

Other groups became involved in the anti-SST campaign. The Conservation Foundation, long concerned with noise pollution, reported extensively on sonic boom hazards in its December 1967 and August 1968 newsletters. The August issue quoted one prominent SST opponent as saying, "The best way of eliminating the threat of sonic boom for once and all in this country would be to have supersonic military planes fly across this capital city for two or three days and nights—when Congress is in session. I am confident that the great SST debate would come to an abrupt and decisive halt."[118]

The Citizens League gained support from other organizations as well. In late May 1967 the Committee for Nuclear Information offered to exchange mailing lists.[119] The World Wildlife Fund and the National Wildlife Federation both expressed an interest in the league's activities.[120] Richard Garwin, a noted physicist at the IBM Watson Laboratory, spoke out against SST development,[121] and the Society for the Psychological Study of Social Issues invited the Citizens League to write an article for its newsletter; Shurcliff sent a statement for publication in January 1968.[122] In May 1968 the National Parks

Association sent Shurcliff one hundred copies of an article on sonic booms that it had published.[123]

In early 1968 two noted scientists, René Dubos and Barry Commoner, expressed their opposition to the SST. Dubos wrote, "The only motivation for the SST is national pride and commercial interest. If we do not have the social courage to stop the SST now . . . it will mean that we are not serious about improving the environment." Commoner called the whole idea of an SST "foolish." In August 1968 Perry Rathbone, director of Boston's Museum of Fine Arts, came out in support of Shurcliff's anti-sonic-boom drive. "Today's occasional sonic booms already present a serious risk of irreparable damage to art objects," he declared. "The possibility that this might become a frequent daily occurrence with the introduction of supersonic transport planes is terrifying and unthinkable."[124] In 1969 the Citizens League received requests for information from the Stanford Conservation Group, the Institute for the Study of Science in Human Affairs at Columbia University, and the National Anti-Pollution Association.[125]

But Shurcliff was not always successful. He failed in the late 1960s to enlist the support of such individuals as Stephen Enke, the economist who headed a Defense Department's SST economic task force that had produced highly negative economic analyses in 1965 and 1966, Luis Alvarez of the Lawrence Radiation Laboratory at Berkeley, and John Gardner, head of the Urban Coalition, or of such organizations as the Democratic National Committee, the World Health Organization, Action on Smoking and Health, the Consumers Union, the Federation of American Scientists, the League of Women Voters, the Environmental Defense Fund, and the National Council on Noise Abatement.[126]

Thus overall public awareness of the SST issue increased considerably from 1967 to 1969. By late 1969 the press was devoting more and more attention to the hazards of the SST and the sonic boom. The Citizens League, which had established a dedicated network of individual SST protesters throughout the country, was the source of much of this public opposition. Moreover, in spite of the FAA's public relations efforts, new organizations joined the opposition, and the anti-SST movement increased in size and power. The SST program was being transformed. No longer was it only a large-scale project with primarily technological, economic, and bureaucratic concerns; it was becoming a potentially widespread public issue.

16 Sonic Boom Three: The Public Debate

The sonic boom debate played a major catalytic role in enlarging the SST conflict and making it a massive public concern. Instead of remaining an integral part of the internal bureaucratic debate (as it had been during the mid-1960s), the controversy over the sonic boom became a spearhead that led the entire SST conflict to grow beyond the confines of an established network of institutions. More specifically, the NAS Committee on SST-Sonic Boom and William Shurcliff's Citizens League Against the Sonic Boom were instrumental in pushing the sonic boom, and hence the whole SST program, into the public view. Ultimately, in 1968, the NAS and Shurcliff squared off into one of the most telling battles in the history of science policy and evaluation.

The NAS Committee's original leadership position in the area of sonic boom studies (as mandated by President Johnson in May 1964) had been usurped during late 1965 by the Office of Science and Technology, which had established the interagency Coordinating Committee for Sonic Boom Studies. Later, in 1967, the Department of Transportation assumed the lead sonic boom policy-making role in the government. But ironically, while the NAS Committee's role concerning the sonic boom in terms of decision-making authority and bureaucratic clout shrunk enormously in the mid-1960s, the committee's actual, albeit unwilling, impact and influence on the evolution of the SST controversy grew.

In retrospect, it is not surprising that the NAS Committee would help extend the sonic boom controversy beyond bureaucratic confines. Many of its members came from outside the government, largely from academic and other research institutions, and the manner in which the committee tended to search for data—by establishing specialized subcommittees—worked to bring in even more outsiders and fields into the controversy. The NAS itself, the committee's parent body, comprised a relatively diverse elite group with allegiances, professional values, and contacts extending beyond the government.

The NAS Committee adapted successfully to its more subordinate role. It would no longer coordinate and guide; instead it became a long-term research and planning arm of the government's sonic boom test program. The most important step taken by the NAS Committee was to establish by the end of 1965 three subcommittees: on public response, on physical effects, and on fundamental research. The public response area was to receive the greatest amount of attention, especially since the NAS Committee decided to maintain the division between public response and physical effects analysis, believing that "there was considerable virtue in keeping the physicists and engineers in one group and the behavioral scientists in another."[1] By creating specialized subunits, the NAS Committee could diversify relatively smoothly and expand the scope and depth of its own investigation. New expertise could be readily incorporated, especially in the increasingly important human response area.

In order to understand the forthcoming struggle between Shurcliff and the NAS Committee, it is necessary to keep in mind the strong de facto alliance between the NAS Committee and the FAA in the mid-1960s. The FAA was helped, for example, by the NAS Committee's tendencies toward greater detail and specialization in sonic boom evaluation. In 1965 and 1966 the NAS Committee increasingly took the position that extensive community overflight sonic boom tests of people's reactions were a waste of time, and instead it supported more thorough, long-range research. As a result, the committee backed the FAA in its fight with Donald Hornig, Nicholas Golovin, and, to a lesser extent, Robert McNamara, to prevent such massive region-wide tests and the quick release of data and conclusions.

Another important factor for understanding the coming battle between Shurcliff and the NAS was the generally pro-SST and pro-FAA inclinations of key NAS Committee members, especially the committee chairman, John Dunning. Dunning, of course, criticized the crash nature of the community overflight plans on technical grounds. He argued, "The immediate initiation of an overflight program, without complete planning and without an adequate theory to support it, may jeopardize the soundness of the sonic boom research program originally proposed." He called for the NAS Committee "to bring to bear calm and considered judgment on [sonic boom] problems."[2] But he also clearly feared the political impact of widespread public reaction to sonic booms. He told the committee in August 1966 that community overflights were "ill-advised," perhaps even dangerous to the entire SST program.[3] By this time Dunning was explicitly backing the FAA's stand on sonic boom testing. He was critical of the Stanford Research Institute's (SRI) research methods (SRI had supported community overflights) and felt that an "engineer manager" should take over the responsibilities of SRI's project director, Karl Kryter.[4] According to Dunning, Kryter was thoroughly knowledgeable in this field, but he was not entirely suitable as project director.[5]

Generally Dunning and his committee backed the FAA fully on the whole issue of community overflights and SRI's reporting of the results of the Edwards Air Force Base tests. Unlike Golovin and Kryter, the committee was skeptical that useful broad findings could be generated quickly. By April 1967 Maxwell had already complained to Dunning of a draft report by Kryter and had detailed the FAA's reasoning behind its stance.[6] Less than a month later Kryter and Dunning clashed on the seemingly technical issue of how to identify and measure the causes of structural damage apparently caused by sonic booms.[7] The chairman of the NAS subcommittee on human response, Raymond Bauer of the Harvard Business School, also disapproved of Kryter's draft and of Kryter's ideas in general, which he felt "would precipitate us prematurely in community studies which would solidify opposition to the SST, even though some future SST might produce a much more benign boom."[8] In late July, just two days before the Edwards report was released, Dunning lashed out against a revised draft in a letter to Donald Hornig; he suggested either releasing the sections of the report separately or drafting a more detailed report. Only when Hornig bluntly rejected Dunning's suggestions, arguing that the report's findings were supported by a great deal of evidence, did Dunning somewhat modify his criticisms.[9]

Dunning's assistance to the FAA clearly went beyond technical advice on designing sonic boom tests. The FAA even felt close enough to Dunning to ask him in September 1966 to take part in the final SST design evaluation.[10] In February 1967 Dunning made it a point to compliment a statement by McKee that defended the SST's profitability even when limited to over-water flight, calling the statement a "masterpiece of condensation."[11] (Dunning's favorable views were reflected at the staff level of the committee. In late 1966 the committee's executive secretary sent the FAA a report analyzing the sonic boom effects on the ability of chickens to produce eggs, which, as he put it, would "hopefully 'lay' this area to rest.")[12] In April 1967 Dunning advised the agency not to release a film that it had produced on the sonic boom,[13] and in June he suggested editorial changes in an FAA sonic boom statement.[14] Maxwell in turn expressed his support for the NAS Committee's research proposals and assured Dunning that "we shall proceed promptly to fund those that you recommend." He also complained of continuing difficulties with Kryter's draft report on the Edwards tests.[15] About two weeks later Dunning praised the FAA and its management of the program, just at the time when the agency was in danger of losing control of the program to the new Department of Transportation. He wrote to the new secretary of transportation, Alan Boyd, to laud McKee's leadership and to urge that the SST program be kept under FAA jurisdiction. He called particular attention to McKee's "skill and foresight" in assembling "key members" of the SST Office and advised Boyd that he would need these people "to carry on the SST program at least until it reverts to the status of a typical industry program requiring normal regulatory supervision."[16] Boyd agreed with Dunning's appraisal of McKee and the SST Office and declared that he would "in no way alter the work of the FAA staff" during the SST prototype construction period.[17]

The flavor of the broad-based supportive relationship between Dunning and the FAA was reflected in a comment by Maxwell to Dunning after Maxwell had informed Dunning that he had visited with the senior editorial staff of the *New York Times*. Maxwell commented, "Let's hope I have made a dent in their position." He also told Dunning that he had met with the aviation committee of the New York Bar Association and that "I believe we did a lot to convince that group" of the soundness of the SST program.[18] The same kind of alliance was in evidence during the spring and early summer of 1968, when the SST program was threatened by possible supersonic flight bans at New York City airports. Maxwell contacted Dunning, who was active in New York City affairs and was chairman of the city's Science and Technology Council. Dunning was very helpful. In a letter to David Rose, vice-chairman and majority leader of the city council, he urged that the anti-SST resolutions be "discouraged," arguing that sonic booms did not "necessarily result in mental and physical injury or damage" and that there was "no evidence at all for direct physical injury." He noted that currently designed American SSTs would not fly supersonically on final approach or directly after takeoff.[19] When the same sort of threat surfaced in the general welfare committee of the

New York City Council, McKee complained to New York's Mayor John Lindsay, making the same technical points that Dunning had used with Rose and urging Lindsay to seek Dunning's advice.[20]

Another important background element for the NAS-Shurcliff battle was the high-level courtship of the NAS Committee that industry and the FAA practiced in the mid-1960s. The pro-SST forces in industry and government made quite sure to keep the NAS Committee well informed on the overall SST program. In 1966 Boeing, Lockheed,[21] General Electric, and Pratt & Whitney[22] briefed the committee periodically. The industry representatives tended to downplay adverse factors. A Boeing spokesman declared that two-thirds of the American SST's potential market involved "non-critical sonic boom routes" and that even with the forthcoming jumbo subsonic jets, there would be "room for all." (But Golovin reminded the committee at the same meeting that studies by the Departments of Commerce and Defense predicted lower markets for a solely transoceanic SST.)[23]

The FAA similarly mounted a campaign to maintain the generally favorable climate of opinion that the committee manifested. In September 1966 the agency agreed to give the committee a "sanitized" (where certain performance data would not be provided) but still rather confidential briefing on the final SST design evaluation that was about to begin.[24]

In January 1967 the FAA was also more than willing to fulfill a request by Dunning for the agency to give the committee a general briefing on the entire program. Dunning wanted this briefing "for prestige purposes."[25] The FAA's subsequent highly revealing performance in mid-February 1967, moreover, clearly was aimed at reinforcing the already favorable view of most committee members. Maxwell opened the meeting with an optimistic overview of the final design competition, which had taken place during the latter half of 1966. He reported that the airlines were generally in agreement with the choice of the Boeing-General Electric configuration, which appeared to have good range and payload features and fewer noise problems than did the Lockheed design. The FAA also appeared considerably impressed with the swing-wing feature of the Boeing model. The agency stressed the Concorde's strength as a formidable competitor, informed the NAS Committee about Senate and White House attitudes, and emphasized the economic stakes. McKee, openly disagreeing with McNamara, declared that the SST would have significant military applications. He mentioned official Defense Department positions in the late 1940s, which had opposed jet aircraft development. If such views had been followed, he said, the rapid growth of commercial jet aviation would have been delayed significantly. Although the FAA's optimism diminished somewhat when the agency turned to sonic boom issues (neither prototype met FAA overpressure objectives, and public acceptability was still unknown), McKee confidently stated that the aircraft was being considered primarily as an over-water vehicle and that transoceanic routes alone would create a demand for five hundred planes, which would generate sufficient revenue to pay for development costs plus 6 percent interest on these costs.

Dunning supported the SST at this meeting, expressing hope that further research would help reduce the effects of the sonic boom. He again appeared skeptical about the SRI methodology used in the Edwards tests, which correlated the effects of jet noise with those of sonic booms. (Certain committee members, including Hallowell Davis and Angus Campbell, strongly defended this methodology and called attention to the rather negative results of the 1964 Oklahoma City tests in spite of their pro-SST bias.)[26]

Therefore, during the 1966–1968 period, the NAS Committee was not merely a strictly objective observer and evaluator of sonic boom phenomena, if it ever had been. Certainly Dunning and some other members favored the development of an SST and supported the FAA. Generally the NAS Committee and the FAA—though often for different reasons—came down on the same side of critical sonic boom issues during this time. (The FAA was not above using NAS Committee support to further its own ends. For example, in answering an inquiry from Senator Frank Moss in February 1967 regarding damage in Bryce Canyon National Park caused by sonic booms, McKee specifically referred to the NAS as participating in the sonic boom studies. Similarly in July Maxwell assured Senator Charles Percy, who had written to the FAA on behalf of a constituent complaining of "air polluted by supersonic boom[s]," that the NAS Committee, "a group of eminently qualified scientists of national reputation," was intensively studying the problem.)[27]

On the other hand, the operating procedures and even the mere existence of the NAS Committee were possible threats to the SST program. The committee members shared a common faith in research, a tendency to look for the complexities in an issue and to avoid simplistic answers, and an inclination to suggest sophisticated problem-solving techniques. The committee's procedures and findings were, at least theoretically, accountable to the high research standards that were exemplified by the scientific elite that made up the NAS as a whole. This approach and the subcommittee structure led to a proliferation of research proposals and an expansion of the committee's interdisciplinary base and scope of research subjects. The NAS Committee was therefore more vulnerable to outside attack than a government agency was because it had allegiances outside the government and because during the mid-1960s certain key committee members, especially the chairman, had become more partisan and were not acting as simply technical advisers on the sonic boom.

The Onslaught

The attack on the NAS came in stages, and not surprisingly its instigator was William Shurcliff, leader of the Citizens League Against the Sonic Boom.

Dunning's contact with Shurcliff began innocuously enough in early December 1966 when Shurcliff wrote requesting information. Dunning informed Shurcliff that first-generation SSTs would not be used on overland flights—a policy "in part due to the recommendations" of the NAS Committee—and confidently predicted that by the mid-1970s many of the SST's problems would be solved as a result of technical research.[28]

In early 1967 the NAS Committee again was involved with Shurcliff's network. The cofounder of the Citizens League and an NAS member himself, John Edsall, wrote President Johnson and Hornig to protest the SST. Objecting to the FAA's involvement in the design and evaluation of the sonic boom tests, he called for the testing of larger populations by an independent agency "with no axe to grind," such as the Public Health Service.[29] Golovin asked the NAS Committee to communicate directly with Edsall on this issue.[30]

The rise of the Citizens League was not unnoticed by committee members. Raymond Bauer was especially sensitive to the league's importance and was urging a thorough comparative historical analysis on the formation of anti-jet-noise and anti-sonic-boom groups.[31] In addition, as a neighbor of Shurcliff's in Cambridge, Massachusetts, Bauer monitored Shurcliff's activities for Dunning. "I will try to keep track of what John Edsall and Bill Shurcliff are doing," Bauer told Dunning in April 1967. "Bill has been very cooperative and has opened up his files to me. I will try and figure out what should be done."[32]

Shurcliff meanwhile increasingly pressured Dunning. In early July 1967 he asked Dunning to notify the press and Congress that the SST could produce sonic booms "eminently capable of adversely affecting people and property on the ground." He chided the NAS for its reluctance to criticize the FAA, a reticence that he claimed was undermining the credibility of the entire scientific community: "Must the entire burden of reputation rest on the shoulders of this tiny league?"[33] Dunning discussed the letter with McKee and passed along a copy of it to the FAA.[34]

Less than two weeks later Dunning telephoned Shurcliff and attempted to calm him. Dunning mentioned that he was fairly certain that the Boeing SST would not be permitted to fly overland at supersonic speeds and that there was no way the sonic boom could be reduced. He stated (rather incorrectly) that he had taken the lead in convincing the other NAS Committee members that overland flights should not be permitted and that he had informed the FAA of this conclusion. Dunning then defended the aircraft, arguing that its sonic boom "should not" cause much physical damage, that it would fly fairly well at subsonic speeds, and that it therefore could fly overland. Dunning stressed that trans-Pacific SST routes would stimulate East-West trade and that the sonic booms from over-water SST flights would be only a small nuisance to ships, given the SST's potentially large overall benefits.[35]

Support for a more skeptical opinion came from the first NAS subcommittee report, *Report on Generation and Propagation of Sonic Booms*, released in October 1967. This was a product of the subcommittee on research chaired by Raymond Bisplinghoff. Although the report recommended further effort in a number of areas, it was generally pessimistic about the prospect of discovering a new design that would drastically minimize the sonic boom.[36]

In addition, Shurcliff was beginning to have an influence on certain NAS Committee members. In mid-December 1967 committee member C. Richard Soderberg sent Shurcliff a particularly poignant letter, which reflected the dilemma facing the committee. Soderberg cautioned the Citizens League to be "guided by a meticulous regard for facts" since much was unknown concerning the sonic boom. But he also deplored "the many other encroachments upon our lives by practically the entire spectrum of technological activities" and sadly observed, "To me this is a shattered dream about the blessings which could have been given to us by science and technology. But this dream was shattered long ago."[37]

Up to this point the NAS Committee had not been a particularly visible part of the sonic boom controversy. Even its positive stance toward the SST had not attracted a great deal of attention. Similarly Shurcliff had annoyed certain members, but he was little more than a nuisance.

All this changed in February 1968 with the release of the report by the NAS subcommittee on the physical effects of the sonic boom. Much of the report was concerned with the design of various tests and with the accurate measurement of test results.[38] One of its conclusions was that continued overflight tests would not help appreciably in explaining the causes of alleged physical damage caused by the sonic boom. Maxwell noted with satisfaction that the report had incorporated all of the editorial changes suggested by the FAA.[39] In fact, the NAS news release that summarized the report stressed that the subcommittee opposed an early resumption of a flight test program because reliable, reproducible data might not be obtainable due to unpredictable atmospheric conditions.[40]

But one statement in particular in the February report proved to be the catalyst for an intense and protracted attack on the NAS Committee by Shurcliff, an attack that proved to be a crucial episode in the whole SST conflict. The controversial phrase, which appeared in the concluding section of the report, read, "The probability of material damage being caused by sonic booms generated by aircraft operating supersonically in a safe, normal manner is very small." Elsewhere in the report, the wording was more careful: "Because the probability of material damage being caused by *a* sonic boom generated by *an* aircraft operating in a safe normal manner is very small, our current inability to provide realistic answers [emphasis added] . . ." At another point, the report read, "The Subcommittee on Physical Effects wishes to emphasize the fact that the probability of *serious* material damage being caused by *a* sonic boom generated by *an* aircraft . . . is very small [emphasis added].''[41] Of all these phrases, the NAS chose to begin its news release with the broadest and least circumspect,[42] and, not surprisingly, the *New York Times* and other papers emphasized this conclusion.[43]

Shurcliff launched his attack in mid-March 1968. Referring to a statement in *Aviation Daily* that the NAS Committee had found that ''there is only a small probability that structures will be damaged by sonic booms from aircraft operating in [a] normal manner,'' Shurcliff argued to Dunning that this report implied that there was little probability that an SST flying across the country ''in some standard manner'' would cause window damage. But surely, Shurcliff exclaimed, such a flight would in fact ''break many windows in several houses.'' The NAS subcommittee statement that was the source of the article, Shurcliff continued, ''allows—almost invites—pro-SST reporters to get almost exactly the wrong impression,'' and he urged Dunning to issue a formal, personal statement to the effect that currently designed SSTs ''would be *likely*, on one trip across [the] U.S.A., to damage windows and plaster walls in several houses.'' He told Dunning, ''You are one of the few people in the country who know the facts, have the prestige, are not under the thumb of aviation, and could at a stroke of the pen dispel the cloud of nonsense such as appears in the . . . paragraph from *Aviation Daily*.''[44]

When Dunning did not follow Shurcliff's suggestions, Shurcliff escalated his attack. In late May 1968 he warned Dunning that the Citizens League was considering a ''major effort'' to counteract the ''central misimpression'' of the report: ''The effort will take a lot of time and effort, and may cause embarrassment to some individuals and to the Academy. We are hoping that you will take some definite and prompt step to correct the situation without need for action by us.'' Shurcliff threatened to take the issue directly to the NAS governing board, to the NAS members, and perhaps to the public at large through the media.[45]

In mid-June Dunning told Shurcliff that the NAS Committee would not issue a clarification, but Shurcliff's demands generated some disagreement within the NAS Committee and were starting to make the NAS as a whole increasingly nervous. A few members suggested a direct counterattack. Donald Weinroth, an NAS Committee consultant, advised that Shurcliff's letter should be quoted in such a way as to be detrimental to his standing in the scientific community.[46] Joseph Zettel, a member of the subcommittee on physical effects, was convinced that Shurcliff would use any information in the "wrong light" and for "his own purposes." He accused the Citizens League of "headline hunting," approved of Dunning's decision not to issue a correcting statement, and thought that Shurcliff's references to past sonic boom damage claims were unwarranted because "so many of the claims are strictly legalized larceny by people who are also trying to make a fast buck."[47]

But other NAS Committee members admitted that Shurcliff had a point. Hallowell Davis pointed out the "unfortunate ambiguity" in the controversial statement, which did not distinguish between "material" and "insignificant or immaterial" damage like a broken windowpane. "I think we have to admit that the sum total of damage," he observed, "even measured in real cost and not by the damage claims, due to broken glass will add up significantly across the nation and across a large number of SST flights, but it is going to be difficult to get across to the public any feeling that this is 'immaterial.' " Davis predicted that the Citizens League would "make things interesting for us for some time to come."[48]

Pressure kept building for clarification by the NAS. Bauer had long urged that the NAS Committee take some action, though he also urged Shurcliff to be more moderate in approach.[49]

But Shurcliff ignored all pleas for compromise. In late June he wrote to the members of the NAS governing board, asking them to issue a correcting statement. He argued that the original wording in the February 1968 report was false and misleading; the *New York Times*, for instance, had reported that the probability of physical damage from an SST's sonic booms was very small but needed further study. If such a clarification were not forthcoming, Shurcliff again threatened to appeal directly to all NAS members and if necessary to bring the issue to the attention of the scientific press, the popular media, and the Congress.[50]

At about the same time Bauer's subcommittee on human response issued its report. In a cover letter, Bauer told Dunning that data from field studies generally were mutually consistent and that even sonic booms with as high an overpressure as 100 psf were not expected to cause hearing damage or direct physiological damage; residents of metropolitan areas would not experience the sonic boom since supersonic flight would not take place within one hundred miles from the airport. The subcommittee was less enthusiastic about psychological response, finding the results of field studies extremely discouraging for the operation of currently designed SSTs over populated areas at supersonic speeds. Curiously, however, the report expressed confidence that "American ingenuity" would produce an SST with an acceptable sonic boom.[51]

The NAS news release on this report was also somewhat strange. Although noting a "growing consensus" that the SST could not be flown over populated areas, the release reported the subcommittee's optimism about the SST's future. Even more astonishing, this release also referred to the earlier report on physical effects, which Shurcliff was attacking, and mentioned that this document had "concluded that the probability of boom damage to structures is small but that additional study is needed to clear up uncertainties and to improve the capability to predict possible damage."[52] The "cautious optimism" expressed by Bauer's subcommittee, however, was undercut by members of the subcommittee on research, who declared in their second report in June 1968 that they did not share such sanguine views: "We are still unaware of any major breakthroughs in the sonic boom problem."[53]

Shurcliff was not particularly pleased with Bauer's subcommittee report, although it contained "no outrageous statement about the acceptability of the boom." Shurcliff complained that it did not live up to its title or to its promise since it primarily recommended only research and its main conclusions concerning human response to the sonic boom were in such vague language that they carried "no clear message." Shurcliff chastised Bauer: "The people of this country have been waiting for years to receive some reliable statement on the human response to the sonic boom. The long-heralded report comes out—and leaves the people empty handed."[54]

Both informally and through correspondence Bauer attempted to convince Shurcliff that his subcommittee report was consistent with the data and appropriate for an advisory body like the NAS Committee. Bauer argued that the NAS Committee could not take the partisan position Shurcliff demanded: policy makers "must be assured that they have received a balanced view of both sides." Shurcliff's position—

"that of an adversary attempting to pile up all possible evidence"—was "a legitimate and established role," Bauer declared, and "our role as nonpartisans should in no way conflict with your partisan role."[55]

Shurcliff softened somewhat, at least temporarily. He agreed that his position as a partisan was much easier to maintain than that of the nonaligned NAS body and acknowledged the vulnerability of Bauer's personal position. Bauer, for instance, was forced to be extremely cautious in his wording, whereas Shurcliff was subject to no such restrictions: "Happily, I am in a less tight spot than you and your colleagues, who felt compelled to avoid gross offense to any one group."[56]

But if his reaction to the report on human response had moderated, his attitude toward the statement in the earlier report on physical effects remained unyielding. In the face of continued NAS inaction, Shurcliff wrote to every NAS member. He called attention to the problems with the controversial statement and asked the members to sign a petition demanding that the statement be corrected. Ultimately 189 NAS members, approximately one-fourth of the academy's membership, signed this petition, and in early September 1968 it was sent to NAS president Frederick Seitz.[57]

By late summer the debate over the controversial statement had expanded far beyond the confines of the NAS Committee. By then Seitz had received several letters from NAS members about this matter, many of which were triggered by contact with Shurcliff. One member told Seitz in mid-August that although he hesitated "in signing a petition asking a committee that has personally studied the matter to reverse its statement," the evidence compiled by the Citizens League appeared "significant," and he was "disturbed to hear that the report speaking for NAS on a matter of general public importance is considered by other responsible scientists to be contrary to available evidence."[58]

Concern over Shurcliff's attack spread to higher levels within the NAS hierarchy, and new pressure was applied to the committee. The NAS Committee on Science and Public Policy was formally responsible for reviewing all NAS reports before release. Its chairman, Harvey Brooks, dean of Harvard University's Division of Engineering and Applied Physics, actually told Dunning in July that the February report "was at least misleading and open to misinterpretation." The report had not been reviewed by his own committee, Brooks explained, because the report was supposed to be preliminary for the sole purpose of recommending necessary research, but the controversial statement "appears to be more or less outside the terms of reference of the subcommittee as I understand them." Brooks added that the media and general public could not be made to understand "these fine distinctions" and that the name of John Edsall, a respected NAS member, appeared on the Citizens League letterhead. Brooks said that his committee took Shurcliff "rather seriously" and would "insist on a thorough airing of the matter, since the good faith of the Academy has been impugned by a presumably responsible group."[59]

Brooks too was a neighbor of Shurcliff in Cambridge and had been in frequent contact with Shurcliff as the controversy over the report's statement grew. In reply to Shurcliff's request (which was sent to all NAS members) that the NAS governing board repudiate the February report, Brooks agreed with Shurcliff that the statement when taken out of context was "open to serious misinterpretation;" but a reading of the full report or even of the conclusions and recommendations alone would clarify the issues; the objectionable statement therefore was not a "misstatement of fact" as Shurcliff alleged. Brooks felt that it would be inappropriate for the governing board to repudiate the report, although the document was "certainly guilty of extreme editorial looseness, and failure to make clear what it means"; still a clarifying statement was being considered. Brooks later noted that only three of the twenty-four press articles on the report, including those in the *New York Times*, called attention to the controversial phrase.[60]

Shurcliff welcomed the suggestion of a clarifying statement, and he urged NAS leaders to issue one. He also later informed Brooks that although the content of most of the press clippings on the report did not emphasize the objectionable phrase, most of the headlines did stress the " 'SSTs-would-do-no-or-practically-no damage' idea." Shurcliff professed to be staggered by the actions of the various government agencies involved with the SST and promised to continue to press his case among NAS members in demanding a correct statement.[61]

By August 1968 the highest levels of the NAS hierarchy were deeply enmeshed in the controversy. That month NAS executive officer John Coleman sent each NAS member a packet of documents on the issue. The executive committee of the NAS council had decided that the offending statement could be interpreted as Shurcliff had suggested and therefore directed Dunning's NAS Committee on SST-Sonic Boom to prepare a formal clarifying statement. The statement that was eventually issued, although misquoting the offending sentence (the new statement read: "The probability of material damage being caused by sonic booms generated by *an* [emphasis added] aircraft operating supersonically in a safe, normal manner is very small"), explained that the February report assumed that it was discussing an aircraft with acceptable "people response" characteristics and that the subcommittee's purpose was to evaluate the cumulative damage resulting from transcontinental flights of currently designed SSTs. It also called attention to the June report on human response to show that the SST would be limited more by human annoyance factors than by material damage. Therefore the statement continued, "To be acceptable to people, the overpressure characteristics of the sonic boom would have to be significantly less in value than those anticipated" for the Boeing SST design. In a short, general epilogue to this statement, the NAS council declared "that the Academy is deeply aware of many threats to our environment and to the health and welfare of our population, including that of noise" and confirmed that the NAS "has not advocated any decision to proceed with the development of an SST transport for overland civilian service."[62]

The next day, in addition to emphasizing the considerable expertise of the subcommittee members and the subcommittee's desire to submit a brief, Dunning informed Seitz that "in all probability, the then current version of the proposed commercial SST would not meet the 'people-response' criteria considered essential for its extensive use over this country at supersonic flight speeds." He told Seitz that "although the members see no reason to change their position on the conclusion about probability of damage," they agreed to an "editorial change" in the controversial statement. The new suggested statement read, "The probability of material damage to a *well maintained* structure by sonic booms generated by a commercial supersonic transport having *acceptable human response characteristics* and operating supersonically in a safe, normal manner would be small [emphasis added]."[63]

Shurcliff was outraged by this statement, which was mailed to all NAS members. He noted that the original controversial sentence had been misquoted so that it appeared to apply only to one SST, not to SSTs generally. He was upset that a later NAS Committee report, the

June report on human response, was used to defend the earlier February report that contained the offending statement. In early September 1968 Shurcliff told Seitz that he wanted a straightforward correction.[64] (Even so, Shurcliff and other Citizens League members were not above using this modified statement in trying to pressure the FAA to change its declarations on the sonic boom's effect.)[65] Bauer and Brooks worked internally for some sort of new statement, and Bauer urged Shurcliff to act "more constructively," but pressure increased.[66] In early October Edsall also wrote to Seitz demanding an unequivocal public NAS statement that the Boeing SST sonic boom would be unacceptable to people and therefore should not be permitted even in sparsely populated areas.[67]

By early October 1968 the NAS still had not formally issued a clarifying statement and, if anything, appeared to become more resistant to Shurcliff's demands. Coleman had ceased to answer Shurcliff's letters and suggested that Dunning's committee do the same. Although he believed that advocates like Shurcliff had a right to their opinion, Coleman called him a "nuisance." He also advised Dunning to tell the NAS governing board that the NAS Committee was doing an effective job and had no need for Shurcliff to identify its goals.[68] Dunning himself vigorously defended his committee before the board in October, stating that some of its most important work lay ahead.[69]

Coleman's unyielding attitude by no means represented that of the entire NAS. Shurcliff's attacks had profoundly shaken the membership, and several NAS participants had acquired a more sober view of the sonic boom. NAS members called for reforms that would enable the organization to maintain its integrity and independence; they criticized the NAS for becoming too closely identified with the contracting federal agencies, recommended more open debate at meetings, and requested better monitoring of NAS reports.[70] In early October the NAS governing board authorized the acceptance of $500,000 by the NAS Committee on SST-Sonic Boom with the proviso that an NAS panel, like Brook's committee, review draft reports.[71]

Bauer publicly declared, first to Senator Clifford Case (R-New Jersey) in late September and later at a meeting of the American Institute of Aeronautics and Astronautics in October, that public reaction to the sonic boom would force the administration to prohibit regular overland SST flights. Given the ambiguity of the June report on human response by Bauer's own subcommittee, his address in October was all the more significant. Basing much of his talk on Kryter's analysis of the Edwards test data, Bauer emphasized the political consequences of the overland SST flight issue: "At this juncture—I repeat—at this juncture—what is crucial is not whether the American public *can* stand the boom of the current generation SST operating overland, or whether they *should*, but whether they *will*." Using Kryter's findings, Bauer concluded that they definitely would not; unlike airport noise, sonic booms from an overland SST route network would affect a vast number of communities. By this time Bauer had also become less enthusiastic about "American ingenuity," and he advised his listeners that "the prospect of an SST which is both economically viable and socially acceptable is quite distant." Still he saw causes for optimism in the sonic boom debate: the issue could well be the best current example of a promising trend called "technology assessment," an approach in which all significant costs and benefits—economic, technological, environmental, and social—are weighed before new technology is introduced.[72]

By late November the NAS still had not replied to Shurcliff's requests for a forthright clarifying statement. In early November Phillip Handler, the NAS president elect, had requested Shurcliff not to take any further action, suggesting that the whole issue would be given close attention. But Shurcliff was impatient: "I have already delayed eight months, and meanwhile the SST programs continue and the threat of the sonic boom increases." In late November he sent Seitz a press release that he said would be distributed to the scientific and popular media and other groups in mid-December. The headline read: "189 members of National Academy of Sciences join in protest against false statement in NAS subcommittee report implying that supersonic transport planes' sonic booms would do virtually no damage to buildings." The text laid bare the protest taking place within the NAS and emphasized the group's resistance to issuing a clarifying statement. The Citizens League offered to provide the press with a list of those who had signed the petition and with excerpts from the protest letters of NAS members.[73]

Shurcliff never directly brought the NAS matter to the attention of the media. Many of his peers in the scientific community, although professing sympathy with his cause, attempted to convince Shurcliff that he had already accomplished a great deal within the NAS and that internal changes had already taken place. Bauer, Brooks, Handler, and others familiar with this issue assured Shurcliff that most NAS members acknowledged that the statement in the February report was at least ambiguous and that to continue pressing the matter would result only in ill will among the scientific community toward the Citizens League.

In a letter to Shurcliff in mid-December, Merle A. Tuve, NAS home secretary, reiterated many of these points; the NAS council generally agreed with Shurcliff regarding the offending sentence, but Tuve advised against portraying the issue as an internal NAS fight. Tuve told Shurcliff that, in effect, he had gone far enough: "We all want very much to achieve the goal of protecting human beings against unwarranted and drastic deteriorations of their environment by repeated sonic booms. Surely this can be done without portraying the Academy as a purveyor of falsehoods and an institution whose officers steadfastly refuse to clean house. Let's be more constructive."[74] Handler congratulated Tuve on the "tone of the letter" and expressed the hope that Shurcliff would direct much of "his ammunition against the Sonic Boom rather than against the Academy."[75]

The urgings of his colleagues finally had some effect.[76] Shurcliff withdrew his threat to issue his press release and he turned his attention elsewhere. Handler and the NAS could breathe a sigh of relief.

But the matter was not finished. A clarification statement of sorts eventually was included in the February 1969 issue of the NAS newsletter. The NAS admitted that the original statement was "by no means clear" and acknowledged that the first generation of SSTs flying supersonically over populated areas would cause "some property damage" and would "pose real problems in terms of human annoyance."[77]

The NAS Committee on SST-Sonic Boom continued to function for another three years; it was formally terminated in June 1971. But after Shurcliff withdrew his threat to go public in December 1968 the committee ceased to be a significant force in the sonic boom or overall SST debates. Until about the end of 1968 the basic deliberations and decisions regarding the SST program had still been internal. But partially as a result of Shurcliff's campaign against the NAS, the focus of controversy began to move beyond government agencies. The NAS membership already included many people from outside the government, and Shurcliff managed to inform a diverse and elite scientific circle of the alleged dangers of the SST's sonic boom. The NAS Committee equivocated, and by the time it was ready to act, under pressure from the entire NAS, opportunity had passed it by. The SST conflict was beginning to grow and to encompass a much larger group of participants. A turning point had been reached.

17 The Changing of the Guard

The new political and bureaucratic context also was a significant factor in enlarging the SST conflict. In particular, Richard M. Nixon's assumption of the presidency in January 1969 introduced an unknown element because Nixon's views on the program were largely a mystery. On the rare occasions when he had mentioned the SST in his 1968 campaign, he had referred to the program merely as a potential target for reduced spending. This changing of the guard clearly would result in some delay as the new administration sorted out its own policies. But SST proponents were primarily concerned about the length of this delay and the priority that the new Nixon administration ultimately would give to the SST effort.[1]

The bureaucratic situation of the SST had also changed significantly by this time. With the Department of Transportation signed into law in October 1966, the FAA became part of the new body, and FAA administrator William McKee soon found himself with a new superior, Secretary of Transportation Alan Boyd. McKee had consistently opposed the absorption of the FAA into a new Transportation Department on the grounds that aviation would not receive the specialized attention it required. This new layer of administration tended to constrain the direct access to the president that McKee had possessed and had used effectively. Indeed it was such direct communication with the White House that had helped protect the SST program from attacks within the government in the mid-1960s, such as over the proposed sonic boom community overflight tests, which Johnson had vetoed at McKee's instigation. But gradually in 1967 and early 1968 the center of SST policy-making deliberations and decisions started to pass from the FAA to the Department of Transportation. Internal bureaucratic rivalry accompanied the early period of the department's history and McKee found his administrative independence considerably diminished. He later recalled being in a constant struggle "with stupid people that [sic] didn't understand one end of an airplane from another [but] who thought they knew all the answers." In July 1968 he resigned as FAA administrator while the SST was in the midst of a critical redesign effort.[2] McKee's colleague in SST affairs, SST program director Jewell Maxwell, stayed at his post, though he now reported to high-level officials at the Department of Transportation, as well as to the acting FAA administrator, David Thomas. The prospect of new Nixon appointees in key positions at the Transportation Department, as well as Nixon's own ambivalence toward the SST, increased the uncertain atmosphere in the program.

Although by early 1969 the Department of Transportation and the FAA had prepared a wide-ranging issue paper on the SST for the incoming administration, Nixon officials clearly did not intend to place a high value on this advice. They were particularly suspicious of any suggestions by the SST Office because the office appeared to be both an administrative and advocacy organization. Outgoing Secretary of Transportation Boyd warned Thomas and Maxwell that his successor, John A. Volpe, wanted his own Office of the Secretary "intimately and continuously involved in the [SST] review process."[3]

The White House soon began to grapple with the SST issue. By the time Nixon held his first cabinet meeting on January 22, a group under presidential counselor Arthur F. Burns had prepared a report outlining proposals for action on various matters, including the SST program. The Burns group recommended establishing a broad SST review committee,[4] and Nixon quickly directed Volpe to create such a body, which was to report to the president in March 1969. Nixon wanted a list of suggested members within two days.[5]

Volpe acted at once, though not with a list of candidates. Instead he described to the president in considerable detail the SST-related activities that were already taking place in the Department of Transportation. He also proposed establishing an Ad Hoc Review Committee under the auspices of the Department of Transportation, which would have senior members from the Departments of Treasury, State, Labor, Commerce, and Interior, as well as from the Council of Economic Advisers and the office of the president's science adviser. This committee would work in cooperation with the Bureau of the Budget and would be concerned primarily with national interest questions, such as balance of payments, employment, and the general technological benefits from SST development. Reflecting his desire to reduce his dependency on the preceding administration, Volpe said that although the committee might use earlier reports from Boeing and the SST Office as "points of departure," he assured Nixon that the group would "come to conclusions of its own." He anticipated that the members would hold "short, informal hearings."[6] Nixon quickly accepted this proposal, though he added to the committee representatives from the Department of Health, Education and Welfare.[7]

Nixon's directive to establish the committee was widely reported in both the aviation and the general press and was perceived as extremely significant. The general opinion was that it indicated a negative view of the SST program on the part of the Nixon administration. One *New York Times* reporter wrote that the administration was seriously considering canceling the project as a result of the SST's design and noise problems and its likely inability to compete with the more economical subsonic 747. As if to corroborate this view, Volpe made no reference to the FAA or the SST program during his confirmation hearings, focusing primarily on highways and high-speed rail transit.[8]

The new undersecretary of transportation, James M. Beggs, was made committee chairman. Beggs had previously been NASA's associate administrator for advanced research and technology, and before that he had worked for Westinghouse. The other committee members were Rocco Siciliano, undersecretary of commerce; Robert Seamans, Jr., secretary of the air force; John Veneman, undersecretary of HEW; Russell Train, undersecretary of the interior; Richard Kleindienst, deputy attorney general; Arnold Weber, assistant secretary of labor; Alexis Johnson, undersecretary of state; Paul Volcker, undersecretary of the treasury; Hendrik Houthakker, member of the Council of Economic Advisers; Lee DuBridge, presidential science adviser; and Charles Harper, deputy associate administrator of NASA.[9]

At about this time, as part of a coordinated effort the SST Office and Boeing had begun to build a detailed case for continuing the SST program. They drafted two separate reports on the SST's national benefits, which they submitted to the Ad Hoc Review Committee in late February and early March. The Boeing piece focused on the possible economic benefits of the SST production phase, which included increased employment and an improved balance of payments. The SST Office document was designed to complement the Boeing report, concentrating on the SST prototype effort and on possible technological benefits. According to the office, the growing divergence in military and civilian aeronautical research needs made a major aircraft program like the SST a necessary "focal point" for civil aviation technology; the SST would prevent the dissipation of "know how" and could well be the "first step" toward commercial hypersonic flight and commercial space flight.[10]

The SST Office had good reason to be concerned. No action had yet been taken on the new B2707-300 fixed-wing design that Boeing had submitted.[11] In addition, Commerce Secretary Maurice H. Stans, although in favor of proceeding with prototype construction, told Volpe in mid-February that he was not opposed to stretching out the prototype phase since the airlines were expending large amounts of capital to outfit their fleets with jumbo jets. Stans was also concerned about the SST's "noise factor."[12] Finally in February Nixon appointed John Shaffer as FAA administrator. Shaffer, a vice-president of the firm TRW who had experience in sales and marketing, said that he did not object to further delay in the SST delivery date to the airlines; the extra time would allow resolution of technical difficulties and would enable the airlines to prepare for the SST. Then in a major blow to the SST Office, Shaffer also recommended that the SST program be formally transferred out of the FAA to the Department of Transportation, where the program would benefit from the "identification with and the shelter of the Secretary's [Volpe's] office."[13]

The first meeting of the Ad Hoc Review Committee, on February 19, 1969, was also not particularly auspicious. In addition to Beggs, the only permanent members who attended were Houthakker and Harper. Other permanent members sent representatives, except for those from the Commerce and Defense departments, who sent no representatives at all. (In contrast, at the even higher-level President's Advisory Committee on Supersonic Transport, which functioned from 1964 to 1966, its chairman Robert McNamara explicitly and successfully prohibited surrogates.) White House Fellow Laurence I. Moss, who would later figure prominently in the SST opposition, was present at this and other Ad Hoc Review Committee meetings. Maxwell, Thomas, and two other SST Office officials also attended.[14]

At the meeting the FAA presented a comprehensive, optimistic overview of the SST program. The agency, for example, emphasized that 73 percent of those polled during the 1964 Oklahoma City sonic boom tests claimed they had not been annoyed (rather than approximately the quarter of the population surveyed who had been annoyed). The FAA also repeated that the estimated SST market was five hundred aircraft and cited the government's evaluation of the B2707-300 that strongly recommended beginning prototype development.

Like the earlier PAC, however, the Ad Hoc Review Committee refused to be limited to information from the FAA. Committee members wanted to talk to airline executives, such as Juan Trippe, Charles Tillinghast, and Najeeb Halaby. Maxwell then also suggested conferring with financial experts Eugene Black and Stanley Osborne, NAS Committee chairman John Dunning, William Allen of Boeing, Jack Parker of General Electric, and FAA technical consultant Raymond Bisplinghoff.[15] Although all of them were outside government, all held generally positive views toward the SST program.

At the committee's second meeting on February 26 there was some improvement in participation. Again, in addition to Beggs, only two full members attended, Houthakker and Harper, but only the Defense Department failed to provide a representative. After this meeting Beggs asked all members to attend personally the remaining sessions.[16]

Beggs established four interagency working panels in the following areas: balance of payments and international relations, with representatives from Treasury, State, and Commerce; technological fallout, with members from the Office of Science and Technology, Defense, and NASA; environmental and sociological impact, with members from HEW, Interior, and OST; and economics, with members from the Council of Economic Advisers, Labor, and Commerce. Requesting that each panel submit a report by March 12, Beggs indicated that these reports would then be reviewed and accepted by the whole committee, which would then "collectively" make its views known to Volpe.

The Transportation Department also attempted to contain the scope of the committee's investigation. It directed the panels to rely principally on information supplied by the Transportation Department, and practically eliminated the sonic boom from consideration by defining the SST as an over-water vehicle. The economics panel was to address itself to areas of domestic impact, such as employment benefits and tax revenues.[17] Houthakker, however, objected, maintaining that the panels should be free to use all relevant information—not just that provided by the Transportation Department—and that the economics panel should also deal with such matters as evaluation of SST demand forecasts, impact on airports and other facilities, and various "financial and organizational problems."[18]

At the committee's third meeting on March 5 a variety of outside individuals presented their views. Among the most provocative was Arnold Moore, director of the Naval Warfare Analysis Group and a member of Stephen Enke's 1965–1966 economics task force in the Defense Department. Curiously, Moore told the committee that he had only had a "couple of hours' notice" before testifying. In any event he was extremely critical of the program, calling it financially and technically risky and tied to overly rigid timetable and performance objectives. Instead he recommended "a flexible R&D program" and cited the Manhattan project and the Polaris program as examples. He also questioned the wisdom of allowing Boeing to switch to a fixed-wing SST after the engine had already been "optimized" for a swing-wing configuration. Both Beggs and Secor Browne, assistant secretary of transportation for research and technology, openly disagreed with Moore, arguing that the B2707-300 design was more advanced and less risky technically than Moore implied.

The other witnesses at this meeting were much more favorably inclined. Gerald Kraft, president of the consulting firm Charles River Associates, which had prepared a 1967 SST economic feasibility report and a similar recent report on the new B2707-300 design, thought that the SST market estimate had not significantly changed between 1967 and 1969. At a base SST sales price of $37 million, a 30 percent rate of return, and a passenger time valued at one and a half times the earning rate, he estimated a 1990 over-water SST market of 479 aircraft, though he pointed out that the market would be quite sensitive to fluctuations in sales price. Najeeb Halaby, then president of Pan American, emphasized the threat of foreign SST competition, the need by American aircraft and aviation industries to maintain market share, the inevitability of supersonic flight, the payback of the government's SST investment, and the probability that solutions to technical problems could be easily found. To Halaby, dealing with such problems as noise and the sonic boom would be "child's play" when compared with solving difficulties associated with the Apollo moon landings. His only negative comment concerned the aircraft's sales price, which he estimated at $50 million. Robert Rummel, TWA vice-president, Harding Lawrence, president of Braniff, and Karl Harr, president of the Aerospace Industries Association, all generally agreed with Halaby's views.[19]

The FAA was quite pleased with the airline and aerospace industry speakers, and breathed a sigh of relief. The next day Maxwell told Shaffer that they had done a "tremendous job and, as a result, I hope we are on the right track. If it hadn't been for them we really would have been in trouble." Maxwell then urged Shaffer to call all the speakers personally to thank them[20] and later wrote to Pan American, TWA, and Braniff, expressing his gratitude for their support.[21]

But as was true throughout the SST program, airline views were not unanimous. In the written comments by airline executives that were submitted to the FAA in February and March, those carriers with primarily domestic routes, such as United and American, minimized the Concorde's impact, while those with overseas routes, such as TWA, Pan American, Northwest, and Braniff, feared competition from the Concorde and, in the case of TWA, the TU-144. Eastern, United, American, Delta, and Braniff were wary about proceeding immediately with a prototype, believing that many technical problems remained to be solved and that the B2707-300 could prove to be unsuitable for domestic operations. Pan American, TWA, Northwest, and Continental, however, saw prototype construction as the best way to find solutions to those problems. They viewed the prototype as an experimental aircraft rather than an economically feasible model ready for production. Pan American and United said that they would not contribute additional risk capital to the SST program.[22] Summarizing airline opinion, Maxwell painted a more favorable "consensus" than really existed, stressing the need to start prototype construction with "nothing further . . . to be gained with additional studies."[23] The FAA then forwarded these airline comments and Maxwell's summary to Volpe in early March, with a strong recommendation to proceed with prototype construction.[24]

The Ad Hoc Review Committee met again on March 12. Two important negative witnesses at this meeting were Elwood Quesada, FAA administrator in the Eisenhower administration, and William Shurcliff. Although acknowledging the tradition of government support for aviation research through military and NASA research and development programs, Quesada felt that the United States had other more pressing aviation needs at this time and that government participation in the SST should be limited to research and development work; arguments concerning national prestige reflected "false values" and carried no real weight. Shurcliff, of course, was totally opposed to any continuation of the SST program.

By this time draft reports by each of the four panels had been submitted, and they ignited an intense debate at the March 12 meeting. The drafts were not particularly reassuring to SST proponents and created a serious problem for SST advocates in the FAA and in the Department of Transportation. (On that same day the *Wall Street Journal* and the *New York Times* and, a few days later, *Aviation Week and Space Technology* reported that Volpe was already strongly inclined to recommend beginning prototype development.) The economics panel emphasized the technical and economic uncertainties connected with the SST. The environmental and sociological impacts panel warned against hazards to passengers, crew, and people on the ground and also discussed the sonic boom, airport noise, and the effects of SST-generated water vapor in the stratosphere. The panel on technological fallout predicted that spinoffs from the SST would be minor and that these should not be used to justify the program. The panel on balance of payments presented two views. The Commerce Department, focusing solely on the aircraft balance-of-payments account, concluded that the SST would have a favorable impact on balance of payments, while the State and Treasury departments, considering the total balance-of-payments effect of the SST (including American aircraft exports and the expenditures of American residents traveling abroad) gave balance-of-payments reasons for delaying the SST program.[25] There was therefore an obvious split in the committee. The Transportation Department clearly favored proceeding with prototype construction, while most members outside the department were more cautious.

The controversy also indicated two very different views of the committee's role. Most of the committee's members, especially those outside the Transportation Department, believed that they were supposed to act as presidential advisers; they expected that their conclusions would reach the president in more or less unadulterated form and that these findings would serve as a basis for the major Nixon administration SST decision. But the Department of Transportation viewed the committee as only one of several important SST evaluation activities in early 1969, which also included the Bisplinghoff group's study of the B2707-300, the interagency redesign evaluation, and the airline reports.

Beggs emphasized that it was Volpe who had been given the responsibility for the overall review.[26]

This confrontation of views in the committee came as no surprise. Houthakker, for example, had been quite suspicious of FAA claims from the beginning.[27] Subsequently there were indications that Houthakker had leaked information to the press of the committee's debate. On March 13 Maxwell met with a *New York Times* reporter who seemed extremely well informed about the recent meeting and whose statements to Maxwell seemed "clearly identifiable" with Houthakker. When questioned, the reporter merely smiled and said that he had spoken with several committee members.[28]

When the reporter's article appeared, the public was informed of the "bitter battle raging" in the Ad Hoc Review Committee and in the Nixon administration as a whole. The major disagreements within the committee were mentioned, in particular the opposition of committee members to Volpe's prototype construction recommendation. It was reported that Beggs would not call another meeting and that Volpe would "bury" the draft panel reports.[29]

The battle escalated even more when on March 19 Beggs distributed a final draft report that allegedly summarized the panels' findings.[30] Many members were astonished at Beggs's highly favorable summary conclusions. Moreover although Beggs requested that the members scrutinize his report carefully, he gave them only one day to prepare comments.[31] Two of the members, Siciliano of Commerce and Harper of NASA, had generally favorable reactions,[32] but the vast majority of the members vehemently disagreed with the summary.

Both Paul Volcker and John Colman of the Treasury Department wrote strongly worded letters of protest emphasizing the many unresolved issues and the need for a more careful review before proceeding to prototype construction, and recommended that another committee meeting be held.[33] Houthakker was even more critical, claiming that the final draft report was extremely distorted. He commented that Beggs's summary contained "primarily the most favorable material, interspersed with editorial comments, and thus distorts the implications and tenor of the [panel] reports." He suggested that the panel reports be submitted directly to Volpe or that each panel chairman draft a separate summary.[34]

Lee DuBridge, the president's science adviser, termed the summary "not acceptable." One of his aides submitted a new review of the findings of the technological fallout panel for inclusion in a final report, which was considerably more negative than Beggs's summary. This statement concluded, "The Panel believes technological fallout should not be considered either wholly or in part as a basis for justifying the SST Program but rather should be considered as a bonus or additional benefit from a program which must depend upon other reasons for its continuation."[35]

Russell Train of the Department of the Interior and representatives from HEW came out against prototype construction on environmental grounds. Train told Beggs, "The growing environmental deterioration in this country and abroad is already the cause of widespread public concern. We believe that the probable adverse environmental impact of the SST is such that the program should not be pursued in the absence of overwhelming evidence of positive advantages."[36]

Arnold Weber of Labor felt that the SST's problems had been understated and the panels' views inadequately represented. Finally a Defense Department representative argued that the issues of the sonic boom and noise had been underemphasized and could not be ignored in the final report.[37]

Pressure for another Ad Hoc Review Committee session intensified. Finally a meeting was arranged for March 25 so that Volpe could hear the members' views and so that an agreement on the format for a final report could be reached. In response to the members' criticisms, Beggs proposed that the report consist of his summary, the four panel reviews, and the letters that he had received from the members. This proposal was accepted at the meeting,[38] and on April 1 Beggs sent the president this entire set of documents.[39]

At about this time another ad hoc committee on the SST was established. At a meeting of the President's Science Advisory Committee in March, Nixon requested a report on the SST from the Office of Science and Technology (OST). Lee DuBridge created an OST Ad Hoc Committee to undertake this review, under the chairmanship of Richard Garwin, a physicist at the IBM Watson Laboratory. During the last half of March this group was quite active. It held discussions with representatives of Boeing, General Electric, and the airlines, and it reviewed detailed material on SST costs and contracts. Copies of its final report were sent to the president and to Volpe on April 2.

The OST Ad Hoc Committee's report was even more negative toward prototype construction than were the Beggs committee's panel reports. The OST group strongly recommended that the SST program be terminated, citing a number of reasons for this stance: SST operating costs would be high; there was doubt about the commercial feasibility of the Concorde and the TU-144; the Concorde could not significantly affect the American balance of payments even if it were successful; American leadership in aviation did not depend on the SST because the United States already flew military supersonic aircraft; and the SST program, ultimately a commercial venture, should look entirely to private capital for financing.

The OST committee's negative findings supported those of the four panel reports and formed a powerful counterpoint to Beggs's more favorable summary report. DuBridge observed that although the findings of the OST group and those of the Ad Hoc Review Committee panels were similar, OST and the Transportation Department had come to widely divergent conclusions.[40]

The submission of two very different major reports on the SST in early April added to the mystery about what the Nixon administration would decide. Some speculated that Nixon would delay the SST and others that he would expedite its development. The administration itself gave contradictory signals. Demands for a quick decision to proceed were made by California Republican governor Ronald Reagan, the Airline Pilots Association, the Air Transport Association, and Senator Howard Cannon (D-Nevada).[41]

It soon became clear that Nixon's decision was not imminent. The budget sent to Congress in April did not contain any new SST funds, though the White House stressed that this absence did not indicate any decision on the president's part. But no one seemed to know when a decision could be expected.[42]

The White House was bothered by the enormous government outlay that was still required to develop the SST under the current financing arrangement. If it decided to proceed in April, it was estimated that an additional $200 million to $247 million would be needed for fiscal 1970 alone in order to keep the project on schedule.[43]

Meanwhile the SST program continued at its current funding rate of $11 million per month, using $90 million to $100 million that remained in carryover funds. At this rate an SST decision could be delayed up to nine months. But the administration's desire to reduce the budget soon increased to such a pitch that even this rate of expenditure began to seem high. During the next two months the SST Office attempted to limit SST spending to $9 million per month, and contractors were asked to submit monthly work plans to the SST Office for approval. In midsummer a Nixon SST decision still appeared to be months away.[44]

Intensifying a new sense of floundering and lack of direction, the SST program in June lost the firm hand of Maxwell, who was reassigned to the air force. On the eve of his departure Maxwell confidently claimed that he was certain that the aircraft, which he had shepherded through some very turbulent periods, would be developed. He added, even more optimistically, that he had not heard a single new argument against the SST in the four years that he had directed the program, that all the charges of the SST critics had been answered, and that the SST would contribute to economic growth. But the program would miss the strong and experienced project manager, and control over the SST program by the Department of Transportation would grow after Maxwell's departure.[45]

Contractor work continued, but the major problem for those working on the SST was the uncertainty caused by Nixon's delay. The Transportation Department attempted energetically to influence the president's decision. In August 1969 Volpe submitted a report to the president that stressed the progress of the Concorde and the TU-144 (the Concorde had just completed its twenty-fourth test flight), the progress of Boeing and General Electric, the growing difficulty of obtaining qualified contractor personnel, and the increasing need for the contractors to make long-term capital investments in order to have a first SST test flight by 1972. Because the Transportation Department had reallocated a number of expensive SST items to the fiscal 1971 budget, Volpe was able to reduce the projected fiscal 1970 SST budget from $210 million to $100 million given a September 1969 go-ahead.

Volpe's lobbying continued into September.[46] He received some assistance from Pan American's president, Najeeb Halaby, who warned that Pan American might even order the Soviet TU-144 if the Concorde proved unsuccessful or if the American SST program did not move ahead. Although he admitted such a possibility was "remote," he argued that Pan American had to stay competitive with the Soviet airline, Aeroflot.[47]

By substantially reducing fiscal 1970 funding needs, Volpe removed a major obstacle to a White House SST go-ahead: Nixon was inclined to support the SST if the funding issue could be resolved. The Department of Transportation's warnings that foreign competitors were advancing and that much of the contractor work was in danger of being dismantled were therefore effective.[48]

On September 23, 1969 Nixon finally announced his commitment to SST prototype construction. Although he acknowledged the "spirited" debate within his administration, he flatly predicted that the SST would be built. The administration, however, requested only about $96 million for fiscal 1970, and Volpe declared that neither the American SST nor the Concorde nor the TU-144 would be allowed to fly supersonically over populated areas unless noise levels were within "acceptable limits." The mood at the White House announcement ceremony, according to the *New York Times,* was "confident and convivial." Washington governor Daniel J. Evans was "beaming" as he stood with Senators Jackson and Magnuson and other members of the Washington congressional delegation.[49]

Nixon's decision triggered a variety of strong reactions. *Aviation Week and Space Technology* naturally approved: "Historically, we would not be surprised if Mr. Nixon's firm decision committing this nation to build a supersonic transport eventually ranks with the late President Kennedy's decision to send Americans to the moon."[50] The *Washington Daily News* termed the decision "wise and inevitable," and the *Washington Post* thought the decision fully justified even though much work was needed in the areas of sonic booms, noise, and airport and runway improvement.[51]

The newspapers that opposed the SST decision were just as vocal. The *New York Times,* stressing the unsolved noise and sonic boom problems, maintained that the money could be better spent elsewhere. The *Washington Evening Star* thought that the project was "a financially dubious investment in luxury travel" for a "handful of eager jetsetters." The *St. Louis Post Dispatch,* in an editorial scornfully titled "From Harlem to Watts in 2 Hours," noted the environmental hazards from the sonic boom, the need to devote funds to worthier social causes, and the possibility that the SST would fly overland despite the sonic boom, given the ever-present desire for increasing profits.[52]

In Congress the reactions revealed a similar pattern. Senators Charles Percy (R-Illinois) and William Brock (R-Tennessee) defended the announcement, stressing the need to take advantage of the SST market. Percy declared, "If there's a market, we have to find a way to build a plane. It's more than a matter of national prestige, it's a matter of dollars and cents." Several members of the House also strongly backed Nixon's SST decision.

At the same time, Nixon's action was harshly criticized on Capitol Hill. Senators Clifford Case (R-New Jersey), William Proxmire (D-Wisconsin), and William Fulbright (D-Arkansas) and Congressman Henry Reuss (D-Wisconsin) denounced the decision. Case and Reuss said they would propose bills to prohibit commercial supersonic flights over the United States until new studies had been done. Proxmire characterized the SST as "little more than a frill . . . a plaything for the jet set."[53]

In early October Nixon formally requested $95.9 million for the SST program for fiscal 1970. This request passed easily in the House. In the Senate, despite Volpe's objections, the Appropriations Committee recommended reducing the $95.9 million request to $80 million. The Senate then rejected a Proxmire amendment to delete all SST funds by a vote of fifty-eight to twenty-two and went on to approve an $80 million fiscal 1970 SST appropriation. Finally on December 19, 1969 the House and Senate agreed to a conference recommendation of $85 million.[54]

But Nixon's decision had unleashed the strongest opposition to date to the SST. Reuss requested a congressional probe of the Transportation Department's refusal to release the four panel reports of the Ad Hoc Review Committee, which were known to be hostile toward the SST program. In late October 1969 Reuss declared that the four panel reports were "a devastating indictment of the SST" and that they directly contradicted Nixon's decision. Then Reuss obtained possession of copies of all the Ad Hoc Review Committee materials that had been sent to Nixon in April and on October 31 released them to the public. Congressman Sidney Yates (D-Illinois), who inserted this material into the *Congressional Record*, commented, "The report of that [Ad Hoc Review] Committee is so unfavorable to the [SST] Program that I am amazed that President Nixon approved the request for the SST."[55]

The release of this material by SST opponents was extremely effective and greatly influenced the press. The *New York Times* cited the documents in an anti-SST editorial and accused the Nixon administration of "ignoring its own report."[56] A widely circulated medical publication, the *Medical Tribune,* ran a full-page feature on the "Beggs Committee Report" and on the committee members' comments.[57]

Reuss continued to press his attack. He soon released the airline letters to the FAA that had been written in late February and early March. Reuss naturally focused on the negative aspects of the airlines' comments and expressed a certain glee that the FAA should have turned to the "friendly airline industry for comments, only to be told that it would give the SST no further support."[58]

The pro-SST forces attempted to counterattack. Boeing's public relations office emphasized that approval of Nixon's decision would reduce Boeing's unemployment rate and would increase employment during the SST production phase.[59] The Defense Department, in a significant departure from its position on the SST in the previous administration, declared that it had great interest in the SST and that it expected a number of technological benefits.[60] In November nine airline presidents publicly called for immediate prototype construction.[61] The FAA prepared public rebuttals to the recently released panel reports of the Ad Hoc Review Committee.[62] In a letter to the *New York Times* in December, for instance, Shaffer objected to the earlier editorial that had charged that the administration was "ignoring its own reports."[63]

Although the SST proponents had easily won the fiscal 1970 SST funding vote in December 1969, Nixon's long delay had seriously weakened support for the program. It was almost as if opposition during most of 1969 had been lying dormant, gathering strength for a renewed battle that would take place after Nixon's decision to proceed. Even the fiscal 1970 congressional funding decision held danger signals for SST supporters. When the Senate Appropriations Committee had approved the fiscal 1970 SST budget, it had stressed the importance of environmental considerations, especially the potentially hazardous effects of the sonic boom. The committee had stated unequivocally that it was "not willing to buy aviation leadership at further cost to the environment."[64]

The tide of public concern over the SST thus rose significantly in late 1969. Although the Transportation Department might look for "one liners" or "short, crisp statements" to defend the SST program (as it had in October)[65] and although presidential counselor Arthur Burns might liken the SST to cathedrals (as he had in November),[66] the floodgates of opposition were opening rapidly. Even as Congress approved the fiscal 1970 SST appropriations request, new forces outside the government were at work attempting to undermine the program, a program to which the administration and the Congress appeared to have just given a vigorous new lease on life.

18 Open War

The Road to Earth Day

Public concern over the SST started increasing in 1967, but the SST did not begin to attract massive public attention until the latter half of 1969. Prior to this time, conflict over the SST had involved mostly members of government agencies, contractors, professional technical associations, some segments of the press, and a few protest groups. President Nixon's decision to proceed in September 1969 radically changed this situation, acting as a catalyst to transform a simmering but fragmented discontent into a widespread effort to make the SST a key target of the emerging environmental movement.

Nixon's action caused a storm of public debate. Anti-SST members of Congress and private citizens deluged the administration with protest letters, and the press too took sides.[1] Newspapers opposing the SST included the *New York Times*, the *Washington Star*, the *Chattanooga Times*, the *Worcester Telegram*, the *Montreal Star*, the *Ann Arbor News*, the *Philadelphia Inquirer*, the *Chicago Sun-Times*, *Newsday*, the *Richmond News-Leader*, the *Louisville Courier-Journal*, and the *Milwaukee Journal*. Newspapers that supported the president's decision included the *Washington Post*, the *Lancaster New Era*, the *Cleveland Plain Dealer*, the *St. Louis Globe-Democrat*, the *Washington Daily News*, the *Kansas City Star*, the *Los Angeles Herald-Examiner*, the *Atlanta Constitution*, the *Boston Globe*, the *Wall Street Journal*, the *Portland Oregonian*, the *Salt Lake City Tribune*, the *Seattle Times*, and the *Journal of Commerce*.[2]

The SST conflict had changed into a widespread, visible battle with both sides actively preparing for combat. Gearing for battle, the SST Office spared no effort to influence congressional and public opinion.[3] Officials in the FAA and the Transportation Department, including Transportation Secretary John Volpe, contacted media representatives and crisscrossed the country in an attempt to promote the program. FAA administrator John Shaffer wrote articles defending the SST and held background briefings in several cities.[4] The major SST contractors also became involved. Boeing officials met with the press, distributed promotional literature, produced film clips on the SST, and supplied Senator Henry Jackson (D-Washington) with information "on the benefits of the SST" for an article that Jackson was writing. Boeing's top SST manager, H. W. Withington, briefed pro-SST witnesses who were to appear on the public television program "The Advocates," and later the company's SST public relations manager told the FAA that he was pleased with the results: "Considering the basic [courtroom] format, our side looked as good, or better, than the opponents."[5] Other pro-SST groups also became more involved: the International Association of Machinists lobbied intensively;[6] the Long Island Association prepared a pro-SST position paper;[7] and the Air Transport Association ran radio commercials in five cities and issued pro-SST press releases to over five hundred U.S. newspapers.[8] The Transportation Department and the SST Office were extremely pleased with these efforts, especially with the way in which the Department of Transportation, the FAA, General Electric, and Boeing were cooperating to present a united front.[9]

The opposition also increased and became more organized. In early October 1969 the Natural Resources Council in Maine announced that it opposed further SST work and urged a reassessment of national priorities.[10] An official of the Conservation Foundation also expressed concern about the SST's environmental impact, calling it potentially worse than the effects of a major oil spill. He told William Shurcliff, "Having lost the ABM battle, I am personally quite anxious to become more involved with the SST problem."[11] The Environmental Defense Fund took a purely legalistic tack, citing the Freedom of Information Act to demand that the FAA release all environmental impact data.[12]

In July 1969 David Brower, former executive director of the Sierra Club, founded Friends of the Earth. Unlike the Sierra Club, Brower advocated a strong anti-SST stance. Friends of the Earth differed from other major conservation organizations, like the Sierra Club, the Wilderness Society, and the Audubon Society, because it did not choose tax-exempt status. Brower wanted his group to lobby freely for legislation and to support actively pro-environment political candidates. Brower himself became president; Gary Soucie, another former Sierra Club officer, became executive director; and George Alderson emerged as chief Washington representative and key lobbyist.

All three were familiar with Shurcliff's activities, and in the fall of 1969 they sponsored a paperback edition of his *SST and Sonic Boom Handbook*, which was published a few months later by Ballantine Books. The *Handbook* came off the presses in February 1970, and the Citizens League ordered 5,000 copies. Shurcliff predicted that the book would benefit from "the great wave of . . . pro-conservation propaganda" currently sweeping the country; by early April 1970 Ballantine had sold 150,000 copies. The SST Office even proposed issuing its own "SST handbook" in retaliation.

Friends of the Earth wrote to newspapers, supplied standard protest letters for its members' use, and supported SST opponents in Congress like Congressman Henry Reuss. Soucie reassured Reuss, "Friends of the Earth intends to devote a considerable share of its attention and resources to the SST, and we are counting on your assistance." The organization also ran full-page advertisements in the *New York Times* and other papers, announcing that the SST "breaks windows, cracks walls, stampedes cattle, and will hasten the end of the American wilderness."[13]

The 85,000-member Sierra Club, despite its earlier reluctance, began to play a more active anti-SST role. In October 1969 the club invited Shurcliff's Citizens League to participate in the first New England Leadership Conference on the Environment, which had adopted the sonic boom as one of its main agenda issues. The club began to distribute Shurcliff's *Handbook* and to include bitter attacks on the SST in its newsletter. In January 1970 at a meeting of the Aerospace Convention, Sierra Club president Phillip Berry urged termination of the entire SST program. The Sierra Club bookstore in San Francisco also distributed the *Handbook*, and by mid-April 1970 the Boston Sierra Club representative had been in close contact with Shurcliff.[14]

Opposition emerged elsewhere. Laurence Moss, a Sierra Club board member and a former White House fellow who had attended virtually all of the Ad Hoc Review Committee meetings in March and April 1969, became a prominent SST opponent in early 1970. Others included Harold Bergan, legislative assistant to Congressman Sidney Yates; James Verdier, legislative assistant to Congressman Reuss; and Gar Kaganowich, executive assistant to Senator Clifford Case. Many independent local groups also began to express interest in the issue.

It was only a matter of time before an organized, national anti-SST campaign was formed. Its catalyst was Kenneth Greif, an independently wealthy thirty-four-year-old English teacher from Baltimore. Greif feared that the large number of environmental groups involved would lead to fragmentation of the anti-SST movement, and he was willing to use his large private fortune to support the cause. In March

1970 Greif's lawyer met with Alderson, Soucie, Moss, Bergan, Verdier, Kaganowich, and W. Lloyd Tupling, the Sierra Club's Washington representative. At this meeting the Coalition Against the SST was born; its office was to be on the first floor of the Sierra Club's Washington headquarters. The coalition banded together fourteen organizations, including the Citizens League (3,000 members), the Consumer Federation of America (37 million members from 150 federated groups), the Sierra Club (95,000 members), the National Wildlife Federation (2.2 million members), the Wilderness Society (60,000 members), Zero Population Growth (10,000 members), and Friends of the Earth (3,000 members).

With the formation of this new consortium, the nature of the anti-SST movement changed. Shurcliff's Citizens League had led the fight for the past three years, but its membership was relatively small and its Cambridge location was outside the political mainstream. Shurcliff's relative role, although still considerable, diminished as the anti-SST campaign became more national in scope and more strictly political in character.[15]

The SST proponents' response to this mounting opposition in early 1970 was basically more of the same. The SST Office's public relations campaign continued, and wherever possible adverse SST media pieces were answered directly. FAA administrator Shaffer, for instance, wrote to the *Oakland Tribune* to protest a critical article by Stanford Professor Karl Ruppenthal,[16] and SST Office acting director B. J. Vierling, as an outraged Stanford alumnus, wrote directly to the university to decry Ruppenthal's "misrepresentation" statements (which, he claimed, were damaging to Stanford's good name).[17] Vierling also met personally with a columnist from the *Washington Evening Star* and complained to the editor of the *Honolulu Star-Bulletin* about an anti-SST editorial.[18] In March the SST Office protested to the *Washington Post* about a piece written by noted ecologist Garrett Hardin, which the office claimed exhibited a "rather shallow understanding" of SST economics.[19] The SST Office also naturally praised favorable reports, such as an article in *Aviation Week and Space Technology*. Vierling told the journal's publisher that the piece was "far beyond anything we had hoped for" and that the SST Office planned to use it in responding to individuals who "criticize our program."[20]

SST supporters were becoming somewhat more edgy over criticism. In January 1970 Matthias Lukens, president of the Airport Operators Council International and deputy executive director of the New York Port Authority, referred to a statement by Shaffer when declaring that operational SSTs would require a whole new system of airports. Lukens was extremely concerned about SST noise levels, and he resisted attempts by Vierling and Shaffer to assuage him on this issue. Shaffer urged Lukens to look at the "big picture" when considering the noise issue, and, exasperated, he finally snapped that the burden of responsibility for quieter airports rested at least as much with airport communities as with the SST.[21]

The SST Office still had to contend with Shurcliff, who constantly wrote letters of protest to Nixon, Volpe, Shaffer, Vierling, and other government officials connected with the SST program. Shaffer became increasingly sarcastic toward Shurcliff and occasionally resorted to name calling. "If someone were to drive a truck through your office," Shaffer wrote, "it could damage the furniture. Similarly, if I were to land an airplane on Harvard, it could damage the roof. The truck that *will not* be driven through your office will not damage your furniture. The airplane that I *am not* going to land on the University will do no damage. Since the SST *is not* going to fly over inhabited land at supersonic speeds (Presidential policy), I suggest that the boom will not be the problem you continuously and monotonously hypothecate." Attacking Shurcliff's extrapolations of the 1964 Oklahoma City sonic boom test results, Shaffer sneered, "[You obviously] didn't take advanced math or science en route to your doctorate." He suggested that Shurcliff might spend his time more profitably being for something rather than against something: "Since you're more a philosopher than a technician," he counseled, "why not identify with something within your field of expertise?" But Shurcliff, of course, ignored this advice. He continued to write letters to government officials urging further definition of policy and formation of regulations regarding sonic booms. One Transportation Department official recommended that the department refrain from all further comment.[22]

During the first five months of 1970 the SST Office responded to numerous congressional inquiries, which resulted in part from Shurcliff's activities and the anti-SST advertisements by Friends of the

Earth. The office went to considerable trouble to demonstrate to Capitol Hill that the SST would not pollute the environment, especially the upper atmosphere, which was a matter of growing concern, and it supported its arguments with references to studies by the Environmental Science Services Administration, the National Academy of Sciences, and an article by George Catham of the Library of Congress in the *Journal of Astronautics and Aeronautics,* all of which minimized the SST's hazards.[23]

SST officials maintained close contact with members of Congress and the congressional staff, not only to placate opposition but also to keep informed of congressional intentions and motives.[24] Volpe himself did a great deal of lobbying within the administration. In April, for example, he met with Russell Train, chairman of the newly formed Council on Environmental Quality, to assure him that the SST program was extremely aware of environmental worries. Volpe proudly declared, "I know of no major technological program that has been more cognizant of the environmental concerns than the SST program."[25]

The program, however, itself lacked effective leadership during this period of mounting attack. Maxwell had resigned as director during the summer of 1969, and Vierling, as acting director, lacked the authority of someone with a permanent appointment. Transportation Department officials like Under Secretary James Beggs and Shaffer were too removed administratively to provide effective guidance. The whole program suffered from a lack of coordination and a sense of purpose.

The situation only improved in April 1970 when William M. Magruder was made director of the SST program. Magruder was a forty-five-year-old aeronautical engineer who had spent many years as a test pilot and designer for the air force and for Douglas Aircraft. He was well known and well liked in the aviation industry, and his appointment was greeted with considerable enthusiasm by colleagues in the United States and abroad. He was already quite familiar with the SST program, having been second in command of Lockheed's SST effort in the mid-1960s. He then went on to supervise the design and development of Lockheed's L-1011 commercial jet transport and all of Lockheed's advanced commercial designs. Magruder was a true believer in aviation generally, and in the SST in particular. He exclaimed to a former colleague after being appointed SST program director, "You can appreciate the thrill of being back in a program that's pushing aviation frontiers as hard as this one is.''[26]

Unfortunately for Magruder not everyone viewed the SST as confidently and as optimistically as he did. Across the country SST critics were marshaling their forces, and conservation groups were alerting citizens to a variety of possible threats to the natural environment. Environmental quality had become an issue of widespread national concern, a concern that expressed itself most dramatically and most effectively on April 22, 1970: the first Earth Day.

Although Earth Day was organized by Environmental Action in Washington, D.C., it was not a Washington-based event; a number of established and ad hoc groups across the country participated. As early as the previous November an *Eco-Directory,* which was to help prepare for Earth Day, had listed potential participants from the northeastern United States, including 6 environmental groups in Connecticut, 23 in Delaware, 6 in Maine, 50 in Massachusetts, 57 in New Hampshire, 6 in New Jersey, 114 in New York, 9 in Pennsylvania, 6 in Rhode Island, and 21 in Vermont. In February, according to the *Boston Globe,* the number in the Boston area alone had grown to 26.

On Earth Day colleges, high schools, libraries, and churches sponsored teach-ins; local groups organized demonstrations; politicians and scholars made speeches; and the media publicized everything. Public awareness of environmental problems was raised to a new, unprecedented level. Some years later one anti-SST activist referred to the 1970 Earth Day as a ''psychological watershed.''

The SST was only one of many issues discussed; others were air pollution, water pollution, weapons technology, pesticides, nuclear explosives, and hunger. But the SST emerged as a major target of protest in many parts of the country. Senator Walter Mondale (D-Minnesota), speaking at the University of Minnesota, angrily compared the $200 million allocated ''to feed hungry children'' with the $290 million fiscal 1971 request for the SST. At Western Illinois University Adlai Stevenson III, Illinois state treasurer, noted that Nixon had requested twice as much money to continue the SST as he had to control air pollution. In Chicago the vice-president of the Amalgamated Meatcutters and Butcher Workmen of North America declared, ''The additional billions we continue to pour into military hardware, the billions our government wants to put in ABMs, MIRVs, and SSTs, must be used instead to protect and reclaim not only the air we breathe and the water we drink, but the human beings who can drink that water and breathe the air." The Citizens League was inundated with requests for information from college groups, high schools, libraries, and local environmental organizations from every state in the Union. Shurcliff's *Handbook* was widely mentioned and was reviewed in several library journals and mass-market publications. In Boston there was a "funeral procession" to Logan Airport and then a "mass die-in" to protest planned SST purchases by TWA, Pan American, and United. In California the UCLA Ecology Action Council prepared a background paper specifically on the SST, terming the aircraft a "gross" national product.[27]

The National Campaign

Magruder's task appeared doubly difficult after Earth Day. Not only was the SST being increasingly attacked on environmental grounds, but it also fell victim to a critical set of hearings held on May 7, 11, and 12 before the Subcommittee on Economy in Government of the Joint Economic Committee. The chairman of this group was none other than Senator William Proxmire, long-time opponent of the SST program. The hearings were orchestrated to a certain extent by the leaders of the Coalition Against the SST, partly in order to fight a $290 million SST budget request for fiscal 1971 that was to be voted on by the House that month and partly to advance the anti-SST campaign in general.

Due to the careful planning by Proxmire and the coalition, the hearings were well timed and the witnesses well chosen. Among those who testified were Henry Reuss, Richard Garwin, Russell Train, Elwood Quesada, William Magruder, Sidney Yates, and Mary Goldring (business editor of the anti-Concorde British journal, the *Economist*). Reuss, the first witness, cited opinion polls to show that over 85 percent of the public was opposed to the SST. Garwin stated that the airport noise of an SST was fifty times that of a 747 and thought this was an unacceptable level for nearby residents to tolerate. Quesada felt that the government was overly involved in the program.

Proxmire saved his heaviest artillery for last. Train declared that the administration would not permit commercial SST development until the "significant environmental problems and uncertainties" were satisfactorily resolved and expressed doubts about the possibility of finding an imminent solution to the SST's severe sideline noise, which he termed the aircraft's "most significant unresolved environmental problem." The greatest bombshell was Train's acknowledgment of the SST's potentially harmful atmospheric effects. He commented that an SST cruising at an altitude of 60,000 to 70,000 feet would discharge into the upper atmosphere "large quantities of water, carbon dioxide, nitrogen oxides and particulate matter" and then noted that a "fleet of 500 American SSTs and Concordes flying in this region of the atmosphere could, over a period of years, increase the water content by as much as 50 to 100 percent." According to Train, this could lead to warmer average surface temperatures and could seriously affect the earth's climate. Train also mentioned that increased water vapor could destroy some of the ozone in the upper atmosphere, which would diminish the ozone's capacity to shield the earth from potentially dangerous ultraviolet radiation. According to a Proxmire aide, what had formerly been a seemingly eccentric scare theory dealing with the SST's atmospheric impacts had been given an aura of legitimacy and "the stamp of seriousness." Proxmire quickly inserted Train's testimony into the *Congressional Record*.[28]

In spite of the best efforts of Magruder, Volpe, other Department of Transportation officials, and executives from Boeing and General Electric to mitigate the hearings' impact by issuing favorable studies, rebuttals, letters, optimistic SST schedules, and statements that the government could legally bar any environmentally hazardous aircraft,[29] the hearings proved extremely damaging to the pro-SST forces. Throughout 1970 and 1971 they were frequently and skillfully cited by conservation groups, by members of Congress, and by the media to support anti-SST arguments.[30]

Magruder's specific task in May 1970 was to defend the SST fiscal 1971 funding request in the House. He reported frequently to White House aide John Ehrlichman and maintained contact with influential congressmen, such as Congressman Edward Boland (D-Massachusetts), chairman of the House Appropriations Subcommittee on Transportation. In a compromise with Russell Train, Magruder backed more research on the SST's environmental effects, recommending a multiagency conference on this and other questions, though he also moved to broaden the scope of this research effort to make it "responsive to the needs of the entire air transportation system, including SST considerations."[31]

Magruder, however, faced the growing power of the anti-SST forces after Earth Day. In April the Massachusetts Committee Against the SST was established to organize statewide SST opposition, to distribute information, and to lobby Massachusetts congressmen such as Boland and Silvio Conte, a Republican. The SST, the organization declared, was "the most dramatic symbol of the worship of technology for its own sake."[32] The Town-Village Aircraft Safety and Noise Abatement Committee near JFK Airport in New York City continued to disseminate anti-SST information in its newsletter.[33] A Seattle housewife, Mrs. Barbara Freeman, collected a quarter of a million signatures from SST protesters and then presented her anti-SST petition to the Washington State congressional delegation.[34] The Coalition Against the SST grew more organized and active. In April George Alderson, who emerged as the effective head of the SST opposition's lobbying efforts, distributed a detailed timetable of expected legislative and procedural events that would precede the House SST appropriations vote.[35]

As the House vote grew closer, the coalition issued strongly worded circulars stressing the SST's hazards and printed a map of the United States showing the country crisscrossed by fifty-mile-wide sonic boom zones. The circulars claimed that the average citizen would bear the cost of a vehicle that only a few "rich people in a hurry" would be able to use; Boeing, it was noted, was unwilling to risk any of its own capital. The coalition urged people to write letters of protest to members of Congress and newspapers and recommended Shurcliff's *Handbook* as a source of information.[36] The Sierra Club too urged its members to take action.[37] As a result of such efforts, the Nixon administration received an ever increasing number of anti-SST letters.[38]

In late May the House Appropriations Committee voted to continue funding for the SST, and thereafter anti-SST lobbying on Capitol Hill intensified, largely coordinated by the coalition. Volunteers traveled to the capital to help in the protest effort. Shurcliff sent the coalition $500 around this same time. Diverse groups joined the effort, including National Tax Action, the National Federation of Social Service Employees, a farmer's organization, labor unions, church groups, and welfare organizations.

The anti-SST forces established a "command post" in the office of a friendly congressman. They focused on influential House Appropriations Committee members and undecided votes and in a highly efficient fashion allocated personnel according to the importance of certain members of Congress.

SST critics were optimistic and expected a close vote in the House; the Department of Transportation seemed worried about the attacks and took pains to refute the opposition's claims to the press.[39] But on May 27 the House defeated by seven votes an amendment that would have eliminated the requested fiscal 1971 SST funding of $290 million.[40] SST foes did not view this event with overwhelming disappointment; the closeness of the vote indicated that congressional opposition to the SST was growing and that previously staunch SST supporters in the House had defected. The vote was a turning point in the conflict and a cause for encouragement among SST critics.[41]

Attention immediately turned to the upcoming Senate vote, and the anti-SST campaign intensified. In a mass mailing the Sierra Club exhorted its members to write letters of protest to the Senate.[42] This and similar appeals by other groups were remarkably effective. Senator Charles Percy, who had voted for SST appropriations the previous year, stated in June that he was convinced he had been wrong and that he would vote against the SST when the funding issue came up again. Senators Harry Byrd, Jr., and William Spong, both Virginia Democrats who had opposed past SST appropriations, said that they would vote to cancel all SST government spending. Senator Mike Mansfield of Montana, Democratic majority leader, remarked that the SST was in "deep trouble."[43]

The coalition continued to coordinate the lobbying and worked closely with congressional aides. The anti-SST effort attracted new recruits. For example, Doug Scott, a young activist who had participated in local Earth Day efforts and who had come to Washington, D.C., in May as a representative of the Wilderness Society, organized during the summer a student intern group of Michigan college students in Washington. Scott, Alderson, and Moss briefed the students thoroughly on the SST at nearby Mount Vernon College. Scott then arranged a meeting on the SST between this group and Senator Robert P. Griffin of Michigan, the Republican whip. The students organized themselves into smaller specialized groups, and by the time they met with Griffin they were virtually experts on the SST. Griffin was clearly impressed, and he told the students that if he decided to support the SST, he would at least meet with them again to give them another chance to dissuade him.[44] At the end of August 1970 Griffin announced his opposition to the SST appropriations request.[45]

During the summer of 1970 the anti-SST campaign grew in size and complexity. The coalition cooperated with other Washington-based organizations, with Capitol Hill staff members, with the Citizens League, and with various state and local anti-SST groups. Shurcliff earned the gratitude of the coalition,[46] Environmental Action,[47] the Sierra Club,[48] Friends of the Earth,[49] and other groups for his contributions of money, informational materials, and advice. Friends of the Earth initiated a mass mailing encouraging its members to contact undecided senators and warning that powerful SST supporters in the Senate, such as Magnuson and Jackson, had not yet begun a serious and active campaign.[50]

In September the prestigious Federation of American Scientists also came out against the SST. In its statement before Congress the organization declared that the SST was a poor financial risk, that the government should invest its funds elsewhere, that the SST would have adverse effects on the environment and balance of payments, and that the race between the American SST and the Concorde was a senseless competition between "white elephants." The federation had also requested that Laurence Moss, currently executive secretary of the Committee on Public Engineering Policy of the National Academy of Engineering, examine SST noise levels at various airports. Moss's study concluded that these levels would be so high that they would result in "vigorous" complaints and "concerted group action" in all or most of the metropolitan areas of New York City, San Francisco, Seattle, Honolulu, Anchorage, Boston, and Los Angeles. Moss's findings were transmitted to members of the Senate in October. The federation also sent each senator a Shurcliff-prepared "boom-zone" map for the particular state that the senator represented.[51]

In July the Airport Operators Council International, which represented executives of all major United States airports, recommended to the Senate that the SST be funded only if the aircraft could meet stringent noise standards. Matthias Lukens, the group's president (on whom the FAA unsuccessfully had already focused a great deal of attention), stated that the public could not be expected to tolerate an SST that would both require larger airports and generate unacceptable noise levels. "The key question of noise" should be settled first, Lukens declared.[52]

August was particularly fruitful for the SST opposition. Early in the month a prestigious group sponsored by MIT, the Study of Critical Environmental Problems (SCEP), warned that a fleet of five hundred SSTs could discharge enough particulate matter into the atmosphere to cause a significant change in the earth's climate. The group compared the SST's possible effects to those of a 1963 volcanic eruption of Mount Agung in Bali, which had increased stratospheric temperatures for several years afterward (initially raising the stratospheric temperature 6° to 7°C, with the temperature remaining at 2° to 3°C above the pre-Agung level for many years). SCEP expressed "a feeling of genuine concern" and forcefully recommended that "the uncertainties" over possible SST atmospheric contamination be resolved before large-scale SST operations began. Although SCEP issued similar warnings for other environmental hazards, such as DDT, ocean oil spills, detergents, and nuclear energy, its SST conclusions received the widest media attention. They were inserted in the *Congressional Record* and subsequently were cited often in anti-SST editorials and circulars. Like Train's earlier testimony in May, in the words of one SST activist, the SCEP findings made the issue of SST contamination of the upper atmosphere "respectable."[53]

With even greater vigor and encouragement than before, a little over two weeks later Proxmire's Subcommittee on Economy in Government called for the cancellation of the SST program because of hidden cost burdens, poor economics, and environmental hazards. The government's SST funding scheme, claimed Proxmire, was an "Uncle Sap" contract, and the SST was nothing but an expensive "frill" for the jet set.[54]

Late in August Congressman Reuss revealed that a top Boeing scientist was unsuccessfully trying to make public a particularly damaging Boeing report on the SST's atmospheric effects. Two days later Richard Garwin, armed with a suitcase full of documents, told the Senate Transportation Appropriations Subcommittee that the prototype's cost would be 40 percent above current estimates and that no imminent technological innovation would solve the SST's environmental problems.[55]

Then in September possibly the most effective of all anti-SST actions occurred: a group of prominent American economists signed individual statements expressing their reservations about the SST. The group included Paul Samuelson, Robert Solow, and C. P. Kindleberger of MIT; Milton Friedman of the University of Chicago; Kenneth Arrow, John Kenneth Galbraith, Wassily Leontief, Francis Bator, and William Capron of Harvard; W. J. Baumol of Princeton; Walter Heller of the University of Minnesota; Arthur Okun, former chairman of the Council of Economic Advisers; and Richard Nelson and James Tobin of Yale. Only Henry Wallich of Yale came out in favor of the aircraft. This attack, skillfully organized by the Coalition Against the SST, had an immediate impact. Senator William Fulbright inserted the statements into the *Congressional Record*, and they were given wide publicity. A *New York Times* editorial less than a week later commented, "Now that most leading economists agree that the plane makes no economic sense for the nation, surely it is time for the Administration to wheel it back to the hangar and turn its attention to more important transportation problems."[56]

Events at this time in Europe also helped the SST opponents. A series of Concorde tests over Scotland and Wales evoked protests from farmers, who complained that, in addition to cracking roofs and shutters, sonic booms caused "hysteria in hens, headaches in humans, and made cows go berserk." An unscheduled Concorde landing at London's Heathrow Airport brought numerous complaints from the nearby community, and the government, worried about possible sonic boom damage to historic buildings, was considering banning all supersonic flights over British soil. An Anti-Concorde Project in Great Britain similar to Shurcliff's Citizens League in the United States was operating actively,[57] and an anti-SST committee had been established in Switzerland.[58] Both groups received support and advice from Shurcliff, and in September it was decided to issue a Concorde SST handbook patterned after Shurcliff's own book.[59]

Also in September an important non-Washington political figure, Mayor John Lindsay of New York City, came out against the SST.[60]

"As mayor of the City of New York," he declared, "I am prepared to do all in my power to prevent any SST from landing at New York's airports until it is proven safe both to our environment and to the health of our citizens."[61]

The anti-SST campaign was spreading. Many new local anti-SST groups were formed at the end of the summer. Although they received information and advice from the coalition, their leadership was usually made up of local or regional officers of national ecological groups, such as the Sierra Club or Friends of the Earth.

The anti-SST campaign in Rhode Island was illustrative of the active and effective grass-roots protest efforts taking place by the late summer of 1970. Ecology Action for Rhode Island, a statewide organization originally founded in 1969 as a Brown University group and the main sponsor of Earth Day events in Rhode Island, throughout the summer of 1970 exerted considerable pressure on the state's pro-SST senators, John Pastore and Claiborne Pell, to oppose the SST appropriations request in the Senate. Pastore, a senior member of the Senate Transportation Appropriations Subcommittee, was especially critical. The anti-SST fight was deftly coordinated by Dwight W. Justice, a nineteen-year-old Elmira College sophomore who was working for Ecology Action during the summer. Justice communicated frequently with both the coalition and the Citizens League. With the help of twenty-five high school students across the state, he spearheaded a petition drive and distributed coalition fact sheets. He obtained sixteen hundred signatures for the petition and presented it to Pastore at the senator's Providence office in early August after alerting the *Providence Evening Bulletin*. That evening, under the heading "Pastore Favors SST Delay" and a picture of Pastore receiving the petition, the newspaper quoted Pastore: "Before we spend larger sums, we must consider what it is doing to the health and welfare of our people. Let's straighten this out before we pour bushels of money into it. Let's make sure we are not adding to the neurosis of our society." Justice immediately sent the article to the coalition in Washington, where Senate opinion was being carefully monitored. Five days later Pell was given a similar petition at his home in Newport. The group presenting the petition reported that Pell leaned toward voting against the SST funding request.

The Rhode Island anti-SST effort received support elsewhere in the state. Later in the summer the City Council of Warwick discussed a resolution that requested the curtailment of SST funding until studies on the environmental effects of the aircraft were completed. In early September the Rhode Island Tuberculosis and Respiratory Disease Association asked Pastore and Pell to help defeat the SST funding. Providence's two important newspapers, the *Journal* and *Evening Bulletin,* were against the SST. In early August the *Journal* emphasized in a major article that both "facts and values" were at issue with the SST. These anti-SST efforts ultimately had the desired effect: both Pastore and Pell voted against the SST.[62]

Iowa was also the scene of strong, organized SST opposition. The state's major newspaper, the *Des Moines Register,* was against the program, and Iowa's Democratic senator, Harold Hughes, had voted against the SST in 1969. Republican Senator Jack Miller, however, had supported the program on national interest grounds and on the basis of a Transportation Department statement maintaining that there was little likelihood that the SST would harm the upper atmosphere.[63]

Other centers of SST resistance appeared around the country. In the Southwest the New Mexico Coalition Against the SST was formed to work on Democratic Senators Joseph Montoya and Clinton Anderson, who had previously supported the SST.[64] In Chicago Theodore Berland, who had recently written *The Fight for Quiet*, founded Citizens Against Noise.[65] In Hawaii Robert Wenkam, Friends of the Earth's Pacific representative, organized citizens groups to lobby pro-SST Senators Republican Hiram Fong and Democrat Daniel Inouye; as of late September, their SST positions were still unclear.[66] In Indiana, whose two Democratic senators, Birch Bayh and Vance Hartke, traditionally supported the SST, an anti-SST campaign began under the direction of Tom Dustin of the Izaak Walton League, a national conservation group.[67]

The SST Office confronted this increasing opposition with a renewed vigor of its own. Magruder's appointment in April had energized the program, and he quickly became the hub of the campaign to push the SST appropriations request through the Senate. His youthful, confident, and knowledgeable appearance, as even many SST opposition leaders acknowledged, impressed Capitol Hill and the media. Magruder favored a new emphasis on positive technological arguments, believing that developments in the commercial and military sectors could be mutually beneficial. With the aid of Boeing and General Electric the SST Office produced informational documents and a brochure on technological fallout.[68] To defend the SST on economic grounds, Magruder asked CAB chairman Secor Browne, a consistent SST supporter, to prepare several studies on airline financial capability. The resulting reports pleased Magruder immensely.[69]

Environmental criticisms, however, appeared to be the hardest to answer. These were gaining more and more credibility as large numbers of experts and scientists joined the opposition. Moreover Reuss and Proxmire accused the FAA in the spring of 1970 of inadequately drafting an environmental impact statement on the SST and thereby not complying with the National Environmental Policy Act of 1969.

Magruder tried to counter anti-SST environmental reasoning by sponsoring a large-scale study on the SST's environmental impact, by forming in July special SST environmental and noise advisory councils, and by funding a $127 million research project on potential SST climatic effects. He coordinated these efforts with Ehrlichman's White House office.[70]

Magruder also publicly attacked the opposition's most prominent arguments. He challenged Garwin's statement that the sideline noise of one SST would equal that of fifty 747s, claiming that SST engine noise was within planned FAA limits and would be confined to airports. At a Senate subcommittee hearing he produced numerous charts and graphs and displayed an impressive knowledge of technical, diplomatic, and economic issues. In response to the SCEP study in August, Magruder conceded that the SST's sulfur dioxide emissions posed a hazard but that these emissions could be easily reduced. He noted with satisfaction that the SCEP group saw no danger of ozone layer damage due to SST pollution and praised the members as a "first-class group of scientists." He later flatly declared before a Senate subcommittee that "the weight of scientific opinion" refuted theories that claimed that the SST caused atmospheric pollution, weather alteration, or increased ultraviolet radiation dangers due to ozone depletion.[71]

In an attempt to enlist airline support and to counteract the impression made by Quesada in May that the airlines did not want the SST, Magruder in June asked American Airlines and United Airlines to make their pro-SST views known to Congress. He also worked with the Air Transport Association to develop a series of pro-SST position papers and persuaded the association's president to support publicly SST funding at Senate hearings.[72]

Magruder responded promptly and carefully to all congressional inquiries. His letters were comprehensive, technically detailed, and often cited specific studies and credible experts. At times, however, his arguments were somewhat inconsistent. In June, for example, he assured Senator Margaret Chase Smith (R-Maine) that proposed FAA regulations would ban all commercial supersonic flights over the United States; later he told Senator Edward Gurney (R-Florida) that these regulations would prohibit such flights only "over populated areas at boom-producing speeds."[73]

The SST Office under Magruder kept in constant touch with congressional aides to provide them with information and to be near sources of congressional opinion. In June the office briefed the staff of the Senate Aeronautical and Space Sciences Committee and learned that the committee was troubled about the success of the Coalition Against the SST.[74] The SST Office also briefed the House Republican Task Force[75] and kept abreast of SST coverage in influential Washington publications such as the *Congressional Quarterly,* often responding directly to anti-SST articles.[76]

Magruder monitored the media closely, supplying reporters and publishers with information and writing numerous letters to protest and to correct anti-SST articles.[77] At the end of June, to increase the SST Office's effectiveness in countering adverse publicity, he met with airport officials and friendly labor leaders and recommended establishing a telephone hot line for the specific purpose of answering press questions.[78]

Magruder was confident and aggressive and liked to confront his opposition directly. In August at a San Francisco press conference he issued a public challenge to SST critics: "I would be perfectly happy to challenge the Council for the Abolition of the SST, the Friends of the Earth, the Sierra Club and everybody else to submit to the Department of Transportation their mailing lists, and I'll mail to them all of the concerns that have been published and the considered opinion of all the major scientific experts of the country on whether these things are really harmful or not."[79]

In September Magruder tangled with George Eads, a young assistant professor of economics at Princeton and adviser to the Coalition

Against the SST, and debated Eads in the office of undecided Senator Robert Packwood (R-Oregon). Eads thought that the prototype should not be built until crucial technical problems had been solved. Magruder attacked Eads on the basis of his youth and his lack of engineering training. Like Najeeb Halaby a half decade earlier, Magruder maintained that prototypes were needed precisely for the purpose of resolving such technical problems. It was generally agreed that Magruder got the best of the debate, and Eads later admitted that Magruder was an excellent salesman: "If I had to find someone to sell Edsels, or iceboxes to Eskimos, he'd be the guy. I liked him. He does his homework. He knows his facts. And he believes in his product."[80]

Volpe and other Transportation Department officials also did their part to promote the program. In June at the unveiling of the Boeing 2707-300 mockup, Volpe compared the threat of foreign SST competition to the inroads that foreign imports had made in the U.S. automobile market: "This could happen to our aircraft industry. There are those who—with ostrich logic—reason that if we don't build a supersonic transport there won't be any supersonic transports."[81]

As the expected Senate vote neared, general anxiety on both sides increased. In late August Nixon had a number of cabinet members and agency heads write pro-SST letters for presentation at Senate hearings. The most significant was written by Russell Train, who denied that two SST prototypes would cause any significant environmental damage. Previously he had been assured by Magruder that the Transportation Department would spend about $27 million during the next three years for research on quieter SST engines and on the SST's atmospheric effects.[82]

But Magruder and other SST proponents were actually unsure whether they had staved off the critics' attacks. Tension heightened. At the end of August the coalition made several last-minute appeals in Washington and urged its grassroots allies to pressure uncommitted senators. Alderson also telegraphed Shurcliff that the Senate vote was expected immediately after Labor Day, but later the SST opponents learned that the vote would not occur before September 24, a delay that gave the opposition more time to organize.[83] Key groups, such as Friends of the Earth, sent out mass mailings that urged continued pressure on senators.[84]

It seemed that the Senate vote would be extremely close. During the summer a number of pro-SST senators had turned into opponents or had become undecided, especially among Republicans and western Democrats, two large blocs of traditional SST supporters. Republicans who came out against the program included Clifford Hansen of Wyoming, Robert Griffin (Republican whip) of Michigan, Winston Prouty of Vermont, Jacob Javits of New York, Robert Packwood of Oregon, Charles Mathias of Maryland, Richard Schweiker of Pennsylvania, Ralph Smith of Illinois, and James Pearson of Kansas. Hansen declared, "When they refer to unpopulated areas, we figure they're talking about Wyoming." Among western Democrats, Gale McGee of Wyoming had become a prominent defector, and Joseph Montoya of New Mexico was wavering. Both were up for reelection in November.[85]

By the second week of September SST opponents were cautiously optimistic and were focusing most of their efforts on uncommitted senators. Gary Soucie of Friends of the Earth, who in early September distributed a confidential tally of thirty-four uncommitted senators, observed that public sentiment and editorial opinion were on their side but that this energy had to be translated into political action; if SST opponents lost in 1970, probably they would not get a second chance.[86]

The coalition continued to be the focal point for anti-SST strategy formulation and for dissemination of literature. Alderson was the main anti-SST operative on Capitol Hill, and he believed that the sonic boom was no longer the central issue. He told Shurcliff that the pro-SST forces were "relying on two things now: the balance of payments, and the deals Magnuson and Jackson can arrange with other senators." Consequently Alderson saw the anti-SST statements by prominent economists in mid-September as "a big boost" for the anti-SST side. According to Alderson, these statements substantially increased the credibility of the opponents' economic arguments and reduced "the issue down to the matter of raw political power, which had little relationship to the facts of the SST."[87]

Senate SST supporters, particularly Magnuson and Jackson, grew increasingly concerned about the growing strength of the anti-SST forces. Majority leader Mike Mansfield pressed for an early vote, but Magnuson and Jackson were powerful enough to engineer further delays. In addition, Magnuson and Jackson agreed, in behind-the-scenes consultation with other SST supporters (including many inside the Nixon administration), that it would be wise to postpone the Senate vote until after the November election. The proponents thought that many senators, especially liberal Democrats, who were running for reelection would be freer to vote for the SST once safely reelected. Jackson himself was up for reelection in Washington and did not want to risk being embarrassed by an SST defeat before the election. Finally Magnuson reasoned that since the Department of Transportation currently operated under a continuing resolution and was permitted to spend funds at the previous year's rate—$18 million per month for the SST program—about $100 million could legally be spent during the first six months of fiscal year 1971. Consequently the real impact of the total $290 million SST request was reduced. Ultimately the acting chairman of the Senate Appropriations Subcommittee, Alan Bible (D-Nevada), postponed the subcommittee's final vote, which made it impossible for the SST appropriations bill to be released to the Senate floor.[88]

Although most SST opponents objected to this postponement and saw it as a signal that the pro-SST forces could not muster a majority vote, others were grateful for the delay, feeling that the SST probably would have passed if the vote had been held before election day. The delay actually injected new energy into the anti-SST drive and made the SST a major senatorial election issue.[89]

In Arizona, for example, where the Republican incumbent, Paul Fannin, was running against Democrat Sam Grossman, members of the League of Conservation Voters (the political arm of Friends of the Earth) ran an intensive media campaign in favor of SST opponent Grossman. Fannin, however, won the election and ultimately voted for the SST.[90]

The anti-SST campaign met with more success elsewhere, such as in Hawaii, where Robert Wenkam of Friends of the Earth led a highly visible organization. Citizens were concerned that a large number of SSTs would land at Honolulu Airport since Hawaii is an island and hence would be less protected by SST overland route restrictions. By forging an effective and intense local anti-SST campaign, Wenkam became well known in Hawaii and among the national anti-SST leaders. He gave several anti-SST speeches at the Honolulu International Airport that were purposely drowned out by the roar of engines. The

Honolulu Star-Bulletin also opposed the SST. In the end, Democrat Daniel Inouye still supported the SST, but Republican Hiram Fong, who was up for reelection, was persuaded to cast an opposing vote.[91]

In Iowa the anti-SST effort focused on Republican Senator Jack Miller, a consistent SST supporter. Anti-SST sentiment in the state ran high. A November poll by the *Des Moines Register* found that 63 percent of all Iowans were opposed to the SST, while only 8 percent favored it. The paper also ran an essay in November by the British historian Arnold Toynbee, who argued against technological innovations that made life worse instead of better. His example: "An airplane carries a negotiator quicker than the fastest ship, but it lands him in the conference room in a state of physical and psychological disorientation in which he is not fit to make decisions." The major anti-SST accomplishments in Iowa were the results of carefully organized work by a number of local groups. This grassroots activity culminated in a three-day Conference on Environmental Action at Iowa State University in Ames, Iowa. David Tranger, an assistant professor at Iowa State, was a key coordinator of the conference and a leader in the Iowa anti-SST campaign. About 250 attended, representing about twenty Iowa organizations, and at this conference the Iowa Confederation of Environmental Organizations was created. The confederation sent letters and telegrams to both Miller and Hughes, both of whom voted against SST funding.[92]

The successful anti-SST drive in Rhode Island led by Dwight Justice of Ecology Action for Rhode Island continued to be active throughout the late autumn.[93]

In California the Bay Anti-Noise Group (BANG) of San Francisco exerted considerable pressure on Republican Senator George Murphy to declare himself against the SST, but Murphy continued to support the program even though he lost his bid for reelection in November.[94]

The Idaho Environmental Council directed its energies at Republican Len Jordan, who had voted for the SST in the past. Idaho's other senator, Democrat Frank Church, was a consistent SST opponent. Ironically Jordan was the one eventually to cast an anti-SST vote; Church was absent from the Senate when the appropriations bill came to the floor. The council's president commended Jordan for his "independence" and then added, "In this courageous stand, made in the face of heavy pressure by SST subsidy supporters, Senator Jordan's vote has a strong claim for the environmental leadership of the Idaho delegation." Although Church's absence, due to an extended stay in Mexico, was expected, the group still characterized it as a "shock."[95]

A number of anti-SST developments were reported in the press during November and early December. In a little noticed but extremely prophetic piece, the *Philadelphia Evening Bulletin* predicted an oil crisis that would trigger a fatal rise in SST fuel costs.[96] In early December the first glimmerings of a new and potentially volatile argument against the SST surfaced when a National Academy of Sciences panel on climate and weather modification warned that depletion of the earth's ozone layer due to SST emissions could result in an increase in skin cancer, and this warning was reported widely in the media.[97] Airline fears about SST economic performance also began to surface. The *Providence Evening Bulletin* reported that both TWA and United Airlines had expressed doubts about the financial viability of the American SST and the Concorde.[98]

Shurcliff continued to be active during the last quarter of 1970. He supplied information, wrote letters to newspapers, including the *New York Times,* and contributed substantial funds to the coalition.[99] Meanwhile the Environmental Defense Fund in late October asked seven major airport authorities to bar future SSTs exceeding federal noise levels. The group warned that these airports ran the risk of damage suits by nearby property owners.[100]

The Senate SST vote was expected in early December, and activists from all over the country flocked to Washington to help the coalition. In late November the Sierra Club, Friends of the Earth, the Federation of American Scientists, Environmental Action, and Zero Population Growth sent a joint letter to each senator, emphasizing the importance of the issue and the expected closeness of the vote. Previous votes on environmental questions had been largely uncontroversial, the letter stated, but the SST question could well depend on a very narrow margin of votes. All senators were urged to be present and to vote against further SST funding.[101]

The SST supporters countered with their own campaign during the fall. Three main groups participated: the administration, led by Magruder; Senate SST proponents, led by Magnuson and Jackson; and industry, led by General Electric and Boeing with assistance from others in the aerospace industry and allied unions.

Magruder characteristically confronted each anti-SST argument directly. For example, a Department of Transportation report submitted in September 1970 to the Council on Environmental Quality declared that the SST's harmful environmental effects would range from "insignificant" to "trivial": the aircraft would be primarily a transoceanic vehicle; takeoff noise would be less than that of subsonics because of the SST's rapid rate of ascent; climatic effects due to emissions would be negligible; and radiation danger to passengers would be insignificant. Unfortunately for Magruder, this report was not well received. It was criticized as being biased and superficial. One scientist called it "pretty miserable."[102]

More credible sources were soon defending the aircraft. In late September Magnuson and Jackson released a report by the Library of Congress that had found little problem with the SST in terms of atmospheric effects, sonic booms, and takeoff and landing noise. Fifteen hundred SSTs, the report claimed, would produce only one twenty-seventh the amount of particulates that the earth attracts from space every day, and a sonic boom generated over water would equal the pressure of a mere three-foot wave. A researcher at the Avco Everett Research Laboratory published an article that came to similar conclusions, though he added that he had not studied the problem of SST-caused "stratospheric smog" that SCEP had found significant. SST supporters were quite pleased, especially with the Library of Congress report. Jackson and Magnuson praised it as being accurate and objective; unlike the Department of Transportation, the Library of Congress obviously had no "axes to grind."

Opposition to SSTs on environmental grounds was also forcing the European proponents to establish lines of defense. Shortly after the Library of Congress report was released, the United States, Britain, and France signed an accord that provided for the exchange of SST environmental data, an action that all parties hoped would help promote public acceptance of SSTs.[103]

Magruder vigorously defended the SST's economics and received some important backing. Henry Wallich of Yale, the only top economist of those surveyed in September 1970 who supported SST funding, declared that the United States could not allow Britain and France to capture the entire SST market.[104] In a study on fuel needs he maintained that American SSTs would require only 15 percent of the world's total consumption of civil aviation jet fuel.[105] In a document attacking anti-SST economists, Magruder accused them of not understanding the airline industry and the importance of keeping up with competition. Magruder's energy and commitment were remarkable; although hospitalized in December because of a heart condition, he was on the phone constantly directing the pro-SST campaign. As a result of his persuasion, Elwood Quesada, currently a director of American Airlines who had spoken against federal SST spending in May, dissociated himself in early December from the environmental critics of the SST. "I want to make it clear," Quesada declared, "that I am not a party to environmentalist suggestions that the SST shouldn't be developed because it'll melt the polar ice cap or things like that. I believe we should research environmental problems. But I think they are tremendously exaggerated and are having an impact I wish didn't exist in the debate."[106]

SST supporters continued to take heart from the successful performances of the SST's foreign-made rivals; the foreign threat appeared more credible. By the end of the second week in October the British Concorde prototype had reached a speed of 1,230 miles per hour and an altitude of 45,000 feet, the prototype's fastest, highest, and longest flight thus far. About three weeks later the French prototype flew at its design speed of Mach 2 for the first time. The Soviet TU-144 flew at Mach 2 as early as May, and former astronaut Neil Armstrong, after returning from a visit to Russia, declared that the TU-144 was a fine-looking aircraft and as good as anything the United States was producing.[107]

The least effective elements of the pro-SST campaign at this time were in industry and labor. The lobbying operations of Boeing and General Electric were fairly weak, though both continued to produce informational materials and to maintain contact with members of Congress. The industry-oriented American Institute of Aeronautics and Astronautics possessed only one Washington lobbyist. The Fairchild-Hiller Corporation, the largest SST subcontractor, was probably the most vigorous in promoting the aircraft on Capitol Hill and in the media. The airlines were publicly supportive, but in private they expressed great concern about the SST's economics. (Friends of the Earth later claimed that it had received underground support from certain airlines.)

Labor was split. The International Association of Machinists and Aerospace Workers, the Airline Pilots Association, the Professional Air Traffic Controllers Organization, and the International Brotherhood of Teamsters all actively promoted the SST. The United Auto Workers joined the Coalition Against the SST when it was formed (but it dropped out of the coalition after its president, Walter Reuther, was killed in a plane crash in May, and it remained neutral for the rest of 1970), along with the International Longshoremen's and Warehousemen's Union, the Oil, Chemical, and Atomic Workers Union, and the National Federation of Social Service Employees. The pro-SST labor groups had little effect. The Transportation Department complained that their lobbying had failed to convince even a single member of Congress to vote in favor of SST funding.[108]

Given the mixed profile of pro-SST, anti-SST, and uncommitted groups in the Senate, an important strength for the pro-SST forces was the enormous political power of Jackson and Magnuson. Washington was the only state outside the South whose two senators were chairmen of standing committees. Seattle moreover was in the midst of a severe recession, with an unemployment rate of about 12 percent. The SST, one congressional aide remarked, was "sort of an anti-poverty project for Seattle"; the vote would not hinge on whether the country needed the plane but on whether Seattle needed the jobs.[109]

In late November the Senate Appropriations Committee approved the $290 million fiscal 1971 SST funding request as part of the overall transportation budget request of $2.7 billion. As a concession to SST opponents, the committee also pledged to support pending legislation that would prohibit SST overland flights at boom-producing speeds. Senator Alan Bible predicted that SST funding would pass in the Senate, but Magnuson felt that the funding bill was in for a tough fight. He was supported in this view by the *New York Times*.[110]

Suddenly on November 30 Magnuson introduced legislation on the Senate floor to ban overland commercial supersonic flights and to reduce takeoff and landing noise. SST opponents were stunned by this move. The bill was referred to the Senate Commerce Committee, which quickly brought it to the floor for immediate consideration by the full Senate; it passed unanimously on December 2. SST supporters now claimed that the environmentalists had no basis on which to oppose the aircraft. Proxmire, however, merely remarked that financial viability was still very much of an issue: the overland SST flight ban guaranteed that the government would not recover its SST investment. In a blistering editorial attack the *New York Times* characterized Magnuson's bill as "the final sales gimmick," observing that such laws could always be repealed or revised.

Pet projects and special interests inevitably became enmeshed in the intensifying Senate battle. Residents in Maryland, for example, wanted the army's Fort Detrick Biological Warfare Center closed, and the Senate had already voted to convert it into a national health center. But both decisions required the help of Jackson, chairman of the Military Construction Subcommittee, and Magnuson, chairman of the Labor-HEW Appropriations subcommittee. It took a certain amount of courage therefore for Senator Joseph D. Tydings (D-Maryland) to vote against the SST; Maryland's other senator, Republican Charles Mathias, voted for further funding.

Similarly Senators Mark Hatfield and Robert Packwood (R-Oregon) supported the expansion of Portland's International Airport, which would involve a "slight intrusion" onto the Columbia River landfill. Magnuson, however, declaring that he was "very jealous of the Columbia River"—the "fastest, cleanest, free-flowing river in the world"—blocked funding until an environmental study could be performed. When Hatfield and Packwood complained, Magnuson said he would consider their arguments after he had disposed of the SST funding request. Hatfield left the Senate chamber minutes before the SST vote, allegedly to catch a flight for a North Carolina speaking engagement; he later angrily denied that he had avoided the SST vote. Packwood voted against the SST.

An increase in CAB subsidies to local airlines was a matter of great concern to certain senators, including Clifford Hansen (R-Wyoming), whose state had limited air service. Hansen objected to spending $290 million on a plane that would shorten travel time on the heavily traveled transatlantic route when subsidies were being cut in areas that had little service to begin with. Magnuson assured him that he would support subsidy increases, but by then it was too late. Hansen had already committed himself to vote against the SST.

The *New York Times* reported in early December that Senator Robert Byrd (D-West Virginia) had offered to support the SST if Jackson and Magnuson would support him as Democratic whip. Byrd reportedly added that he would try to persuade his fellow West Virginia Democrat, Jennings Randolph, to vote for the SST, but Jackson and Magnuson were said to have refused. Both Byrd and Randolph voted for the SST anyway.

The pro-SST forces faced a galvanized and highly adept opposition. On December 1, two days before the vote, the anti-SST side made a number of well-orchestrated moves. Senator Edmund Muskie (D-Maine), early frontrunner for the 1972 presidential nomination, accused the Nixon administration at the confirmation hearings of William D. Ruckelshaus (Nixon's appointee for director of the new Environmental Protection Agency) of violating "the spirit if not the letter of the law" by refusing to release critical reviews of a Department of Transportation's September 1970 SST environmental impact statement. Earlier that morning Muskie and Proxmire had written to the department demanding release of these reviews but had been given the rather lame excuse that not all agency reviews had been received, and it would be "inappropriate" to release them piecemeal. Muskie, one of the Senate's best-known environmentalists, had thus effectively given credibility to Proxmire's environmental allegations and had created doubts in the minds of other senators. Meanwhile SST opponent Senator Gaylord Nelson (D-Wisconsin) released the negative comments of the Department of the Interior, and by this time it was common knowledge that the Council on Environmental Quality and the Department of Health, Education and Welfare had also expressed reservations about the environmental impact statement. On the same day, December 1, the coalition claimed that three formerly uncommitted senators, Alan Cranston (D-California), Len Jordan (R-Idaho), and Jack Miller (R-Iowa), had decided to vote against further SST funding.

Meanwhile Washington had become a mecca for anti-SST activists, who inundated senators with letters and telephone calls, filled the Senate corridors, and besieged the senators' offices. Their lobbying was thorough, enthusiastic, and intense. "We even went after people we thought lost," an activist explained. "We didn't abandon anybody. We thought everyone was eligible for redemption." They monitored the senators' movements carefully and were especially worried that potentially anti-SST senators would try to avoid voting. When Senator Tydings, who was under tremendous pressure from Fairchild-Hiller, left Capitol Hill a few hours before the SST vote, an SST opponent chased him, yelling, "He's trying to duck the vote!" Tydings later returned and voted against the SST. Similarly when SST opponents learned that Democratic Senator Vance Hartke of Indiana, who had promised to vote against the SST during his election campaign, was scheduled to be absent because of a speaking engagement with the United Auto Workers in Indianapolis, they suspected that he was attempting to duck out. They checked and found that his talk was actually slated for the evening and quickly informed his staff. Hartke stayed and cast a negative vote. Later that evening in Indianapolis he apologized for being late, explaining that he had just voted against the SST. To his surprise, the audience broke out in a standing ovation.[111]

Defeat and Resurrection

The milestone Senate SST vote took place late in the afternoon on December 3, 1970. The Senate rejected further SST funding by a surprisingly large margin of fifty-two to forty-one. In the actual vote the anti-SST bloc comprised a diverse group of northern liberals, westerners, southerners, conservatives, Republicans, and Democrats.

Between 1969 and 1970 eighteen senators had switched from a vote in support of the SST to a vote against. There were six Democrats in this group: Birch Bayh of Indiana, Everett Jordan of North Carolina, Gale McGee of Wyoming, Thomas McIntyre of New Hampshire, and John Pastore and Claiborne Pell of Rhode Island. There were twelve Republicans: George Aiken and Winston Prouty of Vermont, Hiram Fong of Hawaii, Robert Griffin of Michigan, Clifford Hansen of Wyoming, Jacob Javits of New York, Len Jordan of Idaho, Jack Miller of Iowa, Robert Packwood of Oregon, Charles Percy of Illinois, Richard Schweiker of Pennsylvania, and John Williams of Delaware.

Various forces worked to defeat the SST in the Senate. The proponents saw tactical reasons for failure. For example, in an editorial *Aviation Week and Space Technology* blamed the pro-SST lobbyists: "There was considerable ineptness and apathy by both the Nixon Administration and the aerospace industry. Boeing might be located on another planet for all the support it provided. The highly paid functionaries of the Aerospace Industries Association daintily avoided the strife until the disastrous vote was recorded." Congressional aides on both sides speculated that delaying the vote may have forced many ambivalent senators to make anti-SST commitments before the election; many northern liberals might have voted for the SST before November because it seemed then that being pro-SST was also prolabor. The delay also gave the media more opportunity to build up the issue. In addition, one conservationist suggested that the Republican Nixon administration did not want to help Democrats Jackson and Magnuson and was happy to "let them sink." Another lobbyist remarked that the White House had not applied true pressure.

But such analysis ignores other more contextual factors: the vigor and intelligence of SST opponents, Muskie's strong opposition, the high visibility of the SST issue, and the SST's close identification with the overall environmental cause. The increasing visibility of the SST issue clearly diminished the effectiveness of Jackson's and Magnuson's political power in this specific affair. On a relatively minor issue, one aide observed, bargaining and vote trading are used readily, but when the issue is one of national importance, members of Congress will often vote on the basis of less parochial considerations.

SST opponents were overjoyed with their victory, the most dramatic environmental success thus far. Proxmire claimed to be "astonished" at the winning margin and said that the "turnaround" in the Senate had been due to a heightened awareness of the SST's potential environmental hazards. The *New York Times* wrote the day after the vote, "Citizens' organizations and individual conservationists collided head-on with an entrenched economic interest group, strongly backed by the political and propaganda resources of the Nixon Administration—and the conservationists triumphed."

SST supporters were extremely bitter. "I personally consider this the most monumental hypocrisy in the history of the Senate," Magruder declared. White House press secretary Ronald Ziegler stated, "The President feels and has felt that the United States should not fall behind in any aspect of aviation. This does not suggest, however, that we intend to overlook environmental questions." At Boeing in Seattle the SST vote was announced over a public-address system by Boeing's president, T. A. Wilson; a Boeing spokesman termed the announcement a "bombshell."[112]

But the anti-SST vote did not necessarily signal total defeat for the program. The SST might survive because the House had already approved the SST appropriations request in May. If the House held firm in the forthcoming Senate-House conference, the Senate would have to reject the entire transportation appropriations bill to terminate the program, and a desire to kill the SST altogether was clearly absent among many senators who had voted against the funding bill. Mansfield called his vote against the SST "about the toughest vote I've ever cast"; Javits said that he would be willing to consider reduced SST appropriations; Griffin agreed that the Senate vote had not permanently killed the SST. Moreover the Senate members of the Senate-House Conference Committee were by and large strong SST supporters, including John Stennis (D-Mississippi), Alan Bible (D-Nevada), Gordon Allott (R-Colorado), and Magnuson. Only three of the seven Senate conferees—Margaret Chase Smith (R-Maine), Clifford Case (R-New Jersey), and John Pastore—had voted against the SST. Similarly only three of the nine House conferees—Sidney Yates (D-Illinois), Silvio Conte (R-Massachusetts), and William Minshall (R-Ohio)—had cast negative SST votes. The House pro-SST group was led by Edward Boland (D-Massachusetts), chairman of the House Appropriations Subcommittee on Transportation, and George Mahon (D-Texas), chairman of the House Appropriations Committee.

Yates, fearing a reversal of the Senate's SST rejection, then made a strategic error and offered a motion to instruct the House conferees to defer to the Senate vote. This move, which resulted in a straight House vote on the SST rather than on the entire $2.7 billion transportation budget, brought to life for the first time in 1970 vigorous and effective pro-SST lobbying on Capitol Hill. Volpe and White House press secretary Ronald Ziegler strongly criticized the Senate's decision. Magruder declared that it would cost $278 million to terminate the SST program, although the Department of Transportation later reduced this figure to about $200 million.

President Nixon came out strongly for the SST on December 5. He called the Senate action a "devastating mistake" and urged Congress to reverse the Senate vote. Among other things, he noted that SST termination would "waste" the $700 million that had already been spent and that the estimated $278 million in termination costs were "only slightly less" than the $290 million requested to continue the program. "It would be like stopping construction of a house when it was time to put in the door,'' Nixon observed. His strong public appeal was particularly significant in view of the limited White House involvement thus far.

Two days later the heads of Pan American, TWA, American, Eastern, and United announced their opposition to cancellation of SST prototype construction. Najeeb Halaby read a joint statement to the press that in effect said that the airlines had misjudged congressional opinion: ''After studying the debate, the chief executives of the airlines represented here today realize that 52 Senators of the United States do not understand the Program and have reversed field in the ninth year.'' Only after a prototype had been constructed, the airlines declared, would they be able to decide whether a production SST should be built. Braniff, Continental, Northwest, and National offered their support, and Boeing lobbyists increased their efforts. The White House also obtained the help of House minority leader Gerald Ford, who gave a strong speech in favor of the SST.[113]

This uncharacteristically vigorous and wide-ranging pro-SST campaign produced results. On December 8 the House voted 213 to 174 to table Yates's motion. For the first time the pro-SST people had actually worked harder than the opposition.[114]

The SST opponents, of course, had not given up. On December 7 Senator Gaylord Nelson proposed a motion that would have effectively barred any SST from operating in or out of American airports. His bill would prohibit all SST operations except for those vehicles that produced no sonic boom on land or water and that produced no stratospheric pollution through engine exhaust emissions.[115]

At the same time, the opportunities for accommodation diminished. Boeing indicated that any significant cut in SST funding would be extremely damaging. ''It is at the stage where it is almost all or nothing,'' Boeing's SST director declared.[116] The ''compromise'' solution finally offered by the conferees on December 10 was to reduce fiscal 1971 funding to $210 million, $80 million less than the original $290 million requested. Several SST critics were strongly opposed to this figure, though they were willing to accept a lesser amount, say $150 million.

Proxmire and Nelson vowed to filibuster the entire conference report on the transportation bill. Because amendments to conference reports were prohibited, there could be no vote solely on SST funding. The anti-SST forces believed that the Senate would reject a straight SST vote but that it might accept the entire $2.7 billion transportation budget. The pro-SST forces strove to keep the SST an integral part of the whole transportation budget and gave their full backing to the $210 million amount. Magnuson called the new funding package a victory, and Nixon announced his support at a news conference. Magruder professed to be delighted, although he had previously argued that anything below $259 million would "destroy" the program.[117]

On December 15 the House reaffirmed its support of SST funding by passing the transportation bill and by voting 205 to 185 to block sending the $210 million compromise figure to conference.[118] Proxmire began his filibuster in the Senate, declaring that he would stop only if a straight SST vote were scheduled for sometime in January 1971. The Senate twice rejected cloture petitions to cut off Proxmire's filibuster.[119] Because the ninety-first Congress's closing date of January 3, 1971 was fast approaching, the filibuster put increasing pressure on the Senate. It created a logjam just when Congress had to deal with such issues as welfare reform, trade quotas, a Nixon veto of a labor bill, and loan terms for Asian and Latin American businesses. In fact, two concurrent filibusters were taking place: a "morning shift" against the SST and a "late shift" against a trade quota provision that had been tacked on to a bill increasing social security payments. Majority leader Mike Mansfield was concerned about the Senate's image, and newspapers began writing about the "breakdown" of the congressional conference system.

After a second Senate cloture vote on December 22 failed to cut off the SST filibuster, closed-door meetings were arranged by Mansfield and minority leader Hugh Scott (R-Pennsylvania) between SST opponents Proxmire and Percy and SST supporters Stennis and Bible. Proxmire demanded and won a straight SST vote in the next Congress. Mansfield and Scott proposed that the SST could only continue to be funded at approximately its current rate until March 31, 1971, after which continuation would require approval by both the House and Senate on a specific SST supplementary appropriations bill for the remainder of fiscal 1971.

By the end of 1970 therefore an interim agreement had been reached. The Senate voted to recommit the original conference report, and the filibuster was broken. Senators Proxmire and Norris Cotton (R-New Hampshire) were added to the conference committee. The

House passed a resolution extending SST funding until the end of March 1971. On January 2, 1971, with a behind-the-scenes agreement already reached, Proxmire extracted a public pledge from the leaders of the Appropriations Committee.[120]

During 1970 the SST had been dramatically transformed as an issue. The topic, which previously had held largely bureaucratic or limited interest group appeal, had exploded upon the national scene in 1970 and had become a matter of great and highly visible concern. At a time, according to a Harris public opinion poll, when Americans were more concerned about the environment than about any other public issue, the SST became a major symbol of environmental danger and evolved into a widespread public issue.[121] The SST was no longer simply a program with bureaucratic, managerial, technological, or limited political problems; by the end of 1970 it was a full-fledged mass issue. During 1970 the anti-SST forces capitalized effectively on the SST's new status and won a crucial Senate vote. But with SST proponents finally mobilizing their considerable resources the question was now whether the SST opposition could maintain its momentum in 1971. The final campaign was about to begin.

19 The Final Confrontation

Prior to 1970 the pro-SST forces had consistently had the upper hand, winning all SST votes in Congress by large margins and preventing the SST from attracting national attention. But in 1970, the year of the first Earth Day, the SST opposition gained on its adversary, becoming more forceful and effective, and it overwhelmed its foe by winning the critical SST vote in the Senate. By early 1971, however, the situation had changed again. SST supporters clearly had been shocked out of their complacency, and for the first time the two sides seemed equally powerful.

Although some SST opponents feared that they were losing momentum after their Senate victory in late 1970, there was still strong anti-SST sentiment throughout the country, especially among students. A nationwide poll released in early February 1971 showed most students opposing SST development.[1]

The Coalition Against the SST remained active and continued to coordinate the anti-SST effort in Washington. It published materials, publicized negative items on the SST, lobbied on Capitol Hill, and provided support for local anti-SST groups. In addition to working for Friends of the Earth, George Alderson formally became coordinator of the coalition. Through a series of memorandums sent out during February and March 1971, he kept member organizations informed on such matters as pro-SST lobbying, new developments in the SST battle, and the legislative schedule related to the SST. The coalition prepared a twelve-page booklet, *New Facts on the SST,* that contained excerpts from the August 1970 SCEP study on climatic effects, the anti-SST statements made by prominent economists in September 1970, a statistical summary showing that forty-four states would lose money on the SST when tax money was deducted from potential SST contractual outlays, and a column from the *Seattle Post-Intelligencer* that claimed that canceling the SST program would actually benefit Boeing by improving its short-term earnings and its ability to compete in the wide-bodied subsonic jet market. Alderson made the booklet available to "leadership people at the grassroots level."[2]

The activities of William Shurcliff's Citizens League complemented those of the coalition. The coalition provided the anti-SST campaign with national leadership and effective political operatives, and Shurcliff's Citizens League provided funds and a network of dedicated activists across the country. Shurcliff maintained a list of about seventy wealthy contributors who regularly sent checks ranging from $50 to $500. He processed thirty letters per day, of which about five were requests from colleges and schools for a notorious map that showed likely sonic boom zones on the Atlantic Ocean. Early in the year Shurcliff prepared a mailing to be sent to some 400 oceanographers; about forty scientists from the Woods Hole Oceanographic Institute had already signed letters protesting the SST, and Shurcliff hoped that another 130 oceanographers would sign a petition expressing concern about the effect of sonic booms on aquatic life. Shurcliff's *SST and Sonic Boom Handbook* remained popular. During 1970 Shurcliff had given away over 9,300 copies, and a similar handbook on the Concorde was due out in February. Shurcliff's son, Charles, traveled to Washington to join the anti-SST campaign armed with sonic boom zone maps, reports, and other material useful to the cause. Shurcliff, meanwhile, offered advice to all who contacted him and worked with the coalition to develop new anti-SST arguments. At one point he suggested publicizing the theory that a "crazed passenger" controlling an SST over, say, New York City could cause about $1 billion worth of damage, about as much as a small atomic bomb.[3]

The coalition and the Citizens League were by no means alone. Friends of the Earth increased the intensity and tempo of its anti-SST appeals after the Senate vote; a January 1971 mailing to its members sounded much like a call to battle. Gary Soucie declared that for the first time SST opponents had the opportunity to win in the House as well as the Senate. He implored members to send more letters to their representatives, no matter how many they had already written, and local chapters responded quickly.[4] The New Mexico chapter, for example, immediately sent out strongly worded brochures, sonic boom zone maps of the United States, and a detailed account of the recent Senate conflict, entitled "Senate SST Vote Shows Environmental Muscle."[5] The Sierra Club, with a membership of some 200,000, also continued its active anti-SST work. Charles Shurcliff, who had previously worked for the Sierra Club, maintained close contact with this group throughout early 1971.[6]

In January 1971 the large grassroots group Common Cause came out against the SST. The following month it distributed a document outlining its arguments against the aircraft and urged its members to contact their congressional representatives and to cooperate with other local anti-SST groups.[7] Another national group, Environment Action, published a leaflet, "No Victory on the SST," which stressed that the December 1970 Senate action had merely delayed the ultimate SST decision and that the critical vote would take place in 1971.[8] The Environmental Defense Fund pursued its legal battle against the SST, filing lawsuits to ban SSTs from seven major airports in the United States because of excessive noise. The Federation of American Scientists continued to campaign vigorously throughout 1971.[9]

One of the newest groups to oppose the SST was Citizens Against Noise, which fought noise pollution. The organization was formed in Chicago by Theodore Berland, who had written an influential antinoise book, *The Fight for Quiet*. It had a prestigious steering committee made up of architects, engineers, lawyers, public relations specialists, and an audiologist. This organization pressured pro-SST senators to change their vote, thanked anti-SST senators such as Charles Percy for their support, and urged passage of a law that would ban supersonic flight over the United States. It exchanged information and ideas with the Citizens League, which it looked to as a model, and Shurcliff forwarded to Berland all the requests he received for noise information. Both men joined the other's group, and Shurcliff sent Berland a "modest check." The activities of Citizens Against Noise were widely reported in the Chicago press, to which the group wrote numerous letters protesting pro-SST columns and editorials.[10]

In early February the SST opposition scored an important victory: Charles Lindbergh, famous aviator and current member of Pan American's board, came out publicly against the SST. In a widely quoted letter written to Congressman Sidney Yates (D-Illinois), Lindbergh declared that the SST's environmental hazards and high economic costs made it too risky to proceed with actual construction. He concluded that exclusive over-water routing was infeasible and saw no "practical way" to avoid sonic boom disturbance; citizens, he felt, were "already subjected to more than enough technological noises."[11]

SST opponents greeted this announcement joyfully. The fact that Lindbergh was a member of Pan American's board made his statement all the more powerful. Alderson speculated that the airlines had been skeptical all along but had presented an optimistic front in response to FAA pressure. He urged that the letter be publicized by sending copies to grassroots leaders and to members of Congress.[12]

Legislative anti-SST activity was growing. By early February bills to limit SST engine noise had been introduced or were planned in several states.[13] In Michigan, for example, anti-SST legislation was supported by a number of environmental groups and by the Republican governor, William Milliken, who had been an SST opponent since 1970. In February 1971 he telegraphed each member of Michigan's congressional delegation to urge them to vote against the SST.[14] Also in February the newsletter of the Student Council on Pollution and Environment, a group focusing on the Great Lakes region, contained a detailed report on a proposed New York State bill to ban noisy aircraft; New York, the group declared, would be an appropriate place to stop the SST.[15]

The Town-Village Aircraft Safety and Noise Abatement Committee, representing communities near JFK Airport, continued its fight against airport noise. In early February it urged its supporters to send anti-SST letters to their representatives. "This is a war," the committee proclaimed, "a war to prevent the loss of our sanity from jet noise pollution. Our troops are the people, their bullets are their letters. Let the bullets fly."[16]

During his reelection campaign New York's Republican governor, Nelson Rockefeller, claimed to be in favor of strict noise legislation, and New York City mayor John Lindsay was a consistent SST opponent. The city's air resources commissioner, Robert Rickles, characterized certain pro-SST arguments made by the head of the Environmental Protection Agency, William Ruckleshaus, as "ridiculous."[17]

In Hawaii the state senate Republican policy leader, Fred Rohlfing, criticized Democratic Senator Daniel Inouye's pro-SST stance. One state representative introduced legislation that would empower the state's health department to ban new commercial aircraft operating above certain noise levels. More than one hundred persons attended a committee hearing on this bill, and Shurcliff supplied the bill's proponents with materials on the SST.[18]

In California as early as April 1970 a bill to limit aircraft noise had been introduced that, according to Shurcliff, would effectively prohibit any SST then under development from landing at California airports. The Sierra Club and Friends of the Earth, both headquartered in San Francisco, kept up their local anti-SST campaigns, along with the Bay Anti-Noise Group. In early February the San Diego Citizens Committee to Ban the SST was organized; it urged all local groups to join in a city-wide coalition against the aircraft.[19]

In New England strong opposition to the SST continued. Shurcliff testified twice before legislative committees at the Massachusetts State House. Peter Koff of the Sierra Club's New England chapter warned SST opponents in the region that Nixon and Volpe would put up a tough fight in 1971. He emphasized the need for a mail campaign to Senators Margaret Chase Smith (R-Maine), John McIntyre (D-New Hampshire), Winston Prouty and George Aiken (R-Vermont), and various New England representatives. Koff encouraged state anti-SST bills and anti-SST positions by local newspapers.[20]

But SST critics faced a stronger and better organized adversary in 1971. Magnuson, Jackson, and Magruder intensified their campaign, supported this time by the White House, the aerospace industry, and labor organizations. SST proponents made use of new technical findings that minimized or negated the alleged environmental and economic weaknesses of the SST. Fred Singer, chairman of the SST environmental advisory committee for the Department of Transportation and deputy assistant secretary of interior for scientific affairs, supported a SCEP finding that SST emissions would not reduce upper-atmosphere ozone or cause a greenhouse effect because of increased carbon dioxide emissions. A Boeing scientist claimed that airplanes accounted for only about 1 percent of urban pollution, compared with some 80 percent caused by ground activities; he maintained that the SST released fewer contaminants into the atmosphere per passenger-mile than any other aircraft.[21] In January Volpe announced that "new evidence" would soon be presented to Congress showing that the SST would have a favorable economic impact and would not be an environmental threat. "The SST is far from dead," Volpe confidently declared.[22] In early February the SST Office reissued a white paper that stressed the SST's benefits to the country in several important areas and the certainty that the Concorde would be a potent competitive force if left unchallenged.[23]

Also at this time, at the suggestion of Magruder, Commerce Secretary Maurice Stans appointed a committee of outside scientists and engineers to study the SST's effects on weather, climate, and atmospheric radiation levels. Heading the committee was Frederick Henriques, chairman of Technical Operations, Inc., and an atmospheric photochemist. Henriques had already questioned the idea that the SST might alter the world's climate. In January he had claimed that particulate emissions from five hundred SSTs would be less environmentally hazardous than the emissions from fifteen hundred Boeing 707s. Another member of this committee, TWA vice-president Robert Rummel, had been an active and consistent SST supporter since the mid-1960s.[24]

SST supporters were especially encouraged by technical findings on aircraft noise. In July 1970 Magruder had established the SST Community Noise Advisory Committee to review SST noise objectives in the light of FAR 36, the strict new regulation that established a maximum level of 108 EPNdB for aircraft noise. This committee was chaired by Leo Beranek, chief scientist of the respected acoustical firm of Bolt, Beranek, and Newman in Cambridge, Massachusetts. In September 1970 the committee had concluded that the allowable noise levels for the production SST should be the same as those stipulated in FAR 36 for new four-engine subsonic commercial aircraft, and Beranek had recommended that Boeing and General Electric give more emphasis to their SST noise programs. During late 1970 the committee monitored the two contractors closely and in turn was supplied with detailed performance data by both firms. In a report to Magruder submitted on February 5 1971 the committee was quite optimistic, declaring that "the level of technology demonstrated by Boeing and General Electric is sufficient to achieve the noise level objectives we recommended;" the production SST would need less engine thrust than previously estimated because it had greater aerodynamic lift capability, it had an improved wing contour that allowed for a steeper takeoff, and it would require a larger redesigned engine to eliminate the need for an afterburner (a noisy device in the engine's rear that creates thrust by burning huge amounts of fuel poured into the exhaust system).[25] The committee was anxious to submit its report before House and Senate SST deliberations began; the findings were released about two weeks after Magruder received the study and were widely reported.[26] In February the SST Office issued its own report paper, based partly on data gathered by the Beranek committee. The accusations of critics regarding SST noise, the report claimed, were not based on fact.[27]

The findings of the Beranek committee clearly stole much of the SST opposition's thunder and momentarily represented a significant victory for the pro-SST forces, though the opposition attempted to respond. The coalition vigorously attacked the notion that a quiet production SST was possible and called such an assertion "a flagrant attempt to mislead the public and the Congress." Alderson emphasized that the engine discussed by the Beranek group existed only on drawing boards and would not be produced for at least two years. Shurcliff commented sarcastically that if General Electric could make quieter SST engines, then presumably it could also make quieter subsonic jet engines, which would still leave SST noise levels "sticking out like a sore thumb."[28]

Soon SST support emerged in other quarters. Organized private efforts to promote the SST shed the low profile they had maintained throughout 1970 and became more vocal. Pro-SST political activity made itself felt in 1971 even in areas that previously had been the scene of significant anti-SST victories. In Hawaii, for example, the state transportation director argued before a state senate committee that the SST should be allowed to land at Honolulu Airport subject to certain controls.[29]

The White House too was taking a more active role, and the *Washington Post* reported that Vice-President Spiro Agnew might be the "quarterback" in a new White House SST drive. The new federal budget submitted in January reflected Nixon's support, calling for the original $290 million in SST funds for fiscal year 1971 and $235 million for fiscal year 1972. In January Boeing chairman William Allen and General Electric's chief executive officer Fred Borch met with White House aide John Ehrlichman to urge the White House to campaign strongly for increased SST funding.

SST contractors and subcontractors were, in the words of one Boeing official, "nervous," feeling that the compromise $210 million in SST funds would cause slippage in the program. Boeing president T. A. Wilson indicated that even the production SST would need strong government support, possibly in the form of loan guarantees.[30] In early February Boeing and General Electric began their most intense SST public relations and lobbying efforts. Shurcliff later reported that Boeing's SST unit had sent out a memorandum to "Suppliers and Subcontractors to the Aerospace Industry," asking these firms to urge Congress, by means of letters from employees, friends, and relatives and in other ways, to vote for the SST; sample letters were enclosed for possible use. The company made it clear that it was determined to support the case for SST prototype construction. Both company executives and employees contributed funds to the pro-SST campaign.

General industry support, which had been weak in 1970, grew in 1971. In February the National Committee for SST was formed, with members from various aviation trade groups. Two pro-SST pamphlets, *SST Fantasy and Fact* and *The Supersonic Transport: Ecology and Economics,* were issued by the Aerospace Industries Association of America. A Seattle-based group, the Committee for an American SST, published a brochure, *Action USA/SST,* which encouraged citizens to write to their senators and representatives in support of the SST. A local radio station owner and a public relations expert headed this group, which by late February had raised about $17,000 to support pro-SST advertising in newspapers and on national television. Other industry groups produced pro-SST decals and bumper stickers. The pro-SST fight in the Los Angeles area was carried by an Aerospace Truth Squad, which had previously campaigned for the new B-1 bomber. In Oregon the Labor and Industry Committee for the SST was formed, and similar groups surfaced in Oklahoma, Maryland, and New York.[31]

In mid-February Donald Strait, vice-president of Fairchild-Hiller, and Floyd Smith, president of the International Association of Machinists and Aerospace Workers, formed the National Committee for an American SST. Strait and Smith established a $350,000 advertising program to counter environmentalist attacks on the SST and to stress the importance of the SST to the national economy. Full-page advertisements appeared in the three daily Washington newspapers, declaring that fears about the SST's environmental hazards and noise levels were unwarranted and emphasizing the threat of foreign competition. A group affiliated with the committee, American Industry and Labor for the SST, sponsored advertisements that stressed the foreign threat in the *New York Times* and other papers.[32] In early March Northrop Corporation urged its suppliers to support the National Committee and to make their support known to Congress. Fairchild-Hiller issued a similar plea to its stockholders and suppliers.[33] At the same time the executive committee of the American Institute of Aeronautics and

Astronautics strongly endorsed a favorable report on the SST that had been prepared by an ad hoc group of that organization.[34]

American SST proponents received a great deal of help from across the ocean. In mid-February British and American SST contractors held a joint news conference in London to announce that their respective SSTs would cause little atmospheric pollution or noise disturbance. *Aviation Week and Space Technology* devoted nearly an entire issue to the Concorde and the technical progress it had made. The journal warned that the Concorde was "the strongest challenger to American domination of the international transport market in history." Later in February the former British minister of technology, Anthony Wedgwood Benn, traveled to New York to speak against a proposed bill to limit aircraft noise. The Russians announced at about this time that the TU-144 would begin making commercial flights between Moscow and Calcutta in the fall of 1971.[35]

With the findings of the Beranek committee, growing industry involvement, active White House support, and the increasing pressure of apparent foreign successes, the pro-SST campaign appeared to have regained the initiative. SST opponents became seriously concerned. Alderson warned that labor organizations at the state and local levels were exerting a great deal of pressure on members of Congress to vote for the SST. He noted that lobbying in Washington was growing more intense and specifically mentioned the committee that had been organized by Smith and Strait. Alderson added that, according to a "captured memo," the SST proponents had failed to obtain astronaut Frank Borman as the chairman of this group and were trying for actor James Stewart; aerospace contractors were making large contributions to the pro-SST effort, apparently viewing the fight as crucial to their attempts to win federal funds for future military and civilian research and development programs.[36]

The active lobbying by both sides in January and February was the prelude for a final confrontation at hearings before the House Appropriations Subcommittee on Transportation and before the Senate Committee on Appropriations. This set of hearings provided the stage for the most comprehensive and wide-ranging public SST debate that had yet been witnessed. Technological, economic, and diplomatic factors were discussed. Moreover the impact of the SST's new meaning as a general symbol of environmental degradation was explicitly articulated and discussed.

The House SST hearings took place on March 1, 2, and 3, and the Senate SST hearings occurred on March 10 and 11. The actual SST vote was expected later in March.[37] The dates of these hearings had been set more than a month in advance, and both sides were well prepared for this final opportunity to present evidence before the critical SST vote. As shown by the success of the SST opposition at the Proxmire subcommittee hearings in May 1970, congressional hearings could have a profound effect on public and congressional opinion.

The pro-SST forces displayed their new vigor at these hearings. As Alderson had warned earlier, the Nixon administration sent several high-level officials to testify before the House, including the head of the Environmental Protection Agency, William Ruckelshaus, and Transportation Secretary Volpe.[38] Ruckelshaus advocated building at least two SST prototypes but not necessarily any production planes. He did not subscribe to the argument that once a large-scale program is funded and involves many jobs, it usually continues. "Technological projects can be stopped if their continuation is found to be environmentally unsound," he declared. Volpe emphasized the economic value of the production aircraft and claimed that terminating the SST program would cost a half-million jobs.

SST supporters presented a battery of endorsements. Magruder submitted pro-SST letters from cabinet members, NASA, the CAB, the National Aeronautics and Space Council, the heads of nine American-flag and three foreign airlines, the Airport Operators Council, the U.S. Chamber of Commerce, the Aerospace Industries Association, the Air Line Pilots Association, and George Meany, president of the AFL-CIO. Magruder produced pro-SST statements from respected technical experts in aviation, including Charles Harper of NASA, Kelly Johnson of Lockheed, Rene Miller of MIT, Wilbur Nelson of the University of Michigan, Maynard Pennell of Boeing, Courtland Perkins, dean of engineering at Princeton, Arthur Raymond of Douglas, and John Stack of Fairchild-Hiller. Floyd Smith, president of the International Association of Machinists and Aerospace Workers, defended the economic prospects of the SST at these hearings.

The most novel and newsworthy aspect of the House hearings, however, was that, in the last days of the SST conflict, they marked the public beginning of an increasingly visible and gripping issue: the SST's impact on the protective layer of ozone in earth's upper atmosphere. James McDonald of the Institute of Atmospheric Physics at the University of Arizona was an atmospheric physicist and a member of a National Academy of Sciences panel on weather and climate modification. He was asked by the panel to study the SST's effect on the upper atmosphere and in November 1970 he had privately reported to the Department of Transportation and the NAS disturbing findings on the relationship involving the SST's water vapor exhausts, depletion of stratospheric ozone, and an increased incidence of skin cancer among humans. At the House hearings in March 1971 he contended that the emissions from five hundred American and foreign SSTs would result in approximately ten thousand additional cases of skin cancer in the United States caused by the depletion of ozone in the upper atmosphere from the aircrafts' water vapor exhaust. McDonald's conclusion was at odds with the SCEP findings, released in August 1970, that depletion of the ozone layer by the SST was not a major problem. Other analysts in the Library of Congress, the Department of Transportation, and Boeing also disagreed with McDonald. But although McDonald's testimony was long, detailed, and in the view of many, tedious, it dealt with an emotional issue and was given widespread publicity.

McDonald's effectiveness might have been even greater had Congressman Silvio Conte (R-Massachusetts) not suddenly asked McDonald about the scientist's theory that power failures in the mid-1960s had been due to "flying saucers." McDonald had testified before a congressional committee in 1968 that there could be a possible correlation between unidentified flying objects and power failures, and Conte's question immediately caused amusement, casting apparent doubt on the credibility of McDonald's SST statements. Later Conte inserted into the *Congressional Record* part of McDonald's testimony on UFOs (which also expressed a belief that contact with humanoid occupants from outer space may have occurred). Conte himself remarked, "A man who comes here and tells me that the SST flying in the stratosphere is going to cause thousands of skin cancers has to back up his theory that there are little men flying around in the sky. I think this is very important." This ridicule of McDonald, however, had no lasting effect, and, to the dismay of SST supporters, McDonald's theories concerning skin cancer continued to be widely reported in the press.

The next day at the House hearings the SST proponents counterattacked. Three respected experts appeared skeptical of charges that the SST represented a large-scale environmental danger. William Kellogg, associate director of the National Center for Atmospheric Research and chairman of the SCEP Working Group on Climatic Change, reiterated SCEP's conclusions that the SST would have an imperceptible effect on the earth's climate, compared to effects caused by natural phenomena. Fred Singer agreed. He rhetorically asked why there was so much "fuss" over the SST, answering that the aircraft's major problem was that it had become a symbol of a general reaction against all technological progress for people who believed that technology was evil. Acording to Singer, the SST's fate should be decided mainly on the specific grounds of economics and national priorities. Leo Beranek repeated the earlier findings of his own committee. All three favored proceeding with prototype construction.[39]

The mass of scientifically credible testimony in favor of the SST proved to be highly effective, though the issue of the SST's effect on the earth's ozone layer would grow in notoriety in the coming years. The House Appropriations Subcommittee on Transportation rejected any compromise and on March 5 approved the Nixon administration's original SST funding request of $290 million. A majority of the subcommittee felt that a funding cut would merely delay the project and ultimately increase its costs. Besides, as one close source observed, the subcommittee "figured that whatever they approved—$210 million, $255 million, $290 million—the opponents would still be against it."[40]

After this pro-SST victory, the SST opponents launched a new offensive of their own. On March 6 Henry Reuss released a letter from George Kistiakowsky, former White House science adviser to President Eisenhower. Kistiakowsky expressed his opposition to government SST funding and claimed that once prototypes were built, the pressure to continue into the production phase would be "irresistible."[41]

At the Senate SST hearings, however, supporters again mustered powerful testimony. Influential individuals such as George Meany, Volpe, and Secretary of the Treasury John Connolly all supported federal funding of the SST primarily on economic grounds. Magruder, Beranek, Kellogg, astronaut Neil Armstrong, and Vice-Admiral Hyman Rickover of the navy also testified in favor of funding prototype development. Armstrong agreed with NASA that two prototypes were needed to determine technical facts, and he praised the TU-144, which he had seen in the Soviet Union. In a lively testimony Rickover minimized the SST's ecological hazards and claimed that public overreaction to environmental problems could affect technological progress just as severely as public indifference could. "An indiscriminately negative approach to technological development, whether by laymen or scientists," he declared, "may well result in needlessly delaying useful and desirable progress." He opposed a suggestion by Proxmire to establish an SST "environmental impact board," telling the senators, "I think you have enough competence right here in Congress to decide the fate of the SST. God spare us from another council, another addition to our bureaucracy. . . . I would rather any time take a technical project and put it up for judgment by Congress than I would to the bureaucracy."

The most forceful counterattack by SST opponents focused on economics. Nobel prize-winning economist Paul Samuelson of MIT, in a widely reported statement, declared that public financial support of the SST would be "clearcut economic folly." He discounted the alleged balance-of-payments and employment benefits of the SST and referred to the Concorde as the "biggest lemon that was ever devised." He urged the government to heed the most important lesson of decision making: cut its losses.

Economist Karl Ruppenthal argued that the SST would make no real difference to the transportation system in the United States and that the real threat to American dominance of commercial aviation lay in the area of short- and medium-range aircraft; the United States would simply be wasting resources that could be better used in more crucial aviation sectors. To Ruppenthal, the SST presented "a clear and present danger to the private enterprise system" because the government would act as banker, marketing agent, and insurer.

The Senate committee received statements from Concorde critics Richard Wiggs, the head of the British Anti-Concorde Project, and from Jean-Jacques Servan-Schreiber, member of the French National Assembly and former publisher of *L'Express*. Servan-Schreiber wrote that the Concorde "looks to us, on this side of the Atlantic, like an industrial Vietnam."[42]

On both sides of the SST debate arguments appeared to center on economic matters, and environmental attacks seemed weakened by the doubt that had been cast on McDonald's credibility. But on March 17, the eve of the House SST vote, Proxmire moved to regain the initiative on the environmental front. At a news conference he released twenty-one written statements from highly respected atmospheric and biological experts, all of whom at least acknowledged that the SST might be a skin cancer hazard. Appearing with Proxmire were J. G. Charney, Sloan Professor of Meteorology at MIT; Thomas Fitzpatrick, chief dermatologist at Massachusetts General Hospital; and Conway Loevy, associate professor of atmospheric sciences at the University of Washington. Charney admitted that McDonald's UFO theories were unorthodox but claimed that his statements showed "honest and fearless pursuit of truth." Proxmire presented other written endorsements of McDonald's work. Some of these statements suggested that McDonald's estimate of 10,000 additional skin cancer cases per year might be too conservative. Dr. Gio Gori of HEW's National Cancer Institute estimated that a fleet of eight hundred SSTs would cause 23,000 to 103,000 additional skin cancer cases. And skin cancer, he warned, was only one of the simpler effects of tampering with the earth's atmospheric defenses; other modifications of the environment might be more destructive.[43]

The House vote finally took place on March 18. By a vote of 215 to 204 representatives decided against continued SST funding. In the House chamber, as it became apparent that the pro-SST campaign had failed, Sidney Yates, who had led so many losing battles against the program, smiled and shook hands with his colleagues. Magruder sat motionless in the gallery.

Only since 1970 had recorded votes been required, and the fact that the House SST vote was recorded demonstrated the enormous impact of this major congressional reform. "The members no longer could duck under some parliamentary guise," Silvio Conte said, and other representatives agreed. The SST opposition captured the votes of 33 of the 51 freshmen representatives; Republicans supported the SST by a vote of 90 to 84, while Democrats voted against the SST, 131 to 114.

SST opponents were overjoyed. Proxmire described himself as "delighted, stunned," and Percy declared that the vote was "a great victory for all people of the country who don't need a flying Edsel."

Michigan environmental organizations praised the sixteen Michigan representatives, out of a total of nineteen, who had voted against the SST. The coordinator of the Michigan Student Environmental Confederation saw the vote as "the biggest victory for environmentalists and conservationists in the House that we ever achieved." The director of the Michigan Confederation of Zero Population Growth observed, "We never expected to achieve victory in the House with this wide a margin. This was one victory we really needed, and believe me, we sweated out every last vote."

SST supporters were stunned by the defeat. Jackson called it a "tragedy." Magnuson vowed to "do all I can to seek approval for the program in the Senate." *Aviation Week and Space Technology,* in a scathing editorial, compared anti-SST arguments to antiquated notions about the danger of railroads, the impossibility of air flight, and the foolishness of envisioning rockets to the moon. Accusations against the SST were just as ridiculous: that the aircraft "would result in the possible death of all fishes, birds, and other life forms in the oceans under its path; or an epidemic of skin cancer from its exhaust emissions; the possible extinction of all life on earth as a result of its flights weakening the protective ozone blanket of the earth; the inundation of all life on earth because its flights might so alter the temperature of the atmosphere that it would melt the polar ice cap."[44]

Attention turned quickly toward the upcoming Senate vote, which was expected on March 24 or 25. On March 21 three Democratic SST opponents in the Senate—William Proxmire, Birch Bayh, and Gaylord Nelson—said that they planned to introduce a resolution to aid workers who would lose jobs as a result of canceling the SST program. The next day Charles Lindbergh told the Minnesota Conservation Society that the SST was unjustified because of potential atmospheric pollution and sonic boom disturbances: "I believe it would be a mistake to become committed to a multi-billion-dollar SST Program without reasonable certainty that SSTs will be practical economically and acceptable environmentally."

Immediately after the House vote, the following senators appeared to be undecided: Hubert Humphrey (D-Minnesota), Margaret Chase Smith (R-Maine), John Sherman Cooper (R-Kentucky), Stuart Symington (D-Missouri), David Gambrell (D-Georgia), Winston Prouty (R-

Vermont), Hiram Fong (R-Hawaii), Marlow Cook (R-Kentucky), Peter Dominick (R-Colorado), James Buckley (Conservative-New York), and Lloyd Bentsen (D-Texas).

In a desperate effort to save the SST, Magnuson and Jackson held a number of last-minute meetings with Ken Belieu, White House congressional relations assistant; John O'Shea, director of the National Committee of Industry and Labor for an American SST; and representatives of the International Machinists Union, the AFL-CIO, the Air Transport Association, Boeing, and General Electric. Each organization was asked to make a final appeal to specified senators. George Meany contacted a number of Democratic senators, including Hubert Humphrey. Meany had been a strong backer of Humphrey in the 1968 presidential election, and Humphrey had supported the SST while he was vice-president. His backing now was essential, and Jackson and Magnuson hoped that Humphrey might call a news conference to announce a pro-SST decision. But their final appeals failed; Humphrey announced that he intended to vote against the SST.

The White House made a number of efforts to influence the Senate vote. One aide, noting that Dick Cavett had not given equal air time to pro-SST views on his television show, forced Cavett to cancel a debate between Magruder and Proxmire and made it possible for Magruder to appear alone. On the air Cavett acknowledged the imbalance but then spoke out against the SST, evoking a burst of applause from the studio audience. In the Senate tension mounted. White House efforts focused on Smith, Prouty, Dominick, Fong, Cooper, and Buckley. Nixon himself met with a number of senators, and after one such meeting Buckley announced that he would vote in favor of the program. Nixon tried to sway Margaret Chase Smith by informing her that a Johnson administration decision to close the Portsmouth Naval Ship Yard near the New Hampshire-Maine border had been reversed, but in the end Smith again voted against the SST. On March 25, the day of the Senate vote, Nixon met with two Republican and one independent senator who had voted against the SST in 1970 but who now appeared to be wavering: Clifford Hansen of Wyoming, Jack Miller of Iowa, and Harry Byrd, Jr., of Virginia. The White House ultimately succeeded in convincing only Fong, Cook, and Dominick to vote in favor of the SST.

TWA meanwhile worked on Symington because it had extensive operations in Missouri. TWA president Charles Tillinghast personally contacted Symington, a respected aviation expert in the Senate, and Symington pledged his vote for the SST.

In contrast, Lockheed and McDonnell-Douglas appeared to cast doubt on the SST. Both firms were then heavily involved in developing large subsonic jets, the L-1011 and the DC-10, and there was some fear that the airlines could not support the additional cost of the SST. It was reported that Lockheed, with large defense plants in Georgia, convinced Gambrell to vote against the SST and that McDonnell-Douglas worked closely with freshman Senator John Tunney (D-California). Tunney allegedly admitted to a Boeing vice-president that McDonnell-Douglas officials had helped him to draft an anti-SST speech.

The Senate SST vote on March 24, 1971 was one of the most dramatic in the Senate's history. Birch Bayh got caught in a Colorado snowstorm, and Mike Mansfield attempted to slow down the roll call to give Bayh a chance to be present. The first surprise came early when Senator Clinton Anderson (D-New Mexico), considered a certain SST supporter, voted against the program. Proxmire called Anderson's vote a "shaker." A key uncommitted senator, Cooper of Kentucky, also voted against the SST; he was said to have been angered by Magnuson's support of strong anticigarette legislation, as were other senators from the tobacco-growing South, including Democrat Sam Ervin (D-North Carolina). The Senate barred funds for the SST for a second time by a vote of fifty-one to forty-six.[45]

Both supporters and opponents acknowledged that this Senate SST defeat signaled the end of the American SST effort. Gloom descended on Boeing in Seattle and on General Electric in Evansdale, Ohio. The Transportation Department ordered Boeing, General Electric, and their subcontractors to stop all SST work. General Electric immediately announced that it would have to terminate about fifteen hundred jobs, and Boeing estimated that about seven thousand workers would be laid off by April 15. Volpe expressed sharp disappointment: "We appear to have no alternative but to disband the teams of scores of experts which have been carrying the program forward and to close down the entire operation."

For a few days after the Senate vote Magruder tried desperately to salvage the SST. He met with James Mitchell, vice-president for aerospace at the Chase Manhattan Bank, with representatives from Boeing, General Electric, and Fairchild-Hiller, and with George Schultz, director of the Office of Management and Budget. But Magruder found no hope. As if to add insult to injury there were hints that a leading Japanese trading firm was interested in buying the assets of the SST program, including patents, blueprints, and completed parts, for about ten cents on the dollar. Edward Uhl, president of Fairchild-Hiller and a consistent SST supporter, tried to create a private consortium to finance the program, but this too failed.

The process of dismantling the SST program continued throughout April 1971. By the middle of the month a detailed termination plan had been worked out between the Department of Transportation, Boeing, and General Electric.[46]

A faint glimmer of reprieve for SST supporters came in mid-May. Immediately after the March Senate vote, Proxmire had knowingly commented, "Nothing is ever dead around here." And less than two months later a bipartisan coalition in the House, led by Gerald Ford, sought to revive the SST program. The SST opposition was not alerted. Ford proposed an amendment to a supplemental appropriations bill in which $85.3 million had been allocated for SST termination costs; the amendment would channel this money into continued SST development rather than termination. In a stunning reversal of its earlier anti-SST vote, the House approved the amendment on May 12 by a vote of 201 to 197.

But this surprising rally was effectively squelched a day later. Thoroughly frustrated, Boeing chairman William Allen declared that reviving the SST program would require a minimum of half a billion dollars in additional funding: "People have been dispersed. Subcontracts have been terminated. We have to pick up the pieces. That in itself will involve an additional expenditure of a very large amount—an expenditure on the top side of about $1 billion, on the low side of at least half a billion." Allen claimed that start-up costs for production would cost the government at least an additional $2 billion.

Allen's comments were the final blow to the SST program since supporters had tried to claim that it would cost more to end the program than to continue it. Volpe said he did not believe that the added costs would be "anywhere near" Allen's figures. The White House promised new estimates from Boeing, but they never appeared. Both Boeing and General Electric were fearful of potentially increased costs due to slowdowns, inflation, and the need to develop a quieter engine. Several administration officials even accused the two contractors of sabotaging the program.

A few days later Harold Johnston, a professor of chemistry at the University of California at Berkeley, declared that a fleet of five hundred SSTs operating seven hours a day would destroy half the earth's protective ozone layer. Johnston's startling conclusion, the most extreme thus far by a responsible scientist, was based on the premise that the nitrogen oxide in the SST's exhaust would be much more destructive than the SST's water vapor, which had previously been identified as the major hazard to the earth's ozone shield.

Given the statement by Allen four days earlier, even Johnston's findings evoked relatively little outcry from SST supporters in the Senate. Magnuson reportedly was convinced by this time that the cause was hopeless. On May 19, the day of the Senate vote on the SST amendment, Magnuson as much as admitted defeat on the Senate floor: "Perhaps the better part of discretion would have been to cut the program down to one prototype, instead of two. I don't know. We might have prevailed with one." The Senate then proceeded to reject the amendment by a vote of fifty-eight to twenty-seven, and a day later the House accepted the Senate's verdict.[47] After struggling for over a decade, the American SST program had finally been laid to rest.

20 Aftermath

Dismay and Euphoria in a New Age

The impact of the final defeat of the SST in Congress had both specific and general dimensions. (In fact, the SST conflict in the United States, in a sense, did not really end in 1971. A new controversy over Concorde landing rights erupted in the mid-1970s.) On the one hand, depending on which side of the debate the participant was on, there was immediate dismay or euphoria among those directly involved in the conflict. The SST's defeat also had a more general meaning, signifying a new view of technology and national decision making.

Public and scientific concern about the SST quickly declined, and the aircraft rapidly lost its status as a prominent, deeply felt issue after the final SST votes in Congress in the spring of 1971. During the early 1970s the conflict appeared to vanish even more abruptly than it had exploded onto the national scene in the late 1960s. As just an indicator of this dramatic decline in public interest, the approximate combined number of general articles listed in the *New York Times Index* on commercial supersonic transports and on the American SST (and primarily excluding articles specifically on the Concorde and the TU-144) for 1969, 1970, 1971, and 1972 was respectively 48, 147, 152, and 11.[1]

But the more general significance of the SST's loss remained clear to nearly all those involved in the conflict: the defeat of the program signified a decline of view that equated technology with progress and an emergence of a new and less enthusiastic attitude toward technological undertakings. The program's cancellation also marked the rise of interested citizen groups as permanent, increasingly institutionalized, and powerful participants in major technological decisions. Such groups clearly had demonstrated a capacity to terminate projects that they opposed, and henceforth these new participants would have to be taken into consideration by decision makers and technological developers in government and industry.

The pro-SST forces were well aware of the SST's symbolic meaning. Senator Henry Jackson called the SST opponents "know-nothings" who had embarked on an "anti-technology" crusade. Senator Warren Magnuson predicted a "technological Appalachia that will create a third-rate nation."[2] *Aviation Week and Space Technology* warned of future assaults on the space shuttle, the B-1 bomber, and "the entire advanced technology base that supports this nation." "The forces that killed the SST development program," it declared, "are out to stamp out technology." Insisting that the SST was a "focal point for some cynically exploited hysteria on the pollution issue," the journal argued that the aerospace industry was more responsive to pollution problems than was, say, the automobile industry.[3] Publisher William Randolph Hearst, Jr., in a March 1971 editorial, feared a new "erosion" of the "adventurous spirit of pioneer leadership which made this country great."[4] The last director of the SST program, William Magruder, who was still with the Department of Transportation, wrote in July 1971 that technology itself was "on trial" and that members of the scientific and technical community had to communicate more effectively with Congress and the public at large. For Magruder, the "SST day is just over the horizon . . . [but] America, for the first time in history, will be left at the starting gates and our aviation industry—second largest in the country—may well go the way of shipping, electronics, steel, autos, movies, and other ones dominating American exports: down, swamped by the advanced products of other nations." Three months later Commerce Secretary Maurice Stans declared that the country's economy needed the SST (as well as the trans-Alaskan oil pipeline and phosphate detergents) more than it needed many environmental regulations.[5]

Even the White House could not forget the SST. At the end of 1971, after personally inspecting the Concorde, President Richard Nixon declared that "one day" the United States would build its own SST, and he again expressed dismay that Congress had terminated the program.[6] During the early phase of the 1972 presidential campaign, Vice-President Spiro Agnew, citing the opposition of leading Democrats to the SST, accused them of being against progress. He labeled them "no-no birds": a no-no bird being a creature who spends most of its time in the same place or flying backward, "because it doesn't want to know where it's going—it just wants to know where it's been," emitting a loud cry, "Don't, don't, don't."[7]

The SST opponents also understood the larger meaning of their victory. The Wilderness Society characterized the SST defeat "as a decision of historic significance" and "the first tangible evidence that the decision-makers of this country are becoming aware of the need for a shift of national priorities—away from the expensive frills that have sapped our resources and endangered our environment, and toward programs that will add to the quality of people's lives in meaningful ways.''[8]

The Battle Continues: The Concorde in the United States

The SST issue did not totally disappear. While the national anti-SST environmental and public interest organizations, flushed with their SST victory, turned their attention to other concerns, William Shurcliff continued to battle the SST. Shurcliff was praised by Michael McCloskey, executive director of the Sierra Club,[9] who told him that it was ''premature'' to dismantle the Citizens League, given its ''pivotal role'' in the anti-SST campaign, and suggested that the Citizens League stay in existence for ''another year or two.''[10] George Alderson, lobbyist for Friends of the Earth and coordinator of the coalition, also lauded Shurcliff as being ''indispensable to the defeat of the SST.''[11]

Immediately after the Senate vote in March 1971 Shurcliff was ''dizzy with pleasure,'' his ''feet nowhere near the ground.''[12] His anti-SST effort was by no means over. The Citizens League continued to function, albeit at a slower pace than before. The organization had been quite active during certain periods—for instance, during the organization's early years (1967 and 1968) and during the period immediately prior to Earth Day in April 1970; but after other larger groups joined the anti-SST campaign in 1970, and especially after the SST's defeat in 1971, the Citizens League's pivotal role, energy level, and correspondence volume atrophied rather rapidly.

The Citizens League, Shurcliff declared, henceforth would aim at the Concorde.[13] In early April 1971 Shurcliff contributed $500 to the London chapter of Friends of the Earth.[14] Energized by Shurcliff's Concorde activity, a representative of this chapter wrote Richard Wiggs, head of the Anti-Concorde Project in Britain, to suggest a meeting of all organizations opposing the Concorde.[15] Shurcliff also provided financial support to the Chicago-based Citizens Against Noise led by Theodore Berland. Between the end of March and the beginning of May 1971 the Citizens League contributed $1,000 to Wiggs's Anti-Concorde Project, $500 to the London chapter of Friends of the Earth, and $200 to Berland's group.[16]

In October 1971 Shurcliff seriously thought of disbanding his group, claiming that Citizens League members had ''lost their pep,'' that press interest in the SST and in the sonic boom had waned, that cash reserves had declined to about $700, and that he was "getting tired and old (62½)." But the league continued to function. In late 1971 Shurcliff published a new *SST Handbook,* and in early 1972 he mailed out an appeal for funds that raised about $6,000. As of mid-1972 the Citizens League still claimed over 5,000 members. By late 1972 Shurcliff was focusing almost exclusively on stopping the Concorde, in close coordination with Richard Wiggs. In February 1973 the Citizens League issued a fact sheet giving "reasons" why almost all airlines had decided not to buy Concordes.

Shurcliff contributed funds periodically to the Sierra Club, Friends of the Earth, Berland's group, an umbrella organization in New York State called the Coalition Against Aircraft Noise, and such foreign anti-SST organizations as the South African chapter of Friends of the Earth. As late as 1975 the Citizens League was still operating.[17]

To a lesser extent, George Alderson remained involved in SST matters and continued to be the key anti-SST leader in Washington. He encouraged vigilance and action against the Concorde and claimed as late as February 1972 that the Coalition Against the SST "still exists." From the latter half of 1971 until at least the end of 1973 Alderson spearheaded a drive to warn SST opponents that the threat of the Concorde was growing each year. In May 1972, after BOAC had formally placed the first firm order for a Concorde, Alderson declared that "the fat is in the fire." He urged the FAA to force the Concorde to meet current federal noise standards and lobbied against the Concorde elsewhere in the government. He and Shurcliff remained in close contact and assisted one another. Alderson was the driving force for promoting anti-SST legislation in various states, especially New York, California, and Hawaii, and was one of the founders of the New York-based Coalition Against Aircraft Noise in early 1973. Like Shurcliff, Alderson communicated frequently with Richard Wiggs.[18]

The Battle Continues: The Concorde Arrives

The SST opposition's continuing fear of the Concorde was well justified. In fact, the Concorde became the center of a new conflict over the SST in the mid-1970s, which, among other things, demonstrated how quickly dormant protest movements can spring back to life.

Although the Concorde's progress was frustratingly slow and difficult, its developers persevered and ultimately overcame the obstacles to at least limited commercialization. The entire SST market outside the socialist bloc belonged to the Concorde after the demise of the American effort, though the exact size of this market was still unknown. In mid-1972 Concorde officials appeared confident. Sixteen airlines, including BOAC and Air France, held options for a total of seventy-four Concordes, and the British were bullishly forecasting sales of 200 to 250 aircraft by 1980.[19]

The Concorde, however, had obvious problems: it was noisy, it had high seat-mile costs, and its range-payload capacity was clearly decreasing. In December 1972 the British government refused to disclose how much of the approximately $3 billion spent on research and development would be recovered through sales. In February 1973 the Concorde suffered a dramatic setback when Pan American and TWA cancelled their options for the aircraft, and other airlines soon followed. By June 1973 only nine orders remained, five for BOAC and four for Air France. Even BOAC's chairman claimed publicly at the time that the Concorde was "too noisy."[20]

Only the British and French state-owned airlines were left as customers. The Concorde developers therefore were forced to demonstrate that the Concorde could meet airline needs before other air carriers would again place orders. Although commercial service would begin in January 1976 with flights from Paris to Rio de Janeiro (via Dakar) by Air France and from London to Bahrain by British Airways (BOAC's new name), the Concorde had to prove its merit on the crucial transatlantic run. Thus by the mid-1970s the United States was again involved with the SST issue. A new full-fledged SST controversy was about to begin.

In summer 1974 the British and French governments informally advised the FAA of their intention to begin regular Concorde service to the United States in early 1976, and Air France and British Airways applied about a year later to the FAA for amendment of their respective operations specifications, which included a list of the type of aircraft to be flown, the airports to be served, and the routes and flight procedures to be followed. In the past, approval had been almost routine because applications usually involved aircraft that either had been produced in the United States and certificated by the FAA, or ones that had been produced abroad, certificated by a foreign regulatory agency, and were quite similar to aircraft already in service in the United States. But the Concorde application was unique because the aircraft was unlike any other in commercial service in the United States, and its environmental impact was alleged to be very different from that of any other commercial aircraft.

The FAA meanwhile began work on a Concorde environmental impact statement in accordance with the National Environmental Policy Act of 1969. The statement, which was released in September 1975, offered four courses of action: refuse to amend the operations specifications, amend them as requested, impose additional restrictions, or take no action (which was not considered permissible by the FAA). The statement appeared to favor approval of the Concorde applications.

On February 4, 1976, after holding public hearings the previous month, Secretary of Transportation William T. Coleman, Jr., in a detailed and extremely important decision, permitted limited scheduled commercial flights of the Concorde into the United States for a trial period of sixteen months. Coleman allowed up to two Concorde flights per day by each carrier into JFK Airport in New York City and one flight per day by each carrier into Dulles Airport near Washington, D.C. He also established data collection and monitoring systems for high-altitude pollution and for noise and emission levels in the vicinity of JFK and Dulles.

It was as if the key environmental issues associated with the American SST had never really been settled. In fact, in order to comprehend fully how much was still unresolved it is necessary to review Coleman's arguments in some detail.

Coleman readily admitted that there were several possible adverse environmental impacts of the Concorde, and he identified four areas that demanded consideration: local air quality, energy impact, excessive noise levels, and ozone depletion in the upper atmosphere. Of the major environmental questions that were raised in the debate over the American SST, only the sonic boom was ignored—not because of its intrinsic unimportance or resolution but because the Concorde would not fly supersonically over land. On the other hand, a new issue, the

Concorde's energy impact, had now become prominent due to rising energy costs and concerns over energy after the 1973–1974 oil embargo.

Coleman did not deny that the Concorde contributed to air pollution. The 1975 environmental impact statement indicated that, compared with the subsonic 747, 707, and DC-8, the Concorde emitted about four to seven times as much carbon monoxide, expelled more unburned hydrocarbons, and except for the 747 produced more nitrogen oxides. Still Coleman rejected the Concorde's impact on local air quality as a reason to ban the plane from American airports. Agreeing with the environmental impact statement, he found the total effect of the proposed limited Concorde operations "negligible."

Similarly Coleman did not find the Concorde's large fuel requirements a compelling argument to reject the British and French applications. Since the 1973–1974 Arab oil embargo and the associated rapid rise in fuel costs, the Concorde looked increasingly like an economic white elephant. Fully loaded, it consumed about 0.063 gallons of fuel per passenger-mile (compared to the 707-300 with only 0.030 gallons per passenger-mile and the even more efficient DC-8, DC-10, and 747 with about 0.020 gallons per passenger-mile); the actual load factors for the Concorde, however, could not be determined without actual commercial service, and given the limited number of Concorde operations, prohibiting such flights would not result in significant overall fuel savings. In answer to demands that the Concorde be banned as a symbolic gesture on grounds of energy conservation, Coleman declared that such a ban would be unfair since the United States did not build the aircraft and did not pay its fuel bills: "It would border on hypocrisy to choose the Concorde as the place to set an example, while ignoring the relative inefficiency of private jets, cabin cruisers or an assortment of energy-profligates of American manufacture."

The most visible and controversial issue in the public debate over the Concorde at this time concerned the aircraft's high noise levels. According to Coleman's figures, although the Concorde's takeoff noise was only 50 percent louder than, and its approach noise actually less than, that of the 707-300, it was twice as noisy as the 747 or the DC-10. The Concorde could not meet the stiff new noise level requirements established in Federal Aviation Regulation 36.

Although acknowledging the seriousness of the noise issue, Coleman argued that the data gave him "no clear direction" and provided "only a descriptive and statistical view of the noise impact"; noise was not "an objective experience," and the same noise could be "a source of irritation to some, but of little to others"; the environmental impact statement indicated, in any case, that the noise impact of six additional daily Concorde flights would be "marginal." Coleman therefore decided to allow a limited number of Concorde flights on a temporary basis in order to collect information on noise levels during actual operations.[21]

The final area of controversy concerned the impact of SST emissions on the upper atmosphere, centering especially on possible depletion of the earth's ozone layer, a problem first publicized in early 1971 at the very end of the congressional battle over the American SST. The focus of the earlier debate had been James McDonald's theory that the SST's water vapor exhausts would reduce the earth's ozone shield. But attention had shifted to the SST's nitrogen oxide emissions, which, it was postulated, would interact with other chemicals in the stratosphere and deplete the earth's ozone shield and in turn would allow more ultraviolet radiation to reach the earth's surface thereby increasing the incidence of skin cancer.

In order to understand Coleman's handling of this issue in 1976 it is necessary to review relevant developments. In late 1971 Harold Johnston of the University of California at Berkeley suggested that the nitrogen oxide emissions of a fleet of five hundred Boeing SSTs flying at an altitude of 20 kilometers could reduce stratospheric ozone by a significant amount—a global average of 10 to 20 percent. About a year later the prestigious National Research Council of the National Academy of Sciences released a study that found Johnston's theories "credible" and that urged intensive research on the SST's probable impact on the environment and climate before beginning full-scale development.

Also in 1971 the Department of Transportation established a climatic impact assessment program to conduct a broad, long-term study on the SST's stratospheric effects. A number of conferences were held and specialized studies were written. The final, long-awaited report of this program was issued in January 1975. Including supporting monographs and appendixes, it came to some 7,200 pages and represented the work of about 550 individuals and numerous agencies and institutions. As with many other official studies associated with the SST, however, in spite of its impressive documentation and apparent thoroughness, this report became the subject of serious and prolonged controversy. For example, at a press conference, the department distributed a twenty-seven-page summary of the whole report that emphasized the report's optimistic conclusions that the currently planned total of 30 Concordes and TU-144s presented no appreciable stratospheric hazard and that there was a high likelihood that low-emission engines and fuels would be developed in the near future. The summary ignored the effect of a large fleet of SSTs. The press mistakenly reported that the director of this study had stated that a full-scale American SST fleet would not weaken the ozone shield. "Scientists clear the SST," the *Christian Science Monitor* declared. Although the report's summary did not explicitly discuss the climatic impact of five hundred Boeing SSTs, the actual report did indicate (using the figures given in tabular form and performing the necessary calculations) that the nitrogen oxides from such a fleet would significantly reduce the ozone shield. Many respected scientists and SST critics charged the Department of Transportation with deception and urged a more accurate packaging of the findings. About a month after the study's release, its director publicly acknowledged the harmful effects of such a fleet. But Johnston and other experts still testified on Capitol Hill in November 1975 that the report was misleading.

Meanwhile this study served as the basis for the conclusions of the FAA's 1975 environmental impact statement on the Concorde. Essentially the statement echoed the earlier study's optimistic findings on the Concorde's "insignificant" stratospheric effect and concluded that the proposed limited number of Concorde flights over a thirty-year period would add annually an average of 200 new cases of nonmelanomic skin cancer (which, according to the FAA and Department of Transportation, is rarely fatal, although it can be expensive to treat and possibly disfiguring), compared to the approximately 250,000 cases already occurring annually in the United States.

The findings of the 1975 climatic impact assessment program study and the 1975 environmental impact statement for the Concorde formed the primary basis for Coleman's Concorde decision in February 1976.

(Coleman also referred to a report, released in April 1975, by the Climatic Impact Committee of the National Academy of Sciences that concluded that a fleet of 100 SSTs would produce 6,000 additional cases of skin cancer; however, with a 10 to 15 percent increase in engine development costs, nitrogen oxide emissions could be eliminated.) Coleman declared that the "slight risk of additional nonfatal skin cancer" did not warrant prohibiting a sixteen-month demonstration effort, especially since extensive monitoring of the stratosphere would take place. He also argued that it would be "unfair" to single out the Concorde without also restricting other, more serious, sources of ozone depletion in the United States. Coleman's decision in this area appeared more reasonable by the late 1970s in the light of much firmer knowledge on atmospheric chemistry. New studies indicated that the impact on the upper atmosphere of the SST's nitrogen oxide emissions was much less severe than Johnston had calculated, though almost all later discussions stressed the necessity for continued research.[22]

Coleman's 1976 ruling seemed to be a major victory for Concorde backers; the Concorde would have the chance to prove itself in the crucial transatlantic market. But Coleman's action also set off a wave of protest among the old anti-SST forces and in communities near the affected airports. The explosive SST conflict briefly flared once again. Various senators and representatives reminded their colleagues that the Environmental Protection Agency had at times opposed the SST, and a number of unsuccessful attempts were made on Capitol Hill to pass a law to ban the Concorde. Clifford Deeds, director of the Town-Village Airport Safety and Noise Abatement Committee near JFK Airport, declared, "The Concorde will never get into JFK if we have to fight it the rest of our lives."

The Concorde had no difficulty in obtaining permission to land at Dulles Airport, which was operated by the FAA. There, on May 24, 1976, amid considerable ceremony, the Concorde made its first commercial landing in the United States.[23] The FAA then issued periodic reports on the findings of its noise and complaint monitoring systems established at Dulles and on its public opinion surveys in the Dulles vicinity. The Concorde's noise characteristics were indeed severe. At takeoff the perceived loudness of the Concorde was about double that of the Boeing 707, four times that of a 747, and eight times that of a DC-10. After the Concorde began operations into Dulles, complaints by private citizens increased enormously. Prior to Concorde operations an average of seventy-seven complaints had been recorded annually. But in the twelve-month period following Concorde flights into Dulles, the "sound complaint center" received 1,387 calls about the Concorde, of which 77 percent dealt with noise. Despite this increase, the FAA found in its first public opinion survey in May 1976 that "more people [in the Dulles vicinity] approved of Coleman's decision than disapproved." In May 1977 the FAA announced that approval of the Concorde by residents near Dulles had increased and opposition to the aircraft had diminished.

In September 1977, however, the FAA's monitoring and testing procedures were severely criticized by the General Accounting Office (GAO), the investigatory arm of Congress. The GAO found the public opinion surveys conducted at Dulles to be unreliable, and criticized the FAA's sampling plan, questionnaire, and response coding and processing procedures. The GAO flatly declared, "We question the validity of the community surveys and would not recommend using them in formulating policy on Concorde operations." In addition, unlike the FAA, the GAO explicitly interpreted the dramatic increase in the number of complaints at Dulles after Concorde operations began as indicating "a generally negative response to the Concorde." The GAO stressed the Concorde's high noise levels and the fact that the Concorde could not be modified to meet current FAA standards. Permitting the introduction of such an aircraft, the GAO declared, "would be a backward step in the national noise abatement program effort."[24]

Still overall prospects for the Concorde appeared to improve after Concorde operations at Dulles began. In September 1976 the House defeated an amendment to the Clean Air Act that would have required modification of the sixteen existing Concordes in order to reduce emissions. (The Environmental Protection Agency had already issued new SST emission rules the previous month for SST engines manufactured after January 1, 1980, and for SST engines certificated after January 1, 1984.)[25]

The British and French also publicized the advantages of the Concorde. In September 1976 British Airways and Air France claimed that the Concorde could serve the United States without New York City, if necessary, by improving its Dulles service. The next month an Air France survey showed that travel fatigue was eliminated by the

Concorde and that 70 percent of all Concorde passengers were business travelers. Both airlines acknowledged that the financial breakeven point for a total system of Concordes (including routes to such cities as New York, Tokyo, and Buenos Aires) was a 65 percent load factor based on an average of 2,750 flying hours annually per aircraft. In 1976 each of the three Air France Concordes flew an average of only 1,190 hours, although the load factors for certain routes were higher than 65 percent (as high as 96.4 percent in September 1976 for the Air France Washington-Paris route and 98.8 percent for the British Airways London-Washington route).[26]

But the Concorde still had not flown into JFK Airport because the New York Port Authority had banned it in March 1976. The Concorde decision was in the hands of the federal courts, where the British and French were suing the Port Authority to rescind its ban. The need for Concorde operations into JFK Airport seemed all the more pressing. Coleman maintained that the Port Authority would have to approve Concorde operations at JFK if New York City wished to maintain its title of "Gateway to Europe."[27] During the latter half of 1976 the British and French organized a multimillion-dollar lobbying campaign in the United States to promote the Concorde.[28] Finally after a series of court rulings and appeals going all the way to the Supreme Court, on October 17, 1977, the court upheld a lower court ruling that the ban was illegal. The way was finally clear for the Concorde to land at JFK Airport. (Throughout this whole process many anti-Concorde activists accused the Port Authority of really staging a charade, claiming that the agency had actually favored the Concorde but out of fear of environmental opponents had used the courts to force Concorde operations into New York.)[29]

On October 19, 1977 the Concorde made its maiden flight into JFK Airport. This test flight was something of an anticlimax after the years of heated opposition by local groups. At various times in 1976 and 1977 opponents had organized anti-Concorde "drive-ins" at JFK, which had tied up roads at the airport and had brought traffic to a standstill. The Port Authority and the FAA soon declared the Concorde's noise levels well within permissible limits, although opponents claimed that the monitoring and flight procedures used did not provide a fair test. With great fanfare, the long-delayed commercial service by the Concorde into JFK began on November 22, 1977; the first Air France and British Airways Concordes landed within minutes of each other, slightly before 9 A.M. The turnout for an anti-Concorde demonstration the day before had been disappointingly low.[30]

The winning of landing rights for the Concorde at JFK was only a partial victory for Concorde proponents. The Concorde's flights to New York apparently did not lose money; at least they covered operating costs. But the entire Concorde system, which consisted of nine flying aircraft, failed to meet the breakeven point of a 65 percent load factor and a flying rate of seven and a half hours per day. In August 1978 Air France and British Airways flew their Concordes roughly four hours and three hours per day, respectively; British Airways reported losing about $33 million on its Concorde operations during the previous fiscal year (about as much as the airline had earned on its other operations), and Air France lost about $65 million annually on its four Concordes. The French government had already assumed about 95 percent of the Concorde losses, and in February 1979 the British government wrote off about $320 million of British Airways' debt. In 1979 only British Airways and Air France, the sole purchasers of the aircraft, flew the Concorde supersonically, with Braniff flying the aircraft subsonically between Dulles Airport and Dallas-Fort Worth. Only nine of the sixteen Concordes already built had been sold, and production facilities were on the verge of being shut down.[31]

The Concorde had already encountered operating problems. In December 1977 an Air France Concorde was forced to break a holding pattern for JFK in bad weather and land at Newark Airport because of a shortage of fuel.[32] At various times between 1976 and 1979 residents along the East Coast of the United States and Canada were subjected to mysterious sonic booms, some of which were probably generated by Concordes flying into JFK or Dulles.[33]

Still Concorde promoters kept pressing. During the spring of 1978 Air France and British Airways launched a major marketing campaign to attract corporate executives to the Concorde.[34] By this time, according to FAA public opinion surveys, anti-Concorde feeling by residents near JFK had declined considerably, and Concorde noise levels were well below official noise limits. Concorde routes had also been extended, including the run by Braniff between Washington and Dallas-Fort Worth and by British Airways (for Singapore Airlines) from Bahrain to Singapore. In 1979, however, under the pressure of rising fuel costs and low load factors on certain runs, there were persistent rumors of taking the Concordes out of service on some, and perhaps all, routes. These rumors turned out to be true. Both the British and French halted Concorde production lines. In 1980 British Airways suspended Concorde service to Singapore, and Braniff terminated Concorde operations out of Dallas-Fort Worth. By the end of 1980 the plane had sustained a $200 million operating loss since beginning commercial service. In early 1980 the French government pledged to cover more than 95 percent of Air France's Concorde operating losses between 1981 and 1984 (previously the French government had pledged only 70 percent of these losses). Air France also announced that it would halt Concorde flights between Paris and Washington in mid-1981. Still claims continued to be heard that at least the transatlantic routes to New York were profitable. But the Concorde market had shrunk drastically.[35]

The Soviet TU-144 (which had beat the Concorde into the air by flying a prototype on December 31, 1968) suffered an even more dismal fate in the 1970s than its Anglo-French rival. A TU-144 prototype tragically crashed at the Paris Air Show in 1973. Although on December 26, 1975 the aircraft began what the Soviet ministry of aviation billed as the world's first supersonic services, these operations were extremely limited, consisting of twice-weekly cargo and mail flights over a 2,000-mile route between Moscow and Alma-Ata in Soviet Central Asia. The TU-144 manifested a whole range of problems, including unreliability, insufficient range, vulnerability to weather conditions, excessive fuel consumption, and high noise levels. The cargo flights in 1976 were reduced to one per week, and at the end of 1977, when scheduled passenger flights to Alma-Ata were supposed to start (at about the same time that British Airways and Air France were beginning Concorde operations at JFK Airport), the TU-144 was able to fly only once in five tries during its first month of passenger operations. Frequently the aircraft, loaded with passengers, sat waiting at the ramp at the Moscow airport until the passengers were transferred to subsonic jets. Technical problems and poor service continued into the late 1970s. In June 1979, after being grounded for eighteen months, the aircraft resumed test flights. The TU-144 proved to be no threat to the Concorde in competing for whatever supersonic market remained after the American SST defeat.[36]

The Impact of the SST Conflict

Long before the 1970s had ended it had become apparent that the vision of SST proponents—the creation of a fleet of several hundred SSTs flying the globe by the 1980s—was a complete fantasy.

In what looked like the beginnings of a second American SST effort, discussions and proposals regarding the development of a cleaner, faster, more advanced SST appeared in the late 1970s, and industry groups urged a "new look" at the aircraft. The U.S. Office of Technology Assessment in 1980 called for governmental "generic" research and development, preproduct definition research and development, for an "advanced supersonic transport." The agency argued that such an aircraft might be more productive in terms of seat-miles generated by an aircraft per unit of time and that new technological advances could possibly remedy some of the problems that had plagued the SSTs designed in the 1960s in such areas as aerodynamics, structures, propulsion, and noise reduction. The agency estimated that the market for an economically viable and environmentally acceptable advanced SST during the period between 1990 and 2010 was about 400 aircraft.[37]

But as even the Office of Technology Assessment admitted, economics and the energy crisis forced aircraft manufacturers to focus almost entirely on developing more efficient subsonics. As one designer said in December 1978, "My vision of the airliner of the future is something big, slow, and ugly, which uses a fuel costing next to nothing, operates without expensive human pilots or crews, and is as reliable as the subway."[38]

As the debate over the SST gradually changed in scope and character throughout the 1960s, the SST dream turned into a nightmare for its advocates. The program became enmeshed in a series of bureaucratic battles over jurisdiction of vital aspects of the program and over the substantive findings of the various studies, the areas of most intense analytical conflict being economic feasibility, financing, and the sonic boom.

The fate of the SST lends support to those who argue against governmental funding of large-scale commercialization technologies.[39] However, there were also important contextual and substantive factors working to defeat the SST that might not exist in all massive civilian projects. Therefore the salient features of a specific project should always be clearly in mind in analyzing the forces behind its evolution, and deriving generalizations from the SST's history for all commercialization programs should be made with considerable caution.

In the SST's case, supporters in government and industry, especially in the FAA, failed to provide the SST with the type of national defense justification that traditionally had been given to most successful military aircraft and weapons systems programs. Also the SST effort was unlike the Apollo program to land an American on the moon. The Apollo effort possessed an enduring, absolute presidential commitment, a continuous identification with national esteem as part of a direct cold war prestige race, and a generally high level of public support. It was also not conceived as ultimately producing a commercial end product. The SST lacked a national defense cloak; it was only partially supported on the basis of prestige; no president gave it a full commitment. It was left fully exposed to hard economic and scientific analysis and later to general public opinion.

The FAA never possessed the bureaucratic power to protect the SST program from opponents or skeptics within the government. Part of the cause for this weakness lay in the particular decision-making structure that emerged in the mid-1960s. By mid-1964 the most important policymaking body had become the interagency President's Advisory Committee on Supersonic Transport (PAC), headed by SST skeptic Robert McNamara and comprising a high-level and prestigious membership. After the assassination of President Kennedy the White House grew less enthusiastic about the effort, although Lyndon Johnson had been a strong SST proponent during his term as vice-president. Early in his presidency Johnson had great faith in McNamara and so heeded his warnings. Johnson also wanted to put his distinctive stamp on what had been a Kennedy program. The doubts of the White House increased when McNamara's key SST aide, Joseph Califano, became a top presidential assistant. Under the direction of the PAC and McNamara, components of the program were parceled out to other agencies or organizations over which the FAA had no control. This multilevel, interagency decision-making network therefore permitted considerable debate, analysis, and delays.

Another factor working to weaken the SST program lay in the area of managerial style and capability. One reason for the formation of the PAC was a lack of faith in the FAA's ability to run such a massive enterprise, a view that was reflected in the important Black-Osborne report in late 1963. Indeed the SST's most outspoken advocate, FAA administrator Najeeb Halaby, though an extremely effective promoter within the Kennedy administration, failed to exhibit the managerial capabilities needed to see the program through. This deficiency was particularly damaging in dealing with McNamara, the most powerful decision maker involved in the SST program in the mid-1960s. McNamara had great faith in sophisticated managerial techniques, and he demanded a level of analysis that Halaby did not meet.

In a very different way, appropriate managerial skills were absent in the later key administrators charged with running the SST program: William McKee, Jewell Maxwell, and William Magruder. Unlike Halaby, all were excellent project managers with extensive experience in directing large technological enterprises. But they did not realize that such endeavors can change radically over time and can exhibit distinct life cycles. The view these individuals held was overly narrow and overly technocratic. The SST managers did not comprehend until it was too late that important societal changes could severely undermine their program, even though early warning signals existed: the large portion of negative respondents in the early sonic boom tests, the initial controversies over industry funding, the consistent criticism of the FAA's managerial ability, and the general decline of FAA authority. Even more important was the fact that early critics like William Shurcliff were basically ignored by program decision makers. Shurcliff, a vigorous, disciplined, and knowledgeable opponent, with his small Citizens League Against the Sonic Boom—the very antithesis of a large bureaucracy like the FAA—seemed unimportant next to the contractors, the White House, and Capitol Hill. Yet Shurcliff was a clear and present danger to the program that went unrecognized, partly because he represented a phenomenon outside the usual experience of successful project managers like McKee, Maxwell, and Magruder.[40]

Substantive problems with the SST also caused ultimately fatal delays. The aircraft's economic feasibility was a vulnerable target, and McNamara assured that this issue was analyzed thoroughly. Economics was the first area of the SST conflict in which experts openly fought with one another. At least three agencies—the FAA, the Department of Commerce, and the Defense Department—examined the subject in great detail. The sonic boom grew into a controversial issue as a host of organizations became involved, including the FAA, the National Academy of Sciences, and the Office of Science and Technology. As the number of experts involved with this issue increased, controversy developed over the design and scheduling of the sonic boom tests. Finally the sonic boom debate expanded beyond the established SST network to include the entire membership of the National Academy of Sciences. Technical problems too did their share of damage: Boeing's swing-wing design proved faulty, causing a crucial two-year delay in the program from 1967 to 1969.

All of these factors resulted in successive postponements that created opportunities for more debate and for more problems to emerge. The anti-SST forces had ample time to form, to organize their campaign, and to take advantage of a conflict that had already begun to spread beyond the confines of the established aerospace industrial, bureaucratic, and congressional network. By the end of the 1960s the SST had been transformed into a national issue and had emerged as the prime target of the growing environmental movement. Already weakened by various defeats within, the SST program soon fell prey to the organized, widespread opposition of a concerned public. The SST advocates and developers were ultimately overwhelmed.

The subsequent experience with the Concorde in the 1970s, an aircraft that also suffered economic failures and interminable delays due to technical problems, certainly confirmed more of the arguments used by the SST opponents than those voiced by the aircraft's advocates. Soon after commercial operations began, it became apparent that there was no hope of recouping the Concorde's huge development costs; even the possibility of covering operating expenses was doubtful. The energy crisis and related rising fuel costs in the late 1970s provided the final blow for at least the first generation of SSTs.

The significance of the SST conflict lies not only in the fact that it involved the rise and fall of a particular aircraft or technology. The episode also both reflected and affected the time in which it occurred and influenced the period that followed.

The original program was the product of a confident and optimistic society that strongly supported technological endeavors and that allowed such decisions to be determined by a relatively small group of advocates from industry, the governmental bureaucracy, various research organizations, and relevant congressional committees. In the early 1960s the SST debate was a mild one and was contained within this supportive network.

But as the decade progressed, society changed. The apparent consensus among experts publicly broke down, leading to broad debates over such issues as SST economic feasibility, sonic booms, and environmental effects. By the late 1960s this concern about technological decisions had spread to the public at large. Various interest groups and self-styled lobbyists involved in the environmental, antiwar, and civil rights movements had become increasingly effective in promoting their causes. Many had taken part in the 1968 presidential campaigns of Eugene McCarthy and Robert F. Kennedy. They knew how to organize at the grassroots and the national levels, how to use the media to their own advantage, and how to work within the system on Capitol Hill. They were dedicated. A new style of participation in technological decision making had evolved by the end of the decade. By the first Earth Day in April 1970 the SST had emerged as a symbol of this new concern and as a prime target of this new kind of opposition. The anti-SST crusade in 1970 and 1971 involved a massive amount of participation by diverse sets of groups: national organizations, ad hoc local groups, cadres of experienced and savvy lobbyists on Capitol Hill, the media, members of Congress and their aides, and a number of economic and technical experts.

The final defeat of the SST in Congress in March 1971 was the environmentalists' most dramatic victory and the most visible manifestation of the new attitude toward technology and decision making. The SST decision sent out powerful, enduring shock waves. Subsequent developments have echoed the pattern that the SST conflict established. The Concorde's entry into the United States was significantly delayed by the SST opposition. A protest movement arose in the 1970s in the United States and elsewhere that effectively stopped the construction of new nuclear power plants. Indeed the first coordinator of the Coalition Against the SST, after living abroad for some years, returned to the United States and became actively involved in the campaign to halt construction of nuclear breeder reactors. In 1978 Russell Train, former director of the Environmental Protection Agency, boldly advocated a worldwide campaign to arrest progressive deterioration of global natural systems.

But after the SST defeat a perceptible change in the environmental movement itself occurred: although it remained strong, it also gradually became institutionalized. The Office of Technology Assessment was established to assist Congress. Many environmentalists assumed high-level positions in the Carter administration. After the 1973–1974 oil embargo, energy also became an important issue. Along with environmental concerns environmentalists increasingly faced the less glamorous task of implementing and enforcing environmental laws and regulations, which was obviously less attractive to the media than, say, the SST debate after the first Earth Day. Even Train reflected this new, less intense mood; in 1978 he spoke out against continuing the abrasive and contentious battles of the late 1960s and early 1970s in fear that they would result in a "largely unproductive" polarization. Instead he favored reaching a common understanding among adversaries. The 1971 SST defeat had ushered in a new, more complex, and perhaps more mundane era.[41]

After the SST conflict key participants moved on to other concerns. Elwood Quesada, FAA administrator under the Eisenhower administration, left the government and went on to develop real estate in Washington, D.C. Najeeb Halaby, enthusiastic SST promoter and FAA administrator from 1961 to 1965, became president of Pan American; after several unsuccessful years he left that airline and eventually started an international consulting firm. General William McKee, who had replaced Halaby at the FAA, returned to the air force and later retired to start a consulting firm in Washington, D.C. General Jewell Maxwell, who ran the SST program from 1965 to 1969, returned to the air force and later became a Boeing executive.[42] William Magruder, who became director of the SST program in 1970, remained in that position until its final defeat in 1971. He then joined the White House staff becoming a special consultant to the president for technology. His chief responsibility was to manage an extremely grandiose exercise called the technologies opportunities program in which a huge number of potential technological initiatives were evaluated and ranked according to cost and attractiveness, particularly political attractiveness. Magruder had lost none of his forceful, if narrow, dynamism. One observer later recalled that this White House effort "took on the distinct aura of a strict military exercise performed according to schedules, PERT charts, and cookbooklike criteria." In retrospect, basically because of its public relations emphasis on visible, politically attractive components, the program's ultimate impact was quite superficial. He later became vice-president of Piedmont Airlines. He died of a heart attack in 1977.[43]

As for the key SST opponents, William Shurcliff continued to run his Citizens League Against the Sonic Boom until the mid-1970s. He then turned to a new cause, solar energy, applying the same techniques and skills that had been so successful in the SST battle. Shurcliff's operating style had not changed. He was still highly organized and was generous in distributing his materials. He became a vigorous proponent of solar home heating and functioned as a one-person clearinghouse and preeminent cataloger on the subject. He published a directory of people, organizations, and designs associated with solar home heating and a number of technical works on solar heating and energy conservation. He also advised a number of would-be SST opponents to direct their attention to other issues, such as alcoholism and cigarette smoking.[44]

Other anti-SST activists remained in the environmental movement. George Alderson stayed on as chief Washington lobbyist for Friends of the Earth. Laurence Moss, one of the founders of the Coalition Against the SST, became president of the Sierra Club, an environmental consultant, and a leader of the National Coal Policy Project—an effort begun in 1977 to foster cooperation between environmentalists and the coal industry.[45]

Significantly SST advocates generally had remained in the aerospace or aviation fields, where they had already worked for so many years. The environmentalists, on the other hand, were busy in diverse ways establishing a new, developing sector in the economy and the government, a new kind of activity that functioned to monitor society and to implement environmental laws and regulations in a host of areas, including technology, energy, transportation, and wildlife. This group was adaptive, taking on new issues and new modes of operating.

A few former proponents eventually admitted that the termination of the SST program had been wise. William Allen, Boeing's chairman emeritus, acknowledged publicly in 1974 that scuttling the SST "could well have been a good decision." The Concorde, he continued, was never competitive, and the spectacular increases in the price of fuel had made all SSTs unable to compete with subsonic jets. Although Allen called ultimate development of an American SST "inevitable," he also added, "I probably won't be around to see it." Allen's age was seventy-three.[46] After the 1971 defeat the SST disappeared as an issue, but its impact remained. The SST conflict was clearly both a catalyst and a harbinger of a new era.

Notes

Most of the information for this book comes from primary documents and from interviews with participants in the SST conflict. I used contemporary journalistic accounts when primary sources were inadequate. I also used some secondary sources. Whenever possible, the location of archival material is given at the end of a listing in the notes. To maintain confidentiality, information obtained from individual interviews is simply listed as "personal interviews."

The primary documents listed in the notes come from the following archives or files:

A. The records of the Office of Supersonic Transport Development of the Federal Aviation Administration, stored in the Federal Records Center, Suitland, Maryland. The accession numbers are:

237-69A-1647. There are 35 boxes in this set. Boxes numbered 1 through 13 were used in this study. Indicated in the references by the symbol, FAA 1647, followed by a slash and the box number.

237-72A-6174. There are 151 boxes in this set. Boxes numbered 1 through 14 and 92 through 97 were used in this study. Indicated in the references by the symbol, FAA 6174, followed by a slash and the box number.

237-70A-905. There are 11 boxes in this set. Boxes number 1 and 2 were used in this study. Indicated in the references by the symbol, FAA 905, followed by a slash and the box number.

B. Documents from the files of the FAA Historian, Washington, D.C.

C. Records of the President's Advisory Committee on Supersonic Transport (PAC). At the time I examined these documents, they had just been transferred from the Department of Defense to the National Archives, Washington, D.C. The accession number of these documents at that time was 69-A-2789, Record Group 330. There are 50 boxes in this set. Boxes numbered 4 through 16, 21, 23 through 25, 28, 29, 32, 34 through 37, 41, and 44 through 50 were used in this study. Indicated in the references by the symbol, PAC, followed by a slash and the box number.

D. Records of the National Academy of Sciences Committee on SST-Sonic Boom, National Academy of Sciences Archives, Washington, D.C. Indicated in the references by the symbol NAS. The newsletter of the NAS Committee, which carried synopses of various SST-related articles, was also used.

E. Various documents stored in the Dwight D. Eisenhower Presidential Library, Abilene, Kansas, from the following collections: White House General Files (Official File), Furnas Papers, and Quesada Papers.

F. Various documents stored in the John F. Kennedy Presidential Library, Boston, Massachusetts, from the following collections: President's Office Files, Sorenson Papers, and Transition Files.

G. Various documents stored in the Lyndon B. Johnson Presidential Library, Austin, Texas, from the following collections: White House Central Files (containers 1 and 2, EX CA; container 398, EX FG 718 and 718A; Miscellaneous EX; containers 2, 3, and 4, Gen. CA; Miscellaneous Gen.); Oversize Attachment (container 2401); WHCF C. F. (CA and FG 718); National Security Files (Supersonic Transport); Administration FAA History (FAA Admin. Hist.—SST and accompanying documents); Administration History of the Office of Science and Technology; and the Office Files of George Reedy.

H. Documents from the records of the Citizens League Against the Sonic Boom, Cambridge, Massachusetts, which are now housed in the Institute Archives and Special Collections, MIT Libraries, Cambridge, Massachusetts. Indicated in the references by the symbol CL.

I. Documents from the personal files of Raymond A. Bauer.

Chapter 1

1. For discussions of governmental funding of large-scale civilian commercialization technologies that cite the SST, see: George Eads and Richard R. Nelson, "Governmental Support of Advanced Civilian Technology: Power Reactors and the Supersonic Transport," *Public Policy,* 19 (1971), pp. 405–427; W. S. Baer, L. L. Johnson, and E. W. Merrow, *Analysis of Federally Funded Demonstration Projects: Executive Summary,* no. R-1925-DOC, *Final Report,* no. R-1926-DOC, and *Supporting Case Studies,* no. R-1927-DOC (Santa Monica, California: Rand Corporation, April 1976). For general discussions of governmental support of large-scale commercialization technologies, see: Richard Schmalensee, "Appropriate Government Policy Toward Commercialization of New Energy Supply Technologies," *The Energy Journal,* vol. 1, no. 2 (1980), pp. 1–40; Otto Keck, *Policymaking in a Nuclear Program* (Lexington, Mass.: Lexington Books, 1981), pp. 1–19. For a more positive general view of such programs, see: Jerome E. Schnee, "Government Programs and the Growth of High Technology Industries," *Research Policy,* 7 (1978), pp. 2–24.

2. For public opinion surveys relating to the public's changing concerns about technology and the environment in the 1960s and beginning of the 1970s, see: G. Ray Funkhouser, "Public Understanding of Science: The Data We Have," *Workshop on Goals and Methods of Assessing the Public's Understanding of Science* (University Park, Pa.: Pennsylvania State University, Materials Research Laboratory, 1973), pp. 26–27; Irene Taviss, "A Survey of Popular Attitudes toward Technology," *Technology and Culture* 13 (1971): 606–621; T. R. LaPorte and D. Metlay, "Technology Observed: Attitudes of a Wary Public," *Science* 4/11/75, pp. 121–127.

3. A comprehensive history and analysis of the social change that occurred during the 1960s still has to be written. However, for an overview of the decade, see: William L. O'Neill, *Coming Apart* (New York: Quadrangle Books, 1971).

4. Large-scale technological endeavors often possess dramatic life cycles that often exhibit an evolution that moves from a narrow and limited set of factors to an increasingly wide range of relevant issues. See: W. H. Lambright, "Government and Technological Innovations: Weather Modification as a Case in Point," *Public Administration Review* (January–February 1972): 1–10; D. Nelkin and S. Fallows, "The Evolution of the Nuclear Debate: The Role of Public Participation," *Annual Energy Review* 3 (1978): 275–312; Mel Horwitch, "Uncontrolled Growth and Unfocused Growth: The United States SST Program and the Attempt to Synthesize Fuels from Coal," *Interdisciplinary Science Reviews* 5 (1980): 231–244; Mel Horwitch and C. K. Prahalad, "Managing Multi-Organization Enterprises: The Emerging Strategic Frontier," *Sloan Management Review* (Winter 1981): 3–16; Peter J. Leahy and Allan Mazur, "The Rise and Fall of Public Opposition in Specific Social Movements," *Social Studies of Science,* 10 (London and Beverly Hills, CA: SAGE, 1980), pp. 259–284.

Chapter 2

1. Robert Hotz, "Supersonic Transport Era; Mach 2 and Mach 3," *Aviation Week and Space Technology*, 5/26/58, p. 21.

2. T. F. Cartaino et al., *Supersonic Transports* (Santa Monica: Rand Corporation, November 1958).

3. Glenn Garrison, "Supersonic Transport May Aim at Mach 3," *Aviation Week and Space Technology*, 2/2/59, pp. 38–40.

4. "Lockheed Studies Mach 3 Vehicle as Next Step in Transport Field," *Aviation Week and Space Technology*, 10/12/59, p. 49; Hall L. Hibbard and Robert A. Bailey, "The Case for the Supersonic Transport," *Interavia*, no. 10 (1959): 1234–1235.

5. William S. Reed, "Study Predicts Mach 2–3 Transport Costs," *Aviation Week and Space Technology*, 11/23/59, pp. 36–38.

6. Robert Hotz, "IATA Girds for Supersonic Airliner Era," *Aviation Week and Space Technology*, 10/26/59, pp. 40–42.

7. For information on the early Concorde program, see: John Davis, *The Concorde Affair: From Drawing Board to Actuality* (Chicago, Ill.: Henry Regnery Co., 1970); Richard Wiggs, *Concorde: The Case against Supersonic Transport* (London: Ballantine, 1971); Andrew Wilson, *The Concorde Fiasco* (Harmondsworth, England: Penguin Books, 1973).

8. For British SST efforts in 1959, see: John Tunstall, "British Weigh Entering Supersonic Race," *Aviation Week and Space Technology*, 5/4/59, pp. 55–57; Peter Masefield, "In the Supersonic Seventies," *Interavia*, no. 9 (1959): 1086; "Supersonic Transport Has Prestige Factor," *Aviation Week and Space Technology*, 5/18/59, p. 41.

9. International Civil Aviation Organization, *The Technical, Economic and Social Consequences of the Introduction into Commercial Service of Supersonic Aircraft: A Preliminary Study* (Montreal: ICAO, August 1960). See especially appendix A, appendix B, pp. 93–104.

10. D. D. Davis to Director, "Audit of the SST Program," 11/20/64, memo, and attached "Chronology of Events from December 1959 through December 1963," p. 1, FAA 6174/96.

11. For the activities of the FAA SST study group and the creation of the FAA SST planning team, see: Davis to Director, "Audit," and attached "Chronology," pp. 1–2; "Things to Discuss with General Preuss (Like Voice)," n.d., and attached "Progress Report, Federal Aviation Agency Supersonic Planning Team," 4/1/60, and "Report on the First Meeting of the Agency-Wide Supersonic Planning Team," FAA 6174/96. For the B-70 program cutback, see J. S. Butz, "Budget Cuts Force Stretchout of B-70," *Aviation Week and Space Technology*, 12/7/59, pp. 26–28. For industry, airline, and CAB views on the SST in early 1960, see Glenn Garrison, "U.S. Supersonic Transport Action Urged," *Aviation Week and Space Technology*, 2/8/60, pp. 46–47; "Supersonic Airliner Development," *Aviation Week and Space Technology*, 2/8/60, p. 46.

12. U.S., Congress, House, *Hearings before the Special Investigating Committee of the Committee on Science and Astronautics*, 86th Cong., 2d sess., 5/17–20, 24/60.

13. U.S., Congress, House, *Report on Supersonic Transports*, Report 2041, 86th Cong., 2d sess., 1960.

14. Davis to Director, "Audit," and attached "Chronology," p. 3; FAA, "Brief History of the United States Supersonic Transport Program," 11/64, p. 20, FAA 6174/14; George B. Kistiakowsky, *A Scientist at the White House* (Cambridge, Mass.: Harvard University Press, 1976), p. 389.

15. "A Program for the Development of a Civil Supersonic Transport," (draft) report, 9/60, FAA 6174/13.

16. For Quesada's setback in the SST area, see: Kistiakowsky, *Scientist,* p. 389; *Aviation Daily,* 10/28/60, p. 358; *New York Times,* 11/1/60, p. 1, 12/11/60, p. 1.

17. FAA, "Commercial Supersonic Transport Aircraft Report," 12/60, FAA 6174/14.

18. For the discussion on Anglo-American SST collaboration during the latter half of 1960, see: *Flight,* 9/6/60, 11/4/60; "National and International Aspects of the Aircraft Business: An Address by Sir George Edwards," *Flight,* 11/4/60, pp. 690–691; "What Price Collaboration," *Flight,* 11/11/60; *Flight,* 12/18/60, p. 960; FAA, "Brief History," pp. 25–30; Preuss to Halaby, "Thorneycroft Correspondence," 3/16/61, memo, FAA 6174/14.

19. "Supersonic Transport Program Launched," *Aviation Week and Space Technology,* 1/16/61, p. 43.

Chapter 3

1. "Report to the President Elect of the Ad Hoc Committee on Space," 1/12/61, pp. 7–8, JFK Transition/1072.

2. U.S., Congress, House, Committee on Science and Astronautics, *Contemporary and Future Aeronautical Research,* Hearings, 87th Cong., 1st sess., 8/1–4, 8/61, p. 3; U.S., Congress, House, Committee on Appropriations, *Independent Office Appropriations, 1962,* Hearings, 87th Cong., 1st sess., 4/17/61, pp. 60–61. Quotations are from: Najeeb Halaby, *Crosswinds: An Airman's Memoir* (Garden City, N.Y.: Doubleday, 1978), pp. 16, 182.

3. Preuss to Halaby, "Letters to Presidents of Engine and Aircraft Manufacturers Requesting Briefings on Supersonic Transport Effort," 3/22/61, memo, and attached Halaby to Allen, 3/24/61, FAA 6174/12.

4. Halaby to Webb, 5/4/61, memo, and attached Halaby to Stern, 5/4/61, FAA 6174/12; Dean to Halaby, "NASA Request for Supersonic Aircraft Funds," 3/22/61, memo, FAA 6174/14; U.S., Congress, House, *Independent Offices Appropriations* (1962), pp. 87–88; U.S., Congress, Senate, Committee on Appropriations, *Independent Offices Appropriations, 1962,* Hearings, 87th Cong., 1st sess., 6/20/61, pp. 622–623.

5. House, *Independent Offices Appropriations, 1962,* Hearings, pp. 80–90.

6. Senate, *Independent Offices Appropriations, 1962,* Hearings, pp. 622–623, "Halaby Urges Congress to Support Supersonic Transport Program," *Aviation Week and Space Technology,* 6/6/61, p. 40.

7. *New York Times,* 7/30/61, p. 40; "Senate Votes Mach 3 Transport Funds," *Aviation Week and Space Technology,* 8/7/61, pp. 38–39; House, *Contemporary and Future Aeronautical Research,* Hearings, pp. 8–9; Davis to Director "Audit of the SST Program," 11/20/64, memo, and attached "Chronology of Events from December 1959 through December 1963," p. 3, FAA 6174/96.

8. Halaby to Kerr, 5/19/61, FAA 6174/12; Senate, *Independent Offices Appropriations, 1962,* Hearings, pp. 623–624; House, *Contemporary and Future Aeronautical Research,* Hearings, pp. 15–16.

9. DOD/NASA/FAA, "Commercial Supersonic Transport Aircraft Report," 6/61, FAA 6174/11.

10. U.S. Task Force on National Aviation Goals, "Report of the Task Force on National Aviation Goals: Project Horizon" (9/61), pp. 14–17, 75–80.

11. *Proceedings of the Eighth Anglo-American Aeronautical Conference*, London, 9/61, pp. 342–345.

12. Davis to Director, "Audit," and attached "Chronology," pp. 3, 6–7; John R. Provan to Halaby, "Establishment of Airline Advisory Group—Supersonic Transport Program," 10/20/61, memo, FAA 6174/14; Halaby, "Establishment of Airline Advisory Group—Supersonic Transport Program," 10/26/61, memo, FAA 6174/14.

13. Davis to Director, "Audit," and attached "Chronology," pp. 3–6; "Brief History of the United States Supersonic Transport Program," 11/64, pp. 21–22, FAA, 6174/14; DOD/NASA/FAA, "Commercial Supersonic Transport Aircraft Report," 6/61, FAA 6174/11.

14. Dean to Halaby, "Management of Supersonic Transport Development," 7/31/61, memo, FAA 6174/14.

15. Davis to Director, "Audit," and attached "Chronology"; FAA, "Brief History," p. 22.

16. "Draft Minutes of a Joint Meeting of the Supersonic Transport Steering Group and the Supersonic Transport Airline Advisory Group," 2/1–2/62, FAA 6174/13.

17. Halaby to Magnuson, 1/22/62, FAA 6174/13; U.S., Congress, House, *Committee on Appropriations, 1963,* Hearings, 87th Cong., 2d sess., 2/1/62, p. 960.

18. Halaby to Webb, circa 1/30/62, draft letter (not sent), FAA 6174/13; Rochte to Halaby, "NASA's FY 1963 Budget Request for the Supersonic Transport," 1/30/62, memo, and handwritten comment by Dean, FAA 6174/13; Dean to Halaby, "Clarification of FAA and NASA Responsibilities for Aircraft Research and Development," 4/27/62, memo, FAA 6174/14; Dryden to Halaby, 2/5/62, FAA 6174/13; Halaby to Magnuson, 1/22/62, FAA 6174/13; "NASA Evolves Two Basic SST Designs," *Aviation Week and Space Technology,* 7/2/62, pp. 214–218; U.S., Congress, Senate, Committee on Appropriations, *Independent Offices Appropriations, 1963,* Hearings, 87th Cong., 2d sess., 8/7–8/62, pp. 630, 901.

19. "Report to the President-Elect of the Ad Hoc Committee on Space," 1/12/61, JFK Transition/1072; Wiesner to Halaby, 6/11/62, note, FAA 6174/12; Golovin to Wiesner, "Supersonic Transport Program," 6/4/62, memo, FAA 6174/12.

20. Rochte to Halaby, "FY 63 Research Program for the Supersonic Transport," 3/16/62, memo, and attached "FY Program Schedule Supersonic Transport, FAA-USAF-NASA," 2/2/62, "A Program to Determine the Technical and Economic Feasibility of a Supersonic Transport," 3/6/62, "Supersonic Transport Program Schedule," and "Techno-Economic Considerations for a Supersonic Transport," FAA 1647/3; "Draft Minutes of the Meetings of the SST Task Group/Airline Advisory Group," 5/8/62, pp. 3, 5, FAA 6174/13; Shank to Halaby, "Supersonic Transport Systems Analysis Program," circa 7/62, memo, and attached "Coordination Sheet for the SST Systems Analysis Pro-

gram,'' FAA 1647/3; Daunt to SST Program Office, ''Supersonic Transport Systems Analysis Program,'' 7/2/62, memo, FAA 1647/3; Loy to Halaby, ''Comments on the Supersonic Transport System's Analysis Program,'' 7/24/62, memo, FAA 1647/3; ''Work Statement for SST Economic and Operational Analysis: SST Systems Analysis Program Chart,'' FAA 1647/3; ''Supersonic Transport Program Division System Analysis of the Supersonic Transport and Work Statement for Technical Analysis: Computer Parametric Program,'' circa 7–8/62, FAA 1647/3.

21. William M. Allen to Roswell L. Gilpatrick, 6/18/62, FAA 6174/12.

22. Welsh to Halaby, ''National Aeronautics and Space Council Meeting,'' 7/30/62, memo, FAA 6174/12; ''National Aeronautics and Space Council: Briefing on Commercial SST Development (August 3, 1962—2:45 P.M. EOB 274),'' FAA 6174/12; ''National Aeronautics and Space Council Meeting—August 3, 1962,'' LBJ roll 2, NASA Microfilm; Rochte to the record, ''Meeting of National Aeronautics and Space Council—August 3, 1962,'' n.d., FAA Historian's files.

23. Senate, *Independent Offices Appropriations, 1963,* Hearings, p. 533; Welsh to Halaby, 10/17/62, FAA 6174/12; Shank to Welsh, 10/22/62, and attached ''SST Cost and Market Analysis,'' 10/22/62, FAA 6174/12; Shank to Halaby, ''Letters of Invitation to Attend Meeting on December 3 in Connection with Rand and Stanford Research Institute Cost and Market Studies of the Supersonic Transport,'' 10/17/62, memo, FAA 6174/12; Bates to Senior Planning Officer, ''Status of Nontechnical Studies on the SST,'' 11/7/62, memo, FAA 1647/3.

24. Listing, ''SST Research and Development Projects Now Underway'' (attached to Director, Aircraft Development Service to Deputy Administrator for Development, ''Proposed Schedule of Events for Preparation and Submission of Supersonic Transport Systems Analysis Report,'' circa 11/62, draft memo), FAA 1647/3; Senate, *Independent Offices Appropriations, 1963,* Hearings, p. 533; Davis to Director, ''Audit,'' and attached ''Chronology,'' p. 7; FAA, ''Brief History,'' p. 17.

25. William Barclay Harding to Halaby, 9/18/62, FAA 6174/12.

26. From Halaby, 9/24/62, memo, FAA 6174/12; J. R. Kennedy to Halaby, 10/22/62, and handwritten comment by Halaby, 11/2/62, FAA 6174/12.

27. Rochte, ''Discussion with Rand-SRI Representatives Regarding the SST Advisory Group's Report on SST Program Planning,'' 12/5/62, memo, FAA 6174/12.

28. Cook to Halaby, 12/14/62, JFK Pres. Office Files; Supersonic Transport Advisory Group (STAG), ''Report to the Chairman Supersonic Transport Steering Group: Supersonic Transport Planning,'' 12/11/62, JFK Pres. Office Files.

29. Imirie to Halaby, 12/20/62, FAA 6174/12.

30. STAG, ''Report to the Chairman,'' p. 21; Dean to Halaby, n.d., note, and attached ''Proposed Budget Message Statements on Supersonic Transport,'' 11/11/62, FAA 6174/12; Halaby to President, ''Race to the Supersonic Transport,'' 11/15/62, memo, JFK Pres. Office Files; F. J. Shank, ''Treatment of Supersonic Transport in FY 64, Budget Message,'' 12/13/62, draft statement, FAA 6174/12; Halaby to President, ''Treatment of Supersonic Transport in FY 1964, Budget Message,'' 12/26/64, memo, JFK Pres. Office Files; BOB director to President, ''1964 Budget for the Federal Aviation Agency,'' 12/14/62, memo, and attached ''Federal Aviation Agency,'' 12/14/62, and ''Federal Aviation Agency, Supersonic Aircraft Development,'' JFK Pres. Office Files; President to

Secretary of Defense et al., "The Civil Supersonic Transport," 1/21/63, press release, JFK Pres. Office Files.

31. IATA, *Symposium on Supersonic Air Transports*, vol. 1: *Report of the Discussions*, 14th Technical Conference, 4/17–21/61, Montreal, pp. 187–207; David A. Anderton, "Industry Shapes Supersonic Design Goals," *Aviation Week and Space Technology*, 4/24/62, pp. 26–27.

32. Richard J. Kent, Jr., *Safe, Separated, and Soaring* (Washington, D.C.: U.S. Government Printing Office, 1980), p. 43.

33. Herbert J. Coleman, "British Stress Mach 2.2 Airliner's Assets," *Aviation Week and Space Technology*, 9/18/61, p. 47; *Proceedings of the Eighth Anglo-American Aeronautical Conference*, 9/61, London, pp. 233–234; *IATA Bulletin*, no. 29 (12/61): 61–62.

34. Ibid., p. 85; IATA, *Symposium*, 1:10–16.

35. David A. Anderton, "Design Problems, Costs Dampen Supersonic Transport Optimism," *Aviation Week and Space Technology*, 4/10/61, p. 31.

36. "Draft Minutes of the Meetings of the Supersonic Transport Advisory Group with U.S. Airlines and Foreign and U.S. Manufacturers," 7/23–25/62, FAA 6174/12.

37. House, *Independent Offices Appropriations, 1962*, Hearings, pp. 85–86; Thorneycroft to Halaby, 3/7/61, FAA 6174/14; "AM" to Halaby, 3/10/61, note, FAA 6174/14; "Summary of 3/7/61 Letter," 3/13/61, FAA 6174/14; Preuss to Halaby, "Thorneycroft Correspondences," 3/16/61, memo, FAA 6174/14; Halaby to Thorneycroft, 3/20/61, FAA 6174/14.

38. "Consortium Urged for SST Development," *Aviation Week and Space Technology*, 10/2/61, p. 43; Maloy, "Discussion Notes of the Ministry of Aviation/Federal Aviation Agency Meeting—Washington, D.C., November 6–8, 1961," 11/21/61, memo, and attached draft notes of MOA/FAA meeting of 11/6–8/61, FAA 6174/14; "Pace of the Supersonic Airliner Development Quickens," *Interavia* (1/62): 108.

39. "Supersonic Transport Progress" (editorial), *Aviation Week and Space Technology*, 11/27/61, p. 17; Rochte to Halaby, 3/28/62, transmittal slip, and attached newspaper article dated 3/23/62, "120 Passenger Supersonic Jet Is Spurred by British and French," FAA 6174/12.

40. "Draft Minutes of a Joint Meeting of the Supersonic Transport Steering Group and the Supersonic Transport Advisory Group," 1/9/62, p. 4, FAA 6174/13.

41. "Supersonic Transport Progress," p. 17; Rochte to Halaby, 3/28/62, transmittal slip, and attached newspaper article dated 3/23/62, "120 Passenger Supersonic Jet"; Allright to Rochte, 2/14/62, FAA 6174/14; Rochte to Allright, circa 3/62, FAA 6174/14.

42. Beall to Allen, "French Supersonic Commercial Transport," 2/15/62, memo, FAA 6174/12; Beall to Halaby, 4/16/62, FAA 6174/12.

43. Kotz to Loy, "Consortium Research, Development and Production of the Supersonic Transport," 7/31/62, memo, FAA 1647/3.

44. Maloy to Halaby, "Supersonic Transport Development—Mr. T. P. Wright's Letter of April 27, 1962," 6/14/62, memo, FAA 6174/12.

45. Maloy to Deputy Director for Development and Loy, "International Cooperation on Supersonic Transport," 8/10/62, memo, FAA 6174/14.

46. "Draft Minutes of the Meetings of the Supersonic Transport Advisory Group with U.S. Airlines and Foreign and U.S. Manufacturers," 7/23–25/62, FAA 6174/12.

47. Carran to Halaby, "Conferences at Sud-Aviation and with General Bonte during Your Visit to Paris," 9/14/62, memo, and attached notes, FAA 6174/12.

48. *Interavia* (newsletter), 11/30/62, no. 5132, FAA 6174/12; U.S. Embassy in London to Department of State, "Civil Aviation—Anglo-French Supersonic Transport," 12/3/62, DOS airgram, FAA 6174/12.

49. Halaby, "Report by the Administrator on the Visit of Sir George Edwards," 12/14/62, memo, JFK Pres. Office Files.

50. House, *Independent Offices Appropriations, 1963,* Hearings, pp. 959–961.

51. Senate, *Independent Offices Appropriations, 1963,* Hearings, pp. 527–30.

52. Halaby to President Kennedy, "Race."

53. Halaby, *Crosswinds,* p. 192.

Chapter 4

1. President to the Secretaries of Defense, Commerce, and the Administrators of NASA and the FAA, the Chairman of CAB and the Director of OST, "The Supersonic Transport," 1/21/63, memo, JFK, Pres. Office Files.

2. "Kennedy Memo Maintains SST Timetable," *Aviation Week and Space Technology,* 1/28/63, p. 38.

3. For Halaby's mood at this time and for Kennedy's appointment of Johnson, see: Halaby to Shank, "Intensification of Supersonic Transport Program," 1/31/63, memo, FAA 6174/12; "Summary, NEH Memo to Mr. Shank (Our #9822)," memo, FAA 6174/12; Halaby to the Vice-President, 1/31/63, FAA 6174/12; President to the Vice-President, 2/19/63, memo, JFK Pres. Office Files; Halaby to the Vice-President, 3/1/63, LBJ V-P.

4. FAA, "Supersonic Transport," 5/14/63, pp. 16–21, charts 9–16, LBJ V-P.

5. Walter W. Heller to the Vice-President, "Supersonic Transport," 3/12/63, memo, FAA 6174/3.

6. For the status of sonic boom thinking during the first half of 1963, see: Office, Director of Development Planning, DCA/Research and Development, HQ USAF, "The Sonic Boom Problem," 3/63, memo, FAA 6174/4; "Fact Sheet on Supersonic Transport Program," circa 3–4/63, LBJ V-P; "A Synopsis of the Research and Study Program of Critical Technical Areas for the Commercial Supersonic Transport: Fiscal Year 1963," 2/8/63, FAA 6174/11.

7. Clarence D. Martin, Jr. to Welsh, "Supersonic Transport," 3/6/63, memo, LBJ V-P.

8. Welsh to Halaby, 3/16/63, FAA 1647/3. Also see: Welsh to Martin, 3/15/63, LBJ V-P.

9. "Fact Sheet on Supersonic Transport Program," circa 3–4/63, LBJ V-P; Rochte, "Memorandum for Record: SST Systems Analysis," memo, 1/28/63, FAA 1647/3; "Supersonic Transport Systems Analysis, Questions of May 1963," circa 3/63, list, FAA 1647/3; F. R. Collbohm to Shank, 3/1/63, FAA 1647/3; Robert C. Seamans to Shank, 2/20/63, FAA 1647/3; "Systems Analysis Team West Coast Trip Itinerary: March 7 through March 15, 1963," FAA 1647/3.

10. "FAA SST Systems Analysis," circa 4/63, FAA 1647/3; D. D. Davis to Director, "Audit of the SST Program," 11/20/64, memo, and attached "Chronology of Events from December 1959 through December 1963" and "Formal Assignment of Personnel to FAA Supersonic Transport Organization," FAA 6174/96; Halaby to Gordon M. Bain, "Supersonic Transport Program," 3/28/63, memo, FAA 6174/12.

11. Loy to Bain, "Attitude of Air Carriers in Early 1950s toward Subsonic Turbo Jet Transports," 5/10/63, memo, 1647/3.

12. Phillip M. Swatek to Halaby, "Beyond SST Speech," 2/22/63, memo, and attached "Beyond the Supersonic Transport," draft speech, FAA 6174/11.

13. Townsend Hoopes to Halaby, 4/8/63, FAA 6174/13; Halaby to Cook, 1/9/63, LBJ V-P.

14. Cook to Halaby, 5/15/63, LBJ V-P; STAG, "Supplemental Report to the Supersonic Transport Steering Group: Supersonic Transport Program Planning," 5/14/63, LBJ V-P.

15. George C. Prill to Halaby, "Concorde," 4/16/63, memo, FAA 6174/13.

16. Ronald A. Brown to the Vice-President, 3/22/63, LBJ V-P; Vice-President to Brown, 4/2/63, LBJ V-P; Robert F. Six to the Vice-President 4/12/63, LBJ V-P; undated and untitled document on North American Aviation, circa 5/1/63, and attached, "Supersonic Transport Design," n.d., LBJ V-P.

17. Colonel Howard L. Burriss to the Vice-President, "Supersonic Transport," 4/25/63, memo, LBJ V-P.

18. Halaby to the Vice-President, 4/12/63, FAA 6174/13. See also the following materials for the 4/26/63 meeting of the Johnson cabinet committee: agenda; attendance list; and "Progress Report—Supersonic Transport Program for the Vice-President, Johnson's Cabinet Committee," list, FAA 6174/11; Welsh to the Vice-President, "Names Suggested for Public Advisory Committee re: SST," 5/3/63, memo, LBJ V-P; Walter Jenkins to the Vice-President, 5/7/63, memo, LBJ V-P.

19. Halaby to the Vice-President, 5/14/63, FAA 6174/12; FAA, "Supersonic Transport," 5/14/63, LBJ V-P; Welsh to the Vice-President, "SST Draft Report," 5/10/63, memo, LBJ V-P.

20. "GER" to the Vice-President, 5/17/63, memo, LBJ V-P.

21. Martin to the Vice-President, 5/2/63, and attached U.S., Department of Commerce, "Potential Effects of the U.S. Supersonic Transport Program on Balance of Payments, Domestic Employment, and Defense Readiness," LBJ V-P; Welsh to the Vice-President, "Supersonic Transport," 5/1/63, memo, LBJ V-P, and "Commerce Meeting of Friday Morning, May 17," 5/20/63, LBJ V-P; U.S., Department of Commerce, "Summary Statement of Position on the Supersonic Transport Problem," "An Analysis of the Factors Influencing the Choice of a Supersonic Transport Program," "The Position of the Department of Commerce with Respect to Long-Range, High-Speed Air Transport," and "Impact of Noise and Sonic Boom on the Success of an SST Program," 5/24/63, FAA 6174/12; Welsh to the Vice-President, "Supersonic Transports," 5/20/63, memo, LBJ V-P.

22. Kermit Gordon to Halaby, 5/23/64, memo, and attached Gordon and Heller to the Vice-President, memo, FAA 6174/12.

23. Welsh to the Vice-President, "Supersonic Transports," 5/25/63, memo, LBJ V-P; Welsh to Halaby, "Supersonic Transports," 5/27/63, memo, LBJ V-P.

24. Golovin to Welsh, "Supersonic Transports," 5/22/63, memo, FAA 6174/12.

25. Welsh to the Vice-President, "Supersonic Transports," 5/25/63, memo, LBJ V-P, and "Supersonic Transports," 5/27/63, memo, LBJ V-P.

26. Ibid., "Supersonic Transports," 5/28/63, memo, LBJ V-P.

27. Edited dictaphone recording, 3/3/63, pp. 24–26, LBJ Office Files of George Reedy box.

28. Welsh to the Vice-President, "SST Letter of May 30, 1963," 6/3/63, memo, LBJ V-P; Vice-President to President, 5/30/63, LBJ V-P.

29. Halaby to the President, "The Commercial Supersonic Transport—The Next Steps," 6/3/63 memo, and attached "Excerpt from Vice-President's Letter Dated May 30, 1963" (tab 1), draft of insert to JFK's speech (tab 2), and to the President, "Commercial Supersonic Transport Program Objectives of the United States," memo (tab 3), JFK Sorenson; "mjdr" to Johnson, 5/31/63, memo, LBJ V-P.

30. Halaby to the President, "Mr. Juan Trippe and the U.S. Government," 6/31/63, memo, FAA Historian's Files; Pan American Airlines press release, 6/4/63, 5 P.M., FAA 6174/12; untitled and undated draft of Pan American press release on Concorde purchase, JFK Pres. Office Files; Warnick to Halaby, 6/4/63, request, FAA 6174/12; L. L. Dity, "Concorde Order Spurs U.S. SST Action," *Aviation Week and Space Technology*, 6/10/63, p. 40, and "U.S. Participation in Concorde Proposal," *Aviation Week and Space Technology*, 6/17/63, p. 40.

31. McNamara to Halaby, 6/1/63, and attached "Summary of Major Specifications for SST Development Contracts, Objectives for SST Development Contract," 5/31/63, FAA 6174/13.

32. FAA draft of JFK Speech (tab 2 of Halaby to the President, "The Commercial Supersonic Transport—The Next Steps," 6/3/63, memo), JFK Sorenson; revisions by Kermit Gordon of FAA draft, JFK Nat. Security; Sorenson draft of JFK speech, 6/4/63, JFK Sorenson; "Remarks of the President at Graduation Ceremonies United States Air Force Academy, Colorado Springs, Colorado," White House press release, JFK, Pres. Office; *New York Times*, 6/6/63, pp. 1, 25.

Chapter 5

1. Halaby to the President, 6/7/63, draft letter, and attached (draft) message to Congress of 6/7/63, FAA 6174/95; Loy to the Vice-President, "Supersonic Transport," 6/11/63, memo, and attached the President to the President of the Senate and the Speaker of the House, 6/11/63, LBJ V-P; "Supersonic Transport," 5/14/63, LBJ V-P; Halaby to the President, "The Commercial Supersonic Transport—The Next Steps," 6/3/63, memo, and attachments, JFK Sorenson.

2. Halaby to Kermit Gordon, 6/17/63, FAA Historian's Files; Communication from the President of the United States, "Amendments to the Budget for the Fiscal Year 1964 in the Amount of $60,000,000 for the Federal Aviation Agency and $310,000 for the Veterans' Administration," 6/24/63, 88th Cong., 1st sess., House of Representatives, Document 126, LBJ V-P.

3. FAA, "Supersonic Transport," 6/19/63, FAA 6174/11.

4. "Statement of N. E. Halaby, Administrator, Federal Aviation Agency, Presented at

Hearing on June 20, 1963, before the House Committee on Interstate and Foreign Commerce on the Supersonic Transport Development Program,'' LBJ V-P; *New York Times*, 6/21/63, p. 59.

5. Halaby's statements come from ''FAA Administrator Halaby Outlines Government-Industry Program to Develop 'Fastest' Supersonic Transport by 1970,'' 6/19/63, FAA news release no. 63-59, LBJ V-P; ''Remarks by N. E. Halaby, Administrator, Federal Aviation Agency,'' 6/19/63, at the American Institute of Aeronautics and Astronautics, Los Angeles, California, LBJ V-P.

6. Handwritten notes on tab B of Vice-President to the President, 5/30/63, FAA 6174/12; Halaby to the President, ''The Commercial Supersonic Transport—The Next Steps,'' 6/3/63, memo, and attachments, JFK Sorenson; *New York Times*, 7/29/64, p. 40, 7/30/63, p. 3, Halaby to the President, ''Status Report on Supersonic Transport Program,'' 7/29/63, (draft) memo, FAA 6174/95.

7. Dean to Halaby, ''Organization and Staffing of Supersonic Transport Development Office,'' 6/12/63, memo, FAA 6174/13; Davis to Director, ''Audit of the SST Program,'' 11/20/64, memo, and ''Chronology of Events'' and ''Federal Assignment of Personnel to FAA Supersonic Transport Organization,'' FAA 6174/96.

8. Bain to Halaby, ''Weekly Activity Report,'' 8/19/63, memo, FAA 6174/12.

9. Bain, ''Weekly Activity[ies] Report,'' 6/28/63, 7/5/63, 7/12/63, 8/5/63, 8/19/63, 8/26/63, memos, FAA 6174/12.

10. For the debate about the need for new SST-authorizing legislation for the FAA, see: Dean to Halaby, ''Establishment and Funding of a Supersonic Transport Office,'' 6/1/63, memo, FAA 6174/13; Dean to Halaby, ''Organization and Staffing,'' and attached ''Alternatives for Requesting Appropriations for the SST Program,'' FAA 6174/13; ''General Counsel's Memorandum Authority of the Federal Aviation Agency to Develop a Supersonic Transport Aircraft,'' 6/28/63, FAA 6174/96; Stimpson to Halaby, 7/14/63, route slip, FAA 6174/12; U.S., Congress, Senate, Committee on Commerce, Aviation Subcommittee, *United States Commercial Supersonic Aircraft Development Program*, Hearings, 88th Cong., 1st sess., October 16, 17, 21, 22, 23, 29, 30, 1963, pp. 1, 87–190; Goodrich to Halaby, ''Supersonic Transport Hearings,'' 10/25/63, memo, FAA 6174/13; Halaby to the President, ''Supersonic Transport Program,'' 10/19/63, memo, FAA Historian's Files; Bain to Orin Harris, 12/31/63, unsigned letter, FAA 6174/96.

11. For Kefauver's SST views and the FAA's reactions to these opinions, see: Kefauver to Halaby, 5/9/63, FAA Historian's Files; Halaby to Kefauver, 5/22/63, FAA Historian's Files; Kefauver to Halaby, 5/29/63, FAA Historian's Files; Davis to the record, ''Senator Kefauver's Request to FAA for SST Report,'' 6/11/63, memo, FAA 6174/13; Kefauver to Halaby, 6/13/63, telegram, FAA Historian's Files; Kefauver to Halaby, 6/19/63, FAA Historian's Files; Stimpson to the record, ''Meeting with Senator Kefauver,'' 6/25/63, memo, FAA 6174/12; Stimpson to Halaby, ''Kefauver Contacts,'' 6/28/63, memo, and handwritten notes by Halaby on memo, FAA Historian's Files.

12. For the NASA-FAA rivalry about the SST during the latter half of 1963, see: Halaby to Shank, 6/3/63, handwritten note, FAA 6174/12; R. F. Goranson to files, ''Discussion of SST Program with A. J. Evans of NASA Hdqts.,'' 6/11/63, memo, FAA 6174/12; Bain, ''Weekly Activity[ies] Report on Supersonic Transport,'' 6/28/63, 7/5/63, 8/5/63, memos, FAA 6174/12; Fred Philips to OA-1, ''NASA Press Briefing on SST and other aeronautical research,'' 7/22/63, draft memo, FAA, Historian's Files; Chester C.

Spurgeon to Halaby, "NASA—SST Program Interest," 8/8/63, memo, FAA 6174/12, "NASA Conference Receives Reports on Supersonic Transport Studies," 9/19/63, NASA news release, FAA 6174/13; Edward H. Kolcum, "NASA Favors Two SCAT Configurations," *Aviation Week and Space Technology*, 9/23/63, p. 26; Robert H. Cook, "FAA Labels SCAT Designs 'Impractical,' " *Aviation Week and Space Technology*, 10/14/63, p. 38; J. Thomas Tidd, to files, "SST Hearings—Monday, October 21, and Tuesday, A.M. October 22, 1963," 10/22/63, memo, FAA 6174/13; Bain to Halaby, "GA-10 Memo, dated 8/8/63—NASA—SST Program," 8/13/63, memo, FAA 6174/12; Harold W. Grant to Webb, 7/31/63, FAA 6174/12; Hugh L. Dryden to Grant, 8/21/63, FAA 6174/12.

13. Webb to Halaby, 12/27/63, and attached "Memo of Understanding," FAA 6174/13.

14. For American attitudes toward SST collaboration with British and French, see: Halaby to the President, "Mr. Juan Trippe and the U.S. Government," 6/13/63, memo, FAA Historian's Files; Chester C. Spurgeon to Halaby, "Report of Discussion between Mr. Alan Greenwook, Mr. Charles Gardner and Myself Regarding BAC-SUD Position on United States SST Development," 6/13/63, memo, FAA 6174/12; L. L. Doty, "Concorde Orders Spur U.S. SST Action," *Aviation Week and Space Technology*, 6/10/63, p. 40; L. L. Doty, "U.S. Shuns Bid to Enter Concorde Effort," *Aviation Week and Space Technology*, 7/8/63, pp. 28–29; L. L. Doty, "U.S. Participation in Concorde Proposal," *Aviation Week and Space Technology*, 6/17/63, p. 40; Bruce to Halaby, 10/16/63, telex, FAA 6174/13.

15. Halaby to Bain, 11/6/63, memo, FAA 6174/13.

16. "ml." to Halaby, 7/16/63, memo, FAA 6174/12; "Russia in SST Race," *Aviation Week and Space Technology*, 10/21/63, p. 38.

17. For the American-flag air carriers' views of the SST, see: Halaby to the President, "Mr. Juan Trippe"; Doty, "Concorde Orders," p. 40; Doty, "U.S. Participation," p. 40; Robert F. Six to E. C. Welsh, 8/14/63, LBJ V-P; "Continental Plans 1970 Service after Order for Three Concordes," *Aviation Week and Space Technology*, 8/5/63, p. 43; C. R. Smith to Johnson, 8/1/63, and attached "Supersonic Transport," LBJ V-P; Smith to Halaby, 8/2/63, FAA 6174/12; "American President Urges SST Action," *Aviation Week and Space Technology*, 8/5/63, p. 45.

18. Bain to Halaby, "Airline Orders for the U.S. Supersonic Transport," 8/20/63, memo, FAA 6174/12; Bain to Halaby, "Meeting with TWA in New York on September 30, 1963," 10/1/63, memo, FAA 6174/13.

19. For the initial airline agreements, see: Halaby to Bain, 10/3/63, note, FAA 6174/13; Tillinghast to Halaby, 10/2/63, draft letter, FAA 6174/13; Smith to Halaby, 10/10/63, telegram, FAA 6174/13; Edward W. Stimpson to Kennedy, 10/14/63, memo, FAA 6174/13; *New York Times*, 10/15/63, pp. 1, 13; Smith to Halaby, 10/16/63, and attached Smith to Dean Rusk, 10/16/63, FAA 6174/13.

20. For further airline agreements, see: Robert H. Cook, "U.S. Carriers Support SST Development," *Aviation Week and Space Technology*, 10/21/63, pp. 38–39; Senate, Aviation Subcommittee, *U.S. Commercial Supersonic Aircraft Development Program*, October 16, 17, 21, 22, 23, 25, 29, 1963, pp. 54, 296, 297; J. Thomas Tidd to files, "SST Hearings—Tuesday Afternoon, October 22, 1963," 10/22/63, memo, FAA 6174/13.

21. "Delivery Positions Established for U.S. Supersonic Transport," 11/14/63, FAA press release, FAA 6174/13.

22. Bain to Halaby, "Delivery Reservations—Northwest Airlines," 11/25/63, memo, FAA 6174/12; Bain to Smith, 12/16/63, FAA 6174/12; Halaby to the President, "Supersonic Transport Development Program," 12/2/63, memo, FAA 6174/95.

23. FAA, "Policy on Reservation of SST Delivery Positions," circa 4/10/64, 6174/96.

24. Handwritten notes on tab B of Halaby to the President, 6/3/63, FAA 6174/12; McNamara to Halaby, 6/1/63, and attached "Summary of Major Specifications for SST Development Contract," 6/1/63, FAA 6174/12.

25. Bain to Halaby, "Weekly Activity[ies] Reports," 6/27/63, 6/28/63, memos, FAA 6174/12; see also Bain to Halaby, "25% Participation in SST Development Program," 7/1/63, memo, FAA 6174/92.

26. Courtlandt S. Gross to Halaby, 7/10/63, FAA 6174/12; "Address by William M. Allen, President, the Boeing Company and Chairman, Aerospace Industries Association, National Editorial Association Luncheon, Boeing Airplane Division—Renton, Washington," 7/20/63, FAA 6174/12; anonymous handwritten notes, circa 8/63, FAA Historian's Files.

27. Bain to Halaby, "Supersonic Transport Development Program," 8/12/63, memo, FAA 6174/12.

28. Halaby to the President, "Status Report on Supersonic Transport Program," 7/29/63, draft memo, FAA 6174/95.

29. Donald W. Douglas, Jr. to Halaby, 8/26/63, and attached Douglas Co. press release, JFK Pres. Office Files; Halaby to the President, "Douglas Aircraft Company Will Not Participate in the Supersonic Transport Development Program," 8/26/63, JFK Pres. Office Files; *New York Times*, 9/10/63, p. 79.

30. *New York Times*, 8/15/63, p. 58; "Six Firms to Enter SST Contest," *Aviation Week and Space Technology*, 9/16/63, p. 49.

31. FAA, "Request for Proposals for Development of a Commercial Supersonic Transport," 8/15/63, FAA 6174/12.

32. Senate, Aviation Subcommittee, *U.S. Commercial Supersonic Aircraft Development Program*, October 16, 17, 21, 22, 23, 25, 29, 1963, pp. 7, 13, 35, 209, 313–315, 362–364, 380, 428, 473, 479–512.

33. Halaby to the President, "Supersonic Transport Program," 10/29/63, memo, FAA Historian's Files.

34. Ibid.; "Financial Advice for Supersonic Transport Program of [sic] Cost Sharing by Industry," 7/25/63, memo, JFK Pres. Office Files; Halaby to the President, "Status Report"; Bain to Halaby, "Weekly Activities Report," 8/5/63, memo, FAA 6174/12; Bain to the record, "Mr. Bain's Telephone Conversation of this date with Secretary Gilpatric, DOD," 8/8/63, memo, FAA 6174/12.

35. Bain to Halaby, "Supersonic Transport Development Program."

36. "Kennedy Names Black as Advisor on Supersonic Transport Program," *Aviation Week and Space Technology*, 8/19/63, p. 37; the President to Eugene Black, 8/13/63, in Eugene Black and Stanley de Osborne, Jr., "Report on the Supersonic Transport," 12/19/63, Baker Library, Harvard Business School.

37. For Halaby's views on JFK and LBJ, see: Najeeb Halaby, *Crosswinds: An Airman's Memoir* (Garden City, N.Y.: Doubleday, 1978), pp. 156, 160; Richard J. Kent, Jr., "In-

terview with N. E. Halaby," 7/30/75, New York, N.Y., pp. 9, 47–50, FAA Historian's Files. For Johnson's pro-SST views, see: Johnson to Keith Kahle, 6/13/63, LBJ V-P; Johnson to S. G. Tipton, 6/11/63, LBJ V-P; Johnson to C. R. Smith, 8/7/63, LBJ V-P; Johnson to Lt. Gen. Ira C. Eaker, USAF (Retired), 6/18/63, LBJ V-P; Edited dictaphone recording, 3/3/63, pp. 24–26, LBJ Office Files of George Reedy box.

38. Black and Osborne, "Report on the Supersonic Transport," p. 6.

39. Halaby to the President, "Black-Osborne Report on Supersonic Transport Program," 12/24/63, memo, FAA 6174/12. See also Halaby to Walter Jenkins, 12/24/63, handwritten note, LBJ Executive Files.

40. McNamara to the President, 1/2/64, memo, and attached the President to Kermit Gordon, "Supersonic Transport Program," 1/2/64, draft memo (also on White House stationery, 1/4/64), LBJ EX CA; Jack Valenti to Bill Moyers, 1/9/64, note, LBJ EX CA.

41. Dean to Halaby, "SST Development Program," 1/3/64, memo, and attached Halaby to the President, "Organization for the Development of a Civil Supersonic Transport Aircraft," 1/3/64, "Examples of Federal Agencies with Promotional and Regulatory Responsibilities," 1/3/64, FAA 6174/92.

42. Halaby to the President, 1/13/64, LBJ EX CA.

43. President to Halaby, 1/20/64, LBJ EX CA.

44. For FAA views and action at this time, see: Halaby to President, "Supersonic Transport Program," 1/29/64, draft memo, FAA 6174/92; Halaby to the President, "Supersonic Transport Program," 2/5/64, draft memo, LBJ EX CA. For Halaby's "report" to the President, see: Halaby to the President, "Interim Report on the Supersonic Transport Program," 2/11/64, memo, FAA 6174/11. See also Halaby to Gordon, 2/11/64, note, FAA 6174/11; Dick Nelson to Walter Jenkins, "Brief of Supersonic Transport Interim Report," 2/12/64, memo, LBJ EX CA.

45. Bain to the President, "Black-Osborne Report on the Supersonic Transport Development Program," 2/28/64, memo, and attached memo, FAA 6174/92.

46. Halaby to Gordon, "Black-Osborne Report on the Supersonic Transport Development Program," 3/4/64, memo, FAA 6174/14.

47. Various anti-SST press pieces were mentioned in or attached with: Phillip M. Swatek to Halaby, 2/11/64, memo, FAA 6174/11; Charles W. Bryan to Senator Spessard Holland, 2/17/64, LBJ EX CA. See also *Aviation Daily*, 2/13/64, pp. 270–272; *New York Times*, 2/22/64, pp. 1, 9.

48. Swatek to Halaby, "Current SST Situation," 2/14/64, memo, FAA 6174/11.

49. Halaby to Gordon, "Black-Osborne Report."

50. For the draft of the executive order to establish the advisory committee, see: Gordon to Halaby et al., 2/3/64, memo, FAA 6174/12; "Bureau of the Budget Proposal on Management of the Supersonic Transport Program," n.d. (after 2/1/64), FAA 6174/92; Gordon to the President, "Assignment of Responsibility for Supersonic Transport Development," 2/12/64, memo, LBJ EX FG; Arthur B. Focke to Myer Feldman, 2/14/64, memo, and attached draft executive order, LBJ EX FG; Focke to Harold Reis, 2/11/64, memo, LBJ EX FG; Halaby to Bain, circa 2/12/64, note, FAA 6174/92; Halaby to Feldman, 2/18/64, note, and attached President to Secretary of Defense et al., and President to FAA Administrator, draft memos, LBJ EX CA; Feldman to the President, "Advisory Committee on Supersonic Transport," 2/27/64, memo, LBJ EX FG.

51. For various agency comments on the SST program, see: Gordon to the President, "Agency Comments and Recommendations on the Black-Osborne Report on the Supersonic Transport Program," 3/25/64, memo, LBJ EX CA; summaries of comments by Department of Commerce, Treasury Department, NASA, AEC, Office of Science and Technology, CAB, Department of State, FAA 6174/92; "Resume of Significant Agency Comments on Black-Osborne Report," 3/6/64, LBJ EX CA; E. C. Welsh to Gordon, "Black-Osborne Report on SST," 2/27/64, memo, 6174/92; Dean to Halaby, "Agency Comments on Black-Osborne Report," 3/5/64, memo, FAA 6174/12; Donald F. Hornig to Gordon, "The Black-Osborne Report," 3/2/64, LBJ Hornig File/1; C. Douglas Dillon to the President, "Supersonic Transport Program," 1/15/64, LBJ EX CA; Halaby to Bain, 3/5/64, route slip, FAA 6174/92.

52. L. L. Doty, "U.S. Supersonic Transport Effort Stalled," *Aviation Week and Space Technology*, 3/9/64, pp. 29–30. See also: *New York Times*, 3/3/64, p. 20; Halaby to the President, "Disclosure of A-11 Technical Data," 3/13/64, memo, FAA 6174/95.

53. To Feldman, 2/13/64, note, and attachments, LBJ Gen. CA; "Government Aviation Experts Convene to Evaluate Design Proposals for Supersonic Transports," 1/6/64, FAA press release, FAA 6174/11; *New York Times*, 1/28/64, pp. 1, 16.

54. FAA, "Supersonic Transport: Program Recommendations," 4/1/64, p. 1, FAA 6174/96.

55. Dean to Halaby, "SST Status Report," 3/17/64, memo, FAA 6174/14; Halaby to the President, "Status Report on SST," 3/18/64, memo, LBJ EX CA; Halaby to Feldman, 3/25/64, FAA 6174/95.

56. Gordon to the President, "Agency Comments and Recommendations"; ibid., "Current Presidential Decisions to Be Made on the Supersonic Transport Program," 3/25/64, memo, LBJ EX CA.

57. Feldman to the President, "Supersonic Transport," 3/26/64, memo, LBJ EX CA.

58. *New York Times*, 3/12/64, p. 68.

59. Halaby to Feldman, circa 3/26/54, handwritten note, LBJ EX CA; Gordon to the President, "Current Presidential Decisions."

60. "Executive Order Establishing the President's Advisory Committee on Supersonic Transport," 4/1/64, LBJ EX FG.

61. President to Black, the President to Osborne, the President to McCone, 4/3/64, LBJ EX FG.

62. Robert H. Cook, "Heavy SST Financial Role Seen for DOD," *Aviation Week and Space Technology*, 4/6/64, pp. 28–29.

Chapter 6

1. For the discussion on the sonic boom, see: William Baxter, "The SST: From Watts to Harlem in Two Hours," *Stanford Law Review* 21 (11/68):1–57; Boeing Co., *Aircraft Noise Reduction Status Report*, D6-23482, 9/68; Boeing Co., *Sonic Boom: A Review of Current Knowledge and Developments*, D6A1959801, 1/67; William Shurcliff, *S/S/T and Sonic Boom Handbook* (New York: Ballantine Books, 1970); U.S. Department of Transportation, *Second Federal Noise Abatement Plan: FY 1970–71* (Washington, D.C.:

GPO, 1971); Wyle Laboratories Research Staff, *Noise Primer for the Supersonic Transport* (El Segundo, Cal.: Wyle Laboratories, 1971).

2. FAA document presenting overview of past U.S. sonic boom studies and description of Oklahoma City sonic boom tests, circa 4/64, FAA 6174/96.

3. Memo, John R. Provan to Dean, "Response to BOB Letter on Clearance of the Sonic Boom Survey," 5/14/64, memo, FAA 6174/14; Dean to Halaby, "Bureau of the Budget Clearance of Sonic Boom Survey," 5/16/64, memo, FAA 6174/14.

4. Dean to David E. Cohn, 5/18/64, FAA 6174/14.

5. William M. Jackson to Bain, "Sonic Boom Studies—Progress Review," 2/25/64, memo, FAA 6174/14.

6. Bain to Halaby, "Status Report—Supersonic Transport Program," 5/21/64, memo, FAA 6174/97; Halaby to the President, "Progress Report on the Supersonic Transport Development Program," 6/1/64, memo, LBJ EX CA.

7. Frederick Seitz to the President, 6/25/64, LBJ CF CA.

8. Oklahoma City Chamber of Commerce to "SST Booster," 6/3/64, CL.

9. Halaby to Norman Cousins, circa 6/12/64 (never sent), draft letter, FAA 6174/11; "The Era of Supersonic Morality," *Saturday Review*, 6/6/64, p. 49.

10. Bain to Halaby, "Sonic Boom Test Program," 6/16/64, memo, LBJ FAA Hist. Doc.

11. C. P. Wolle to Halaby, "Early Impressions—Oklahoma City Boom Program," 6/23/64, memo, FAA 6174/97.

12. George H. Shirk to Bain, 6/16/64, FAA 6174/11; Bain to Shirk, 7/13/64, FAA 6174/11.

13. Bo Lundberg, "SST Sooner Boomer," circa 6/64, FAA 6174/11.

14. *Washington Post*, 7/15/64, p. A7.

15. For the Citizens Advisory Committee, see: Halaby to Jo-Ann McGinnis, 8/6/64, FAA 6174/11; Bain to Halaby, "Sonic Boom Claims Advisory Committee," 8/6/64, memo, 6174/11; Martin Menter to the files, "Appointment of Jo-Ann McGinnis as Special Consultant," 8/6/64, memo, FAA 6174/11; Halaby to A. Mark Eudaley, 8/6/64, FAA 6174/11; Sonic Boom Advisory Committee, "Initial Report," 8/15/64, "First Interim Report," 8/24/64, "Summary Report," 9/14/64, and "Terminal Report," 10/5/64, FAA 6174/11; "Comments on the Sonic Boom Advisory Committee Summary Report," n.d., FAA 6174/11.

16. For Monroney's view of the damage claims, see: Halaby to Bain, "OKC Sonic Boom Advisory Committee and Secretary Zuckert," 10/13/64, handwritten memo, FAA 6174/11; Dave R. McKown to Monroney, 10/7/64, FAA 6174/11; Monroney to McKown, 10/17/64, FAA 6174/11.

17. Monroney to Bain, 12/30/64, FAA 6174/11.

18. Bain to Halaby, "Public Opinion Poll, Oklahoma City," 12/17/64, memo, and attached Paul N. Borsky, "Preliminary Report on Community Reactions to Sonic Booms, Oklahoma City Area, February–July 1964," National Opinion Research Center, University of Chicago, 11/64, FAA 6174/11.

19. McNamara to the President, "First Interim Report of President's Advisory Committee on Supersonic Transport," 5/14/67, memo, LBJ FAA Hist. Doc.; LBJ to the NAS President, FAA Administrator, and Secretary of Commerce, 5/20/64, memo, NAS.

20. Minutes of the first, second, and fourth NAS Committee meetings, 7/29/64, 8/26/64, 10/19/64; NAS; *Who's Who* (Chicago: Marquis Who's Who, 1965, 1975).

21. Minutes of the Special Panel on Structures meeting in New York City, 9/18/64, NAS.

22. Bain to Dunning, 9/19/64, NAS; Minutes of the third NAS Committee meeting, 9/21/64, NAS.

23. Minutes of the fourth NAS Committee meeting, 10/19/64, NAS; Seitz to Halaby, 11/4/64, and attached "Comments on SST-Sonic Boom," 11/3/64, FAA 6174/11.

24. H. Davis to Dunning, 9/24/64, NAS; Dunning to Bain, 9/30/64, NAS.

25. Kenneth Youel and Robert Harper, "Public Opinion Aspects of Sonic Boom Problem and Public Relations Recommendations: A Report to Committee on Sonic Boom, NAS" (attachment to Youel to Park, 10/15/64), NAS.

26. Fourth NAS Committee meeting minutes, NAS; "Comments of the Committee on SST-Sonic Boom," 11/3/64, FAA 6174/11; Fourth NAS Committee meeting charts: "Analysis of Oklahoma City Damage Claims," "Preliminary Conclusions," and "By-Products of Our Economic Study Having Social/Political Importance," NAS.

27. Park to Whitcomb, 10/26/74, NAS; Whitcomb to Park, 11/18/74, NAS.

28. Minutes of the fifth NAS Committee meeting, 11/17/64, NAS.

29. BRAB, *Interim Report No. 1 to the NAS SST-Sonic Boom Committee* (Washington, D.C.: NAS-NRC, November 10, 1964), NAS; Seitz to Halaby, 11/23/64, FAA 6174/11.

30. BRAB, *Interim Report No. 2 to the NAS SST-Sonic Boom Committee* (Washington, D.C.: NAS-NRC, 1964), NAS.

31. Bain to Dunning, 11/14/64, NAS.

32. Bain to Halaby, "Press Visit to Sonic Boom Structural Response Test Site at White Sands, New Mexico," 12/10/64, memo, and attached John H. Wiggins to Bain, 12/7/64, FAA 6174/11; minutes of sixth NAS Committee meeting, 12/22/64, NAS; personal interviews.

33. Dillon to Dunning, 12/31/64, NAS; Dunning to Bain, 1/4/65, LBJ FAA Hist. Doc.

34. Minutes of the sixth NAS Committee meeting, 6/22/64, p. 3, NAS.

35. Halaby, "Memorandum of Conversation: Dean John Dunning," 1/12/65, FAA.

36. NAS Committee on SST-Sonic Boom, *Status Report* (Washington, D.C.: NAS, 1/27/65), NAS.

37. "Summary Conclusion" from Draft Status Report of the NAS Committee on SST-Sonic Boom, 12/18/64, NAS; Seitz to McNamara, 1/28/65, NAS; Dunning to Seitz, 1/27/65, NAS.

38. NAS Committee, *Status Report*, 1/27/65.

39. Minutes of the seventh NAS Committee meeting, 2/4/65, NAS; "First Two Volumes of Report on Oklahoma City Sonic Boom Study Present Data on Boom Overpressure Levels, Meteorological Effects," 2/11/65, FAA news release no. 65-15, LBJ FAA Hist. Doc.; "Reports on Sonic Boom Studies in Oklahoma City and New Mexico Released by FAA," 4/25/65, FAA news release, no. 65-34, LBJ FAA Hist. Doc.; *New York Times*, 4/25/65, p. 62.

40. Kingsley Davis, "A Statement on Public Reaction to the Sonic Boom," 3/15/65, NAS.

41. BRAB, *Interim Report No. 3 to the NAS SST-Sonic Boom Committee: Long-Range Structural Response Research and Testing Program* (Washington, D.C.: NAS-NRC, 7/2/65), NAS.

42. Dunning to Seitz, 7/13/65, NAS.

43. NAS Committee on SST-Sonic Boom, *Status Report* (Washington, D.C.: NAS, 7/21/65), NAS.

44. Phillip M. Boffey, *The Brain Bank of America* (New York, N.Y.: McGraw-Hill Book Co., 1975), p. 119.

Chapter 7

1. This overview of McNamara is based on the following: David Halberstam, *The Best and the Brightest* (New York: Fawcett Crest, 1971), pp. 263–325; William W. Kaufmann, *The McNamara Strategy* (New York: Harper & Row, 1964), pp. 168–274; Robert J. Art, *The TFX: McNamara and the Military* (Boston: Little, Brown, 1968); *TFX: The Commonality Decision*, ICCH Case No. 9-375-035 (Boston: Intercollegiate Case Clearing House, 1974); Najeeb Halaby: *Crosswinds: An Airman's Memoir* (Garden City, N.Y.: Doubleday, 1978), p. 207; Richard J. Kent, Jr., "An Interview with N. E. Halaby," 7/30/75, New York, N.Y., p. 35, FAA Historian's Files.

2. Gordon Bain, "Supersonic Transport: Program Recommendations," 4/1/64, FAA 6174/96.

3. FAA, "Comments on the Memorandum to the Advisory Committee," 4/10/64, FAA 6174/96; Bain to Califano, 4/9/64, and attached to Califano, "Comments on Department of Defense Draft Memorandum," 4/9/64, memo, FAA 6174/96.

4. Halaby to the President, "Recommendations Regarding the Supersonic Transport Program," 4/6/64, revised 4/9/64, draft memos, FAA 6174/96.

5. "Draft Proceedings of the President's Advisory Committee on Supersonic Transport," 4/13/64, PAC/32.

6. Ibid., 4/18/64, PAC/32.

7. President to Halaby, 4/23/64, memo, FAA 6174/96.

8. "Draft Proceedings of the President's Advisory Committee on Supersonic Transport," 5/1/64, PAC/32. For the engine designs, see: George N. Catham and Franklin P. Huddle, *The Supersonic Transport* (Washington, D.C.: Congressional Research Service, 5/18/71), pp. 9–11.

9. Halaby to the President, "Press Speculation about Supersonic Transport Program," 5/4/74, memo, LBJ EX CA.

10. L. L. Doty, "Prospect of SST Shift to Pentagon Grows," *Aviation Week and Space Technology*, 4/27/64, p. 38; *Aviation Daily*, 4/28/64, p. 366; *Business Week*, 5/2/64, p. 36.

11. For the disagreements in May 1964 over continuing the SST design competition and SST financing policy, see: PAC to the President, "First Interim Report of the President's Advisory Committee on Supersonic Transport," 5/5/64, draft memo, FAA 6174/96; "Comments—First Interim Report of the President's Advisory Committee on SST," 5/7/64, rough draft, FAA 6174/96; Halaby to PAC, "Comments on Advisory Committee's Draft Report," 5/8/64, memo, FAA 6174/96; "Draft Proceedings of the President's Advisory Committee on Supersonic Transport," 5/8/64, PAC/32; "Memorandum for the

President: First Interim Report of the President's Advisory Committee on Supersonic Transport," 5/14/64, PAC/32.

12. For the disagreements in May 1964 over performing SST economic analysis, see: PAC to the President, "First Interim Report," 5/5/64; Califano to Halaby, 5/5/64, memo, and attachment; Dean to Halaby, "Commerce Department Transportation Economics Resources," 5/7/64, memo, FAA 6174/96; "Draft Proceedings of the President's Advisory Committee on Supersonic Transport," 5/8/64, PAC/32; Halaby to PAC, "Comments on Advisory Committee's Draft Report," 5/8/64, memo, FAA 6174/96.

13. For the disagreements in May 1964 over the control of sonic boom studies, see: PAC to the President, "First Interim Report," 5/5/64; Bain to the PAC, "Sonic Boom Memorandum Prepared by Dr. Brown," 5/1/64, memo, FAA 6174/96; Dean to Halaby, "Approaching the National Academy of Sciences," 5/7/64, memo, FAA 6174/96; Halaby to PAC, "Comments on Advisory Committee's Draft Report"; "Draft PAC Proceedings," 5/8/64; "Memorandum for the President: First Interim Report of the President's Advisory Committee on Supersonic Transport," 5/14/64, PAC/32.

14. Office of the Secretary of Defense to Lawrence F. O'Brien, 5/12/64, memo, LBJ EX FG.

15. "Memorandum for the President: First Interim Report"; Halaby to the President, "Supersonic Transport Development Program," 5/15/64, memo, FAA 6174/95.

16. Bill Moyers to the President, circa 5/15/64, note, LBJ CF CA.

17. President to Halaby, Luther Hodges, and Frederick Seitz, 5/20/64, memo, LBJ CF CA.

18. For the friction in late May and early June 1964 between the Commerce Department and the FAA over performing the SST economc analyses, see: Hodges to Halaby, 5/25/64, FAA 6174/11; Hollomon to Halaby, 5/28/64, FAA 6174/92; Halaby to Hollomon, 6/8/64, FAA 6174/92; Acting Secretary of Commerce to the President, 6/1/64, LBJ CF CA; Halaby to Hodges, 6/8/64, FAA 6174/92; *New York Times*, 6/17/64, p. 85.

19. For the continued general Commerce-FAA bickering, see: Hodges, "Memorandum on SST Economic Analysis," 7/2/64, memo, FAA 6174/92; Hodges to Halaby, 7/2/64, FAA 6174/92; Katz to SST Study Advisory Group, "SST Study Advisory Group Meeting," 7/7/64, memo, and attached distribution list, FAA 6174/92; Hodges to Halaby, 7/7/64, FAA 6174/92; Dale Davis to the record, "FAA/Department of Commerce Relationships," 7/16/64, memo, FAA 6174/92; Katz to Bain, "Procedures to Be Followed by Department of Commerce Personnel in Contacting Companies and Organizations under Contract to the Federal Aviation Agency," 7/24/64, memo, FAA 6174/92; "Memorandum of Understanding on Transfer of Technical and Economic Data from the Federal Aviation Agency to the Department of Commerce," 8/5/64, FAA 6174/95. For the Commerce-FAA conflict over funding the SST economic studies, see: Gordon to Halaby, 6/6/64, FAA 6174/11; H. B. Alexander to Halaby, 6/15/64, route slip, and handwritten notation by Dean, FAA 6174/11; Hodges to Halaby, 6/24/64, FAA 6174/92; Alexander, "Telephone Call Record," 7/7/64, FAA 6174/11; Hodges to Gordon, 6/16/64, and attached cost estimates of contractual services, FAA 6174/12; Gordon to Halaby, 7/22/64, FAA 6174/92; Harold R. Grant to Gordon, 7/23/64, FAA 6174/92; "Chronology of Actions Regarding Transfer of Funds to Commerce for SST Economic Study," 7/25/64 (attached to telephone call records of 7/24/64), FAA 6174/11; Bain to Halaby, "Department of Commerce's Request for Funds," 7/24/64, memo, FAA 6174/92; Alexander to

Halaby, "SST Economic Study," 7/27/64, memo, FAA 6174/92; Halaby to Hodges, 7/27/64, FAA 6174/92. For the Commerce-FAA conflict over sonic boom studies, see: Katz to Dr. John Dunning, 8/13/64, FAA 6174/92; Katz to Bain, "Study of Economic Effects of Sonic Boom," 9/11/64, memo FAA 6174/92; Bain to Katz, 9/11/64, FAA 6174/92; Bain to Halaby, 9/14/64, note, and Halaby to Bain, 9/15/64, handwritten note, FAA 6174/11; Hollomon to Bain, 9/15/64, FAA 6174/92; Bain to Hollomon, 9/16/64, FAA 6174/92; Hollomon to Bain, 9/18/64, FAA 6174/92; Bain to Hollomon, 9/22/64, FAA 6174/92; Douglas Doil to Special Assistant to Deputy Administrator for SST Development, "Sonic Boom Data Requested by IDA," 10/13/64, memo, FAA 6174/92.

20. For Commerce Department activity, see: Hodges, "Memorandum on SST Economic Analysis," 8/3/64, memo, FAA 6174/92; Katz to Bain, "Report on Visits to Airframe Manufacturers, August 14–18, 1964," 8/21/64, memo, 6174/92; Hodges to Halaby, 8/10/64, and Halaby to Bain, 8/12/64, handwritten note, FAA 6174/11; Bain to Hollomon, 8/24/64, memo, FAA 6174/11; Hollomon to Bain, 8/26/64, FAA 6174/11.

21. Halaby to Bain and PO-1, 9/4/64, memo, FAA 6174/92.

22. Bain to Halaby, "Department of Commerce Economic Study Program—Your Memorandum of September 4, 1964," 9/14/64, memo, FAA 6174/11.

23. Bain to Halaby, "Commerce Economic Study and National Academy of Sciences Sonic Boom Committee Report Dates," 1/7/65, memo, FAA 6174/11.

24. Charles G. Warnick to Halaby, 5/11/64, memo, and Bain, 5/22/64, note, FAA 6174/97.

25. For the Guggenheim Aviation Center's SST resolution and for Halaby's response, see: Elwood Quesada to McNamara, 5/25/64, and attached "Resolution on Supersonic Transport Development," FAA 6174/12; Quesada to Halaby, 6/3/64, note, and attached Guggenheim Foundation press release, 6/1/64, FAA 6174/12; Halaby to Quesada, 6/11/64, FAA 6174/12; G. Piel to Moyers, 6/8/64, LBJ Gen CA.

26. Warnick to Halaby, "Lundberg Speech," 6/3/64, memo, and attached Lundberg, "Aviation Safety, Supersonic Transports and Their Effects on Society," 5/22/64, speech, FAA 6174/11.

27. R. L. Bisplinghoff, "The Supersonic Transport," *Scientific American* (June 1964):25.

28. *New York Times*, 7/2/64, p. 32.

29. Halaby to Warnick, 8/24/64, handwritten note, FAA 6174/11.

30. Warnick to Halaby, "SST," 9/11/64, memo, and "SST Article," 12/16/64, memo, FAA 6174/11.

31. Halaby to Edward Weeks, 12/21/64, FAA 6174/11.

32. For the beginning and the structure of the SST design competition in June 1964, see: Halaby to the President, "Progress Report on the Supersonic Transport Development Program," 6/1/64, memo, FAA 6174/12; "Design Contracts Awarded to Boeing, Lockheed, General Electric, Pratt and Whitney in U.S. Supersonic Transport Development Program," 6/2/64, FAA press release, FAA 6174/12.

33. For the disagreements between the Defense Department and the FAA about the SST design competition, see: Bain to Califano, 6/14/64, FAA 6174/96; Gordon to Halaby, 6/15/64, FAA 6174/12; Halaby to Gordon, 6/16/64, handwritten note, and attached Halaby to Boeing, Lockheed, General Electric, and Pratt & Whitney, 6/16/64, draft letter, FAA 6174/12; Halaby to Califano and Gordon, 6/23/64, memo, and attached Halaby

to SST contractors, 6/23/64, draft letter, FAA 6174/12; Halaby to (individually) William Allen (Boeing), William P. Gwinn (Pratt & Whitney), Courtlandt S. Gross (Lockheed), and Jack S. Parker (G.E.), 6/26/64, FAA 6174/12; Bain to Halaby, "Contracts for Phase II-B," 10/31/64, memo, FAA 6174/11; Califano to McNamara, "Supersonic Transport Advisory Committee," 11/6/64, memo, PAC/10; "SST Program since May 14, 1964," 1/18/65, PAC/10; "Review and Comparison of Recommendations by the President's Advisory Committee on Supersonic Transport and Contractual Actions by FAA since 14 May 1964" (attachment to Califano to Moyers, 1/13/65, memo), PAC/10.

34. For Defense Department control of classified information in the evaluation and for general Defense Department authority in this process, see: Halaby to McNamara and McCone, "Security Classification—SST Engine Development," 5/27/64, memo, FAA 6174/97; Califano to Halaby, "Security Classification—SST Engine Development," 6/17/64, memo, FAA 6174/97; Halaby to McNamara, 6/22/64, FAA 6174/97; Halaby to McNamara, 6/8/64, FAA 6174/97; D. D. Davis to Halaby, "SST Phase II-A Contracts; Consent to Release Subcontracts," 6/26/64, memo, FAA 6174/96; Bain to Halaby, "DOD Contract Administration Assistance on SST Contracts," 6/4/64, memo, and attached "Memorandum of Agreement," FAA 6174/97; Halaby to McNamara, 6/4/64, FAA 6174/97; Califano to Halaby, 6/25/64, memo, and attached (unsigned) "Memorandum of Agreement," FAA 6174/96; Halaby to Califano, 7/9/64, FAA 6174/96. For FAA discussions and for the creation of systems and models for the interagency evaluation, see: FAA, "Proposed Supersonic Transport Economic Model Ground Rules," 6/5/64, and attached letter, FAA 1647/3; "Summary of Comments on SST Economic Model Ground Rules," 7/7/64, and attached mailing list, FAA 1647/4; "Agenda Outline" of 7/7/64 meeting on SST economic analysis ground rules, and attached attendance list, FAA 1647/4; Bain to Halaby, 10/13/64, note, and attached Bain to Tillinghast, draft letter, FAA 1647/4. For the FAA official's quote, see: *New York Times*, 9/29/64, p. 19.

35. For summaries of the manufacturers' proposals, see: Bain to (individually) Hodges, Webb, McCone, Osborne, and Black, 11/6/64, FAA 6174/96; Bain to Robert Prestemon, 11/6/64, FAA 6174/96; Bain to Halaby, "Manufacturers' SST Proposals," 11/2/64, memo, FAA 6174/97; T. C. Muse to Special Assistant to the Secretary and Deputy Secretary of Defense, "Summary of the Supersonic Transport Phase II-A Proposals," 11/4/64, memo, PAC/10; FAA, "Summary of Information Presented in Manufacturers' Proposals Submitted November 1, 1964," circa 11/2/64, PAC/10.

36. Halaby to McNamara, 11/4/64, handwritten letter, LBJ CF CA; Halaby to the President, 11/4/64, handwritten letter, LBJ CF CA; Halaby to Gordon, 11/4/64, handwritten letter, FAA 6174/97.

37. Califano to McNamara, "Supersonic Transport Advisory Committee," 11/6/64, memo, PAC/10.

38. Black and Osborne to McNamara, 11/16/64, LBJ FAA Hist. Doc.

39. McNamara to the President, 11/21/64, memo, and attached President to Halaby, draft memo, and draft press release, PAC/10.

40. Califano to Moyers, 11/21/64, memo, PAC/10.

41. Gordon to McNamara, 11/28/64, memo, and attached Gordon to the files, "Presidential Letter to Halaby on SST Program," 11/28/64, memo, PAC/10.

42. For the evaluation results, see: "Summary of the Evaluation," circa 12/8/64, LBJ FAA Hist. Doc.; Lawrence E. Levinson to Califano, "Al Flax and the SST," 12/3/64,

memo, PAC/10; "SST Program since May 14, 1964," 1/18/65, PAC/10; Halaby to Bain, 12/12/64, handwritten note, FAA 6174/97.

43. Bain to Halaby, "Supersonic Transport—Program Recommendations," 12/19/64, memo, FAA 6174/97. For Bain's view of Lockheed's submission, see also: Bain to McKee, "Summary of Lockheed SST History," 7/15/64, memo, LBJ FAA Hist. Doc.

44. "Minutes of the Meeting: Lockheed Aircraft Corporation/Federal Aviation Agency—Discussion of Phase II-A Evaluation Results," 12/18/64, FAA 1647/41.

45. Halaby, "Memorandum of Conversation with Secretary of Defense McNamara," 12/21/64, draft memo, LBJ FAA Hist. Doc. For McNamara's view on including Lockheed and Pratt & Whitney in the SST competition, see: "SST Program since May 14, 1964," 1/18/65, PAC/10; *New York Times*, 12/23/64, p. 1; "Boeing's SST Given Aerodynamic Edge," *Aviation Week and Space Technology*, 12/21/64, p. 29; Halaby, 12/23/64, handwritten note, FAA 6174/96.

46. Halaby to the President, "Supersonic Transport Program," 12/23/64, memo, FAA 6174/11.

47. "SST Program since May 14, 1964," 1/18/65, PAC/10.

48. Halaby, "Memorandum of Conversation with Secretary of Defense McNamara," 1/5/65, draft memo, LBJ FAA Hist. Doc.

49. Califano to Bill Moyers, 1/13/65, memo, and attached President to Halaby, "Supersonic Transport Program," 1/13/65, draft memo, FAA 6174/11; President to Halaby, 1/21/65, memo, FAA 6174/11.

50. Alan L. Dean to Halaby, "New Look at SST Program," 12/2/64, FAA 6174/11; Dean and Packard Wolle to Halaby, "Proceeding with the SST Development Program," 12/2/64, memo, FAA 6174/11.

51. For Bain's complaints, see: Bain to Halaby, "Supersonic Transport Development Program," 12/10/64, memo, FAA 6174/97; Halaby to Bain, "Black-Osborne letter of 11/16/64," 12/7/64, handwritten memo, FAA 6174/97; Bain to Halaby, "Black and Osborne Letter Dated November 16, [1964] to Secretary McNamara," 12/10/64, memo, FAA 6174/96.

52. Bain to Halaby, "Lockheed Letter—Memorandum of Conversation with Secretary McNamara," 1/6/65, memo, LBJ FAA Hist. Doc.

53. *New York Times*, 3/7/65, p. 6.

54. Califano to Halaby, 2/25/65, FAA 6174/95.

55. *New York Times*, 3/7/65, p. 6.

56. Halaby to Bain, Dean, and Mackey, 2/10/65, memo, FAA 6174/95.

57. Katz to Bain, 2/3/65, FAA 6174/92.

58. Halaby to Bain, 3/5/65, handwritten note, FAA 6174/95; U.S. Department of Commerce, *SST: An Economic Analysis Part I—Executive Summary (Preliminary)*, 3/9/65, NTIS no. 655603; DOD, "Memorandum for the President's Advisory Committee on Supersonic Transport," 3/24/65, PAC/10.

59. Halaby to the President, 3/16/65, FAA 6174/95; Halaby to McNamara, 3/16/65, FAA 6174/95. See also: Bain to Halaby, "Comments on Draft 3/16/65 of Letter to Secretary McNamara," 3/16/65, memo, FAA 6174/96.

60. DOD, "Memorandum for the President's Advisory Committee on Supersonic Transport," 3/24/65, PAC/10; Califano to C. Douglas Dillion et al., 3/29/65, PAC/10.

61. Daniel J. Edwards, through J. Stockfisch, to the Secretary of Treasury, "Supersonic Transport Study of the Department of Commerce," "FAA's Brown Book Dated March 16, 1965," and "DOD's Memorandum for the President's Advisory Committee on Supersonic Transport, dated March 24, 1965," 3/26/65, memos, PAC/11.

62. "Draft Proceedings of the President's Advisory Committee on Supersonic Transport," 3/30/65, pp. 40–41, PAC/32.

63. Bain, "Comment—Department of Defense Memorandum for the President's Advisory Committee on Supersonic Transport, Dated March 24, 1965," 4/5/65, draft memo, FAA 6174/95; Halaby to McNamara, 4/23/65, PAC/16; FAA to the PAC, "Comment—Department of Defense Memorandum for the President's Advisory Committee on Supersonic Transport, Dated March 24, 1965," 4/23/65, memo, PAC/16.

64. "Draft Proceedings of the President's Advisory Committee," 3/30/65; personal interviews; Halaby to McNamara, 4/16/65, PAC/10; Halaby to McNamara, 4/23/65, LBJ FAA Hist. Doc.

65. For pressures to keep Lockheed and Pratt & Whitney in the SST design competition, see: H. M. Horner to Halaby, 3/15/65, LBJ FAA Hist. Doc.; H. M. Horner to McNamara, 4/2/65, FAA 6174/95; McCone to the record, "Discussion with Mr. Horner, Mr. Gorton and Mr. Jordan of Pratt & Whitney Concerning SST Engine Development and Problems, 1 April 1965," 4/8/65, memo, PAC/11; McCone to the record, "Discussion with Courtlandt Gross, Dr. Hall L. Hibbard, Director and Senior Corporation Advisor, Mr. Robert A. Bailey, Project Manager of the SST Program and Mr. Kelly Johnson, concerning SST Project at Lockheed," 4/7/65, memo, PAC/11; McCone to the record, "Discussion with . . . Pennell and Mr. Lloyd Goodmanson, Who Are Directing SST Project at Boeing," 4/7/65, memo, PAC/11; James F. Alexander to the record, "SST Engine Proposed by the General Electric Company," 5/5/65, memo, PAC/11.

66. U.S. Department of Commerce, *SST: An Economic Analysis—Part I, Executive Summary, Preliminary, Supplement I,* 4/30/65, NTIS no. 655604; U.S. Department of Commerce, *SST: An Economic Analysis—Part I, Executive Summary, Preliminary, Supplement II,* circa 5/4/65; John T. Connor to Bill Moyers, 5/4/65, LBJ CF CA.

67. Enke to Charles J. Hitch, Califano, and Alain C. Enthoven, 4/9/65, memo, and attached "Major Strategy Issues and Overall Strategy Affecting the SST Program," PAC/41; Enke, "The SST Venture—Issues and Questions," 4/12/65, SST Task Force Internal Paper no. 1, PAC/41.

68. Arnold Moore, "Profitability of Subsonic and Supersonic Aircraft—Contrast and Conundrum," 4/30/65, SST 1965 Task Force Internal Paper no. 14, PAC/41; Arnold Moore, "Over-Investments in SSTs," 5/4/65, SST 1965 Task Force Internal Paper no. 15, PAC/41.

69. Daniel James Edwards, "Some Aspects of SST Development," 4/29/65, SST 1965 Task Force Internal Paper no. 9, PAC/41.

70. For the Enke group's criticism of Commerce's SST economic studies, see: Enke, "Reasons Why the Cost-Benefit Model of Commerce Needs Considerable Amendment," 4/26/65, SST 1965 Task Force Internal Paper no. 8, PAC/41; Enke, "Inadequacies of Commerce Cost-Benefit Analysis of U.S. SSTs," 4/30/65, SST 1965 Task

Force Internal Paper no. 12, PAC/41; Enke, "Evolving a Supply-Demand Model for SST," 4/27/65, SST 1965 Task Force Internal Paper no. 7, PAC/41.

71. Enke, "The Profitability and Place of Supersonics in Air Transport (An Interim Statement)," 4/29/65, revised 5/6/65, SST 1965 Task Force Internal Paper no. 13, PAC/41; Enke, "Economic Studies Needed," 5/17/65, SST 1965 Task Force Internal Paper no. 19, PAC/41.

72. Halaby to McNamara, 4/16/65, PAC/10.

73. "Memorandum for the President, Second Interim Report of the President's Advisory Committee on Supersonic Transport," 5/8/65, PAC/32. For McNamara's views and the PAC's decision on government-industry cost sharing in the SST program, see: Califano to Halaby, 4/16/65, FAA 6174/96; "Draft Proceedings of the President's Advisory Committee on Supersonic Transport," 5/5/65, PAC/11; PAC, "Executive Session," PAC/11.

74. For Halaby's decline of power and resignation, see: DOD, "Memorandum for the President's Advisory Committee on Supersonic Transport," 5/1/65, PAC/11; Halaby to McNamara, circa 4/27/65, handwritten note, FAA 6174/95; White House "Contact Cards" for Halaby, LBJ.

Chapter 8

1. Interview of General William F. McKee by Dorothy Pierce, tape no. 1, 10/28/65, LBJ Oral History; White House "Contact Cards" for McKee, LBJ; personal interviews.

2. Califano to Fowler et al., n.d., and attached "Suggested Issues for Consideration at the May 21, 1965 Meeting of the President's Advisory Committee on Supersonic Transport," PAC/11.

3. Defense Department SST Economics Task Force, "Summary of Findings to Date," 5/17/65, PAC/11.

4. J. A. Stockfisch to Stone, "Request for Comments on Airline Expensing of Supersonic Transport Research & Development," 5/11/65, memo, PAC/11; Stockfisch to Fowler, "Materials for Tomorrow's SST Advisory Committee Meeting," 5/20/65, memo, and attached "Some General Questions Facing the Committee," 5/20/65, and "Comments on the 'List of Issues' for the May 21st Meeting," PAC/11.

5. Califano to McNamara, 5/19/65, memo, PAC/11; "Suggested Issues for Consideration at the May 21, 1965 Meeting of the President's Advisory Committee on Supersonic Transport," n.d., and handwritten notes on this paper, PAC/11.

6. "Draft Proceedings of the President's Advisory Committee on Supersonic Transport," 5/21/65, PAC/32. For the PAC decisions made at the 5/21/65 PAC meeting, see also: Califano to McCone, 5/26/65, and attached "Schedule of Work Agreed to at the May 21, 1965 Meeting of the President's Advisory Committee on Supersonic Transport," PAC/12; Califano to Black, 5/24/65, PAC/12; Califano to Halaby, 5/24/65, PAC/12; Califano to Hornig, 5/24/65, PAC/12; Califano to General Carroll, 5/26/65, memo, PAC/12.

7. For the conflict over McKee's confirmation, see: James R. Ashlock, "McKee Issue May Swell SST Opposition," *Aviation Week and Space Technology*, 6/28/65, p. 31; Bain to Halaby, "July 1965 Programming for Supersonic Transport Development," 6/14/65, memo, LBJ FAA Hist. Doc.; "McKee Confirmation Follows Quiet Hearing," *Aviation Week and Space Technology*, 7/5/65, p. 25.

8. "Remarks of the President at the Swearing-In Ceremony of General McKee," White House release, 7/1/65, FAA 6174/95.

9. Halaby to the President, 7/1/65, LBJ EX Misc. Files; Halaby to the President, 7/1/65, memo, LBJ, EX Misc.

10. The President to Halaby, 7/9/76, LBJ EX Misc.

11. For the beginning of the extended design competition (phase II-C) and for industry complaints about the cost-sharing provisions of the SST contracts, see: Bain to McKee, "Supersonic Transport Development Program," 7/1/65, memo, LBJ FAA Hist. Doc.; Bain to Leonard C. Mallet, 7/1/65, FAA 6174/95; Bain to McKee, various "Daily Highlight Reports" (nos. 1, 3, 5, 6, 7, 9, 11–15, 19–25, 27) beginning 7/6/65 and ending 7/26/65, memos, FAA 6174/95; Bain to McKee, "Exhibition of Supersonic Transport Development Hardware," 7/6/65, memo, FAA 6174/95; Bain to McKee, "Daily Highlight Report No. 6," 7/13/65, memo, and attached "Reasons for Twenty-Five Percent Financial Participation by Contractors," FAA 6174/95; "Companies Balk at SST Contract Clauses," *Aviation Week and Space Technology*, 7/12/65, p. 25; Bain to McKee, "Summary of Lockheed SST History," 7/15/65, memo, FAA 6174/95; Allen to Bain 7/23/65, FAA 6174/95; Bain to Executive Secretary, "Report of FAA Activities for the White House," 7/26/65, memo, FAA 6174/95; "Industry-Government Talks Continue," *Aviation Week and Space Technology*, 7/26/65, p. 30; Bain to McKee, "Lockheed's Financial Proposal," 7/27/65, memo, FAA 6174/95; "SST Bidders Given 2 Concessions as Negotiations Approach Deadline," 8/2/65, *Aviation Week and Space Technology*, p. 37; Bain to McKee, "Execution of SST Phase II-C Contract with Lockheed Aircraft Corporation," 8/10/65, memo, FAA 6174/95; Bain to Steadman, 8/11/65, FAA 6174/96; Bain to McKee, "Execution of SST Phase II-C Contract with the Boeing Company," 8/12/65, FAA 6174/95; Allen to McKee, 8/13/65, FAA 6174/95; *New York Times*, 8/14/65, p. 46; Davis to Executive Secretary, "Report of FAA Activities for the White House," 8/24/65, memo, FAA 6174/95; personal interviews.

12. For the decline of Bain's role in the SST effort, see: Bain to McKee, "Status of SST Office," 8/26/65, memo, FAA 6174/95; H. M. Horner to McKee, 7/21/65, FAA 6174/95; "Departure of Bain Spurs Concern over Future FAA Direction of SST," *Aviation Week and Space Technology*, 8/30/65, p. 36; personal interviews.

13. For the formalization and upgrading of the SST Office under Maxwell, see: Maxwell to McKee, "Delegation of Contracting Authority for the SST Program," 9/24/65, memo, FAA 6174/96; Director of Audit to Dean, "Proposed Delegation of Contracting Authority to the Director, Supersonic Transport Development," 9/27/65, memo, FAA 6174/14; Dean to Executive Secretary, "Delegation of Contracting Authority for the SST Program," 9/27/65, memo, FAA 6174/14; McKee to Maxwell, "Delegation of Authority," 9/30/65, memo, FAA 6174/92; Maxwell to McKee, "Delegation of Contracting Authority for the SST Program," 10/12/65, memo, FAA 6174/96; Director of Management Services to Dean, "SST Delegations of Contracting Authority," 10/16/65, memo, FAA 6174/96; Maxwell to SST Contracts Manager, "Delegation of Authority," 10/20/65, memo, FAA 6174/96; Maxwell to McKee, "Delegation of Contracting Authority to Aeronautical Center for the Supersonic Transport Program," 10/22/65, memo, FAA 6174/96.

14. For the Enke group's continual criticism of FAA economic analyses and of the Commerce Department's SST economic studies, for the FAA and Commerce counterattacks, and for the final decision by McNamara to distribute simultaneously all SST economic studies to the manufacturers, see: Enke to Hollomon, "Continuing Economic Analysis,"

6/10/65, memo, PAC/12; Connor to McNamara, 6/8/65, PAC/12; Bain to McKee, "DOD Presentation—Commerce/FAA Supersonic Transport Economic Viability," 7/2/65, memo, FAA 6174/95; Enke to Hitch, Flax, Califano, Enthoven, and Levinson, "Transmittal of Draft Report of Chairman's SST Economics Task Force," 7/14/65, memo, PAC/41; Connor to McNamara, 7/21/65, LBJ Oversize; Enke to Steadman, "Reply by McNamara to Connor's Letter of July 21," 7/29/65, memo, PAC/12; Steadman to McNamara, "Economic Studies of SST," 7/29/65, memo, PAC/12; Enke to Steadman, 7/30/65, memo, PAC/41; Enke to Enthoven, "The SST Study," 7/30/65, memo, PAC/41; McNamara to Connor, 7/31/65, PAC/12; Enke to Steadman, 7/30/65, memo, PAC/41.

15. Enke to Enthoven, "The SST Study," 7/30/65, memo, PAC/41; Enke to Steadman, 7/30/65, memo, PAC/41.

16. SST Economics Task Force, "The Economics of a U.S. Supersonic Transport (Report on Findings to Date)," 8/20/65, PAC/25.

17. Enthoven to Steadman, 8/21/65, memo, PAC/41; Walgreen and Mayo to Enthoven, "The Economics of a U.S. SST," 8/25/65, memo, PAC/41.

18. Katz et al. to Boyd, "Comments on Report, 'The Economics of a U.S. Supersonic Transport,' by an O.S.D. Task Force, Date 8/20/65," 9/20/65, memo, PAC/41.

19. Ralph E. Parsons to SS-1, "OSD Study—SST Economics," 9/13/65, memo, FAA 1647/3.

20. For the public support of the SST program by the airlines, see: Bollech to SS-3, "Review of Industry Comments on SST Economic Studies Prepared by Department of Commerce and OSD," 9/23/65, memo, FAA 1647/3; James R. Ashlock, "SST Contracts Complicate Airlines Plans," *Aviation Week and Space Technology*, 7/12/65, pp. 26–27; Erwin J. Bulban, "Possible Market for 500 SST's Forecast," *Aviation Week and Space Technology*, 5/24/65, pp. 27–28; "Industry-Government SST Talks Continue," *Aviation Week and Space Technology*, 7/26/65, p. 30; C. R. Smith to Bill Littlewood, 7/13/65, LBJ EX CA; Smith to McKee, 7/20/65, LBJ FAA Hist. Doc.; McKee to Bain, 7/22/65, handwritten note, LBJ FAA Hist. Doc. For the airline comments on the SST economic studies by the Commerce and the Defense departments, see: William J. Hogan to M. R. Monahan, 9/20/65, PAC/41; C. R. Smith to Boyd, 9/23/65, PAC/41, Tillinghast to Boyd, 9/20/65, PAC/41; Chabon to Steadman, 12/17/65, handwritten note, and attached "Pan Am Comments," 10/6/65, PAC/41; S. B. Kauffman to M. R. Monahan, 9/17/65, PAC/41; W. A. Patterson to Boyd, 9/20/65, PAC/41.

21. For optimistic public statements by Boeing and Lockheed, see: Erwin J. Bulban, "Possible Market for 500 SST's Forecast," *Aviation Week and Space Technology*, 5/24/65, pp. 27–28. For the airframe manufacturers' comments on the SST economic studies by the Commerce and Defense departments, see: Courtlandt S. Gross to M. R. Monahan, 9/17/65, PAC/41; Edward L. Wells to Boyd, 9/15/65, PAC/41. For the engine manufacturers' comments on the SST economic studies by the Commerce and Defense departments, see: J. S. Parker to Boyd, 9/27/65, PAC/41; L. C. Mallet to Monahan, 9/17/65, PAC/41.

22. For the new, calmer atmosphere in the SST economic debate, see: James R. Ashlock, "Johnson Advisors Split on SST Timing," *Aviation Week and Space Technology*, 10/11/65, p. 36; McKee to Califano, 7/21/65, handwritten note, LBJ Oversize; T. V. Bollech to Special Assistant to the Director, "Review of Industry Comments on SST Economic Studies Prepared by Department of Commerce and OSD," 9/23/65, FAA 1647/3.

23. Enke to Steadman, 10/7/65, note, and attached "Organizing Economic Studies," PAC/28.

24. Maxwell to McKee, "Meeting with Secretary McNamara—October 8, 1965," 10/8/65, memo, LBJ FAA Hist. Doc.

25. "Draft Proceedings of the President's Advisory Committee on Supersonic Transport," 10/9/65, PAC/32.

26. For the beginning of SST financing work by Black, Osborne, and McCone, see: Califano to McNamara, "SST: Black/Osborne/McCone Working Group," 6/1/65, memo, PAC/12; Califano to Halaby, 6/10/65, FAA 6174/95.

27. For the status of and the FAA's position on airline SST delivery positions and further airline financing, see: Bain to McKee, "Daily Highlight Report No. 8," 7/15/65, memo, FAA 6174/95; McKee to McNamara, 8/13/65, FAA 6174/92, McKee to McNamara, 9/1/65, FAA 6174/96; McKee to Steadman, 9/3/65, FAA 6174/96; Bain to Steadman, 9/3/65, FAA 6174/96; "Draft Proceedings of the President's Advisory Committee on Supersonic Transport," 10/9/65, PAC/32.

28. For the airline positions on further airline SST financing, see: Tillinghast to McKee, 8/23/65, LBJ FAA Hist. Doc.; C. R. Smith to McKee, 8/19/65, and attached "Comments—Program for Financing the SST," 8/18/65, FAA 6174/92; W. A. Patterson to McKee, 8/23/65, 6174/92; Curtis Barkes to W. A. Patterson, "Supersonic Transport Financial Plan," 10/2/65, memo, FAA 6174/92.

29. For the manufacturers' positions on further airline financing, see: D. J. Haughton to McKee, 5/19/65, FAA 6174/92; Lockheed Aircraft Corp. to Bain, 8/23/65, 6174/92; comments of GE on FAA financial plan, circa 8/65, FAA 6174/92; Allen to McKee, 8/23/65, FAA 6174/92; H. M. Horner to McKee, 8/19/65, FAA 6174/92; McKee to McNamara, 9/1/65, FAA 6174/92.

30. Maxwell to McKee, "Meeting with Secretary McNamara—October 8, 1965."

31. For the PAC discussion at the 10/9/65 meeting and for the PAC decisions made at this meeting, see: Steadman to McKee, 10/16/65, draft letter, FAA 6174/96; "Draft Proceedings of the President's Advisory Committee on Supersonic Transport," 10/9/65, PAC/32. For earlier recommendations on airline SST financing by the Black-Osborne-McCone working group, see: McCone to Steadman, 8/27/65, FAA 6174/96; Steadman to McCone, 8/31/65, FAA 6174/96. For the private discussion with McNamara, see: Maxwell to McKee, "Meeting with Secretary McNamara—October 8, 1965."

32. For the history of the debate and passage of fiscal 1966 SST supplementary funding a number of documents were reviewed. Among the most useful were: Henry Rowen to Califano, "SST Appropriation Estimate," 7/8/65, memo, and attached "Supplemental Estimate of Appropriation, Fiscal Year 1966: Civil Supersonic Aircraft Development," PAC/12; Dean to McKee, "SST Supplemental Appropriations Request and Related Matters," 7/27/65, LBJ FAA Hist. Doc.; Dean to McKee, "Notification of Appropriations Committees Regarding Use of SST Contingency Reserve Funds," 8/16/65, memo, FAA 6174/14; Edward W. Stimpson to the record, "Discussions Regarding Use of Contingency Reserves Pending Congressional Action," 8/31/65, memo, LBJ FAA Hist. Doc.; McKee to Lee C. White, "Funding of SST Contracts," 8/31/65, LBJ FAA Hist. Doc.; McKee to Califano, 8/19/65, memo, LBJ Oversize; Monroney to Connor, 10/21/65, and attached, James R. Ashlock, "Johnson Advisors Split on SST Financing," *Aviation Week and Space Technology*, 10/11/65, p. 36, FAA 6174/96; Stimpson to McKee, "Sta-

tus of SST Appropriations," 10/8/65, memo, LBJ FAA Hist. Doc.; "McKee Calls SST Next Step in Transport," *Aviation Week and Space Technology*, 10/18/65, pp. 28–29; U.S. Senate, *Supplemental Appropriations Bill, 1966*, Report No. 912, 89th Cong., 1st sess., 10/19/65, FAA 1647/2; Migdon Segal, *The Supersonic Transport—A Legislative History*, no. 71-137 SP (Washington, D.C.: Congressional Research Service, 5/26/71).

33. "Draft Proceedings of the President's Advisory Committee on Supersonic Transport," 11/6/65, PAC/32; "Memorandum for the President: Third Interim Report of the President's Advisory Committee on Supersonic Transport," 11/15/65, PAC/13.

Chapter 9

1. For Maxwell's systematic preparation, see: Maxwell to FAA Deputy Administrator, "Major Objectives for 1966," 1/10/66, memo, FAA 6174/95. The complexity of the evaluation was apparent to all in early 1966. See: Enke to Maxwell, 4/21/66, and attached Enke, "Major SST Program Events During 3/30/66 to 1/1/67," 4/20/66, FAA 6174/96.

2. For selection of Edmund Learned as FAA consultant, see: Department of State to Beirut et al., "Civil Aviation: Information on U.S. Supersonic Transport Program," DOS airgram, 4/20/66, FAA 1647/6.

3. McKee to Secretaries of Defense, Treasury, and Commerce, Director, Bureau of the Budget, and Chairman, Civil Aeronautics Board, "Establishment of the SST Inter-Agency Economics Advisory Group," 12/10/65, memo, PAC/28.

4. For the elaborate CAB proposal, see: Murphy to McKee, 12/3/65, FAA 6174/92; Irving Roth to Skaggs, 12/13/65, and attached "CAB Program for Supersonic Transport Economic Analysis," 12/10/65, FAA 6174/92.

5. For the FAA's response to the CAB's proposal and for the ultimate CAB role in the economic studies, see: Skaggs to Maxwell, "Proposed CAB Economic Research Support," 12/15/65, memo, FAA 6174/92; Skaggs to Maxwell, "Possible CAB Economic Research Support," 12/21/65, memo, FAA 6174/92; Murphy to McKee, 12/16/65, FAA 6174/92; Maxwell to McKee, "CAB Support on Economic Studies," 12/22/65, memo, FAA 6174/92; Dean to Maxwell, "Financing Proposed CAB Program for Economic Studies Relating to the SST," 12/16/65, memo, FAA 6174/92; Allen H. Skaggs, "A Plan for Economic Research Studies: Supersonic Transport Development," 1/12/66, FAA 6174/92; "Draft Proceedings of the President's Advisory Committee on Supersonic Transport," 5/6/66, PAC/32; "Agenda Item IV: SST Economic Studies" (from "FAA Back-up Material" for PAC meeting of 5/6/66), FAA 6174/96.

6. For the decline of the era of good feeling between Enke and the FAA under McKee and Maxwell and for the increasing criticism by Enke, see: Steadman to McKee, 12/15/65, PAC/28; McKee to Steadman, 12/22/65, PAC/28; Maxwell to Enke, 10/29/65, and attached FAA briefing document on SST financing, FAA 6174/96; Enke to McKee, 11/1/65, PAC/28; Enke to Steadman, 1/28/66, memo, PAC/28.

7. Enke to members of the SST Interagency Economics Advisory Group, 2/7/66, memo, PAC/28.

8. Maxwell to Skaggs and members of the Interagency Economics Advisory Group, 2/9/66, memo, PAC/28.

9. Enke to McNamara, "FAA Resistance to Economic Studies Recommended by PAC-

SST," 2/18/66, memo, PAC/28; Enke to Steadman, 2/21/66, memo, PAC/28; McNamara to McKee, 2/24/66, LBJ FAA Hist. Doc.

10. McKee to Maxwell, 2/25/66, handwritten note, LBJ FAA Hist. Doc.

11. For continued complaints to the FAA by Enke, see: Enke to Skaggs, 3/9/66, 3/18/66, PAC/28; Enke to Maxwell, 3/18/66, PAC/28; Enke to Skaggs, "Economic Studies: Contents, Contractors, and Schedules," 3/25/66, memo, PAC/28.

12. Enke to Skaggs, 4/18/66, and attached "Comments on Draft Description of Economic Studies," PAC/28; Enke to Maxwell, 4/27/66, PAC/28.

13. For Enke's demand for passenger surveys, see: Enke to Skaggs, 3/25/66, PAC/28; Enke to Maxwell, 4/8/66, FAA 6174/92.

14. R. W. Rummel to Skaggs, 4/15/66, FAA 6174/92.

15. Maxwell to Enke, 4/20/66, PAC/28.

16. Enke to Maxwell, 4/25/66, PAC/28.

17. FAA, "Status Report: Supersonic Transport Economic Research Program for Presidential Advisory Committee," 6/66, PAC/14.

18. Enke, "Comment on Major SST Program Events" (briefing document for McNamara prior to PAC meeting of 7/9/66), PAC/14.

19. "Draft Proceedings of President's Advisory Committee on Supersonic Transport," 7/9/66, pp. 30–43, PAC/32.

20. Steadman to McNamara, 8/23/66, memo, and attached Enke, "Proposed Strategy for PAC between Now and End of CY 1966," PAC/13.

21. Skaggs to Enke, 8/31/66, FAA 1647/7.

22. For the disagreements between Enke and the FAA, see: Steadman to Maxwell, 8/23/66, FAA 6174/95; Maxwell to Steadman, 9/10/66, FAA 6174/95; Steadman to Maxwell, 9/8/66, FAA 1647/7; Walgreen to Skaggs, 9/8/66, 9/21/66, FAA 1647/7; Steadman to McNamara, 9/26/66, memo, PAC/14.

23. Maxwell to Special Assistant for Supersonic Transport Development, "Comments on Dr. Enke's and Dr. Walgreen's Proposed SST Paper," 11/2/66, memo, FAA 1647/8; Chief, Engineering Division, to Chief, Analysis and Control Division, "Review of Dr. Enke and Walgreen's Paper," 11/2/66, memo, FAA 1647/8.

24. For IDA's findings, see: Norman J. Asher et al., *Demand Analysis for Air Travel by Supersonic Transport*, NTIS no. AD652309 (Arlington, Va.: Institute for Defense Analyses, December 1966), vol. 1; FAA briefing materials for the PAC meeting of 10/6/66, FAA 6174/95; "Draft Proceedings of the President's Advisory Committee on Supersonic Transport," 10/6/66, pp. 10–11, PAC/32.

25. Skaggs to IEAG, "Minutes of Sixth Meeting of the IEAG—September 19, 1966," 9/23/66, memo, FAA 1647/7; Skaggs to Niskanen, 9/26/66, FAA 1647/7.

26. R. V. Radcliffe to Skaggs, 11/1/66, FAA 1647/8; Skaggs to D. J. Lloyd-Jones, 10/22/66, FAA 1647/7.

27. Skaggs to Learned, 10/15/66, FAA 1647/7; Learned to Skaggs, 10/21/66, FAA 1647/7.

28. Maxwell to Steadman, 11/4/66, FAA 1647/8.

29. Skaggs to the PAC, "Preliminary Economic Feasibility Study," 11/7/66, memo, FAA 1647/8.

30. Skaggs to Learned, 11/7/66, FAA 1647/8.

31. For Learned's briefings and his views, see: Learned to McKee and McNamara, 11/14/66, FAA 6174/95; Learned to Skaggs, 11/15/66, FAA 1647/8.

32. Learned to McKee, 11/22/66, FAA 6174/95.

33. Maxwell to Walgreen, 11/9/66, FAA 1647/8; Walgreen to Skaggs, 11/9/66, PAC/15; Steadman to McNamara, 11/16/66, memo, PAC/15; Captain Chabon to Steadman, 11/16/66, memo, PAC/15; Walgreen and Rastatter to Steadman, "Pre-PAC Meeting Economic Items," 11/15/66, memo, PAC/15; Walgreen and Rastatter to Steadman, "Pre-PAC Meeting Economic Items," 11/16/66, memo, PAC/15.

34. Edwards to Skaggs, "Comments on FAA's 'Summary Economic Feasibility Report of the United States Supersonic Transport' Dated November 4, 1966," 11/9/66, memo, PAC/15.

35. FAA, "Preliminary Summary Economic Feasibility Report: United States Supersonic Transport," 11/66, PAC/4.

36. Walter B. Boehner to Skaggs, 11/9/66, PAC/15.

37. "Draft Proceedings of the President's Advisory Committee on Supersonic Transport," 11/17/66, PAC/32.

38. Maxwell to Steadman, 12/1/66, PAC/4; Skaggs to Learned, 12/2/66, FAA 1647/8.

39. FAA, "Summary Economic Feasibility Report: United States Supersonic Transport," 12/66, PAC/15.

40. Skaggs to SS-20, "Memo for the Record—Appointment with DOD Consultants," 12/5/66, memo, FAA 1647/8.

41. "Comments on FAA Summary Economic Feasibility Report," 12/66, PAC/5.

42. For the Bureau of the Budget's review of the FAA report, see: Skaggs to Learned, 12/2/66, FAA 1647/8; Steadman to McKee, 12/6/66, and attached Bureau of the Budget, "Evaluation of Estimates Presented in FAA's Economic Evaluation of the SST," FAA 6174/96; Bureau of the Budget, "SST Economic Viability: Summary and Conclusions," 12/9/66, FAA 6174/96; "Draft Proceedings of the President's Advisory Committee on Supersonic Transport," 12/7/66, p. 78, PAC/32.

43. "Draft Proceedings of the President's Advisory Committee on Supersonic Transport," 12/10/66, pp. 22–28, PAC/32.

44. FAA, "Summary Economic Feasibility Report: United States Supersonic Transport," 12/66, PAC/15.

45. "Draft Proceedings of the President's Advisory Committee on Supersonic Transport," 12/7/66, pp. 40–48, PAC/32; "Draft Proceedings of the President's Advisory Committee on Supersonic Transport," 12/10/66, pp. 2–20, PAC/32.

46. For Lockheed's criticism of IDA's findings, see: William P. Kennedy to Skaggs, 11/23/66, FAA 1647/8; McCone to McKee, 1/5/67, FAA 1647/8; Lindon U. Cockroft to Skaggs, "Letter to General McKee from Mr. J. A. McCone covering letter from Mr. A. Carl Kotchian, Executive V.P. of Lockheed-California Corp. (12/30/66) Which Critiques the Institute for Defense Analyses Work on SST Balance of Payments Considerations," 1/12/67, memo, FAA 1647/8.

47. O. E. Melby to Skaggs, 1/25/67, telex, FAA 1647/8.

48. Charles W. Trippe to Skaggs, 2/15/67, FAA 1647/9.

49. William A. Niskanen to Skaggs, 12/19/66, FAA 1647/8; Norman L. Christeller to J. W. Locraft, 12/29/66, FAA 1647/5.

50. For the FAA's selection of outside economic reviewers and for its delight with the results, see: Skaggs to Dr. William F. Pounds, 1/9/67, FAA 1647/8; Pounds to Skaggs, 12/24/67, FAA 1647/8; Skaggs to C. E. Jackson, 1/26/67, FAA 1647/8; Skaggs to Walter Lederer, 1/13/67, FAA 1647/8; Ira Dye to Lederer, 1/23/67, FAA 6174/3; Skaggs to Gerhard Colm, 2/1/67, FAA 1647/9; Skaggs to Jackson, 1/30/67, FAA 1647/8; Skaggs to the record, "Balance of Payments Review," 1/12/67, memo, FAA 1647/8; Maxwell to McKee, "Balance of Payments Paper," 1/25/67, memo, and attached "Balance of Payments," FAA 6174/92; Skaggs to Learned, 2/14/67, FAA 1647/9.

51. FAA, "Economic Feasibility Report: United States Supersonic Transport," 4/67, chap. 8, FAA 6174/92.

52. Osborne to McKee, 2/9/67, FAA 1647/9.

53. FAA, "Economic Feasibility Report: United States Supersonic Transport," 4/67, FAA 6174/92.

54. FAA, "[Draft] Financial Plan" and "[Draft] Guidelines for Preparation of a Financial Plan," 11/23/65, FAA 6174/96.

55. Enke to Maxwell, Financial Committee, "SST Financial Plan," 12/6/65, memo, and attached "SST Financing Plan," FAA 6174/96. For retrospective summaries by pooling advocates of the SST financing schemes, see: Stephen Enke, "Government-Industry Development of a Commercial Supersonic Transport," *American Economic Review*, 57 (May, 1967): 71–75; J. A. Stockfisch and D. J. Edwards, "The Blending of Public and Private Enterprise: The SST as a Case in Point," *Public Interest*, Winter, 1969, pp. 108–117; John A. Walgreen, E. H. Rastatter, and Arnold B. Moore, "Economics of Supersonic Transport," *Journal of Transport Economics and Policy*, 7 (May, 1973), pp. 1–8.

56. For the FAA's financial plan, see: Maxwell, "Finalization of Phase III Finance Plan," circa 1/66, FAA 6174/96; Maxwell to McKee, "SST Finance Plan for PAC," 1/15/66, memo, FAA 6174/96; FAA, "SST Financial Plan," 1/21/66, PAC/13; "Cost-Sharing Options," circa 1–2/66, FAA 6174/96; Maxwell to Financial Advisory Committee members, "SST Financial Plan," 1/26/66, memo, FAA 6174/96. For the PAC recommendations about SST financing, see: "Memorandum for the President: Third Interim Report of the President's Advisory Committee on Supersonic Transport," 11/15/65, PAC/13.

57. McKee to McNamara, 1/25/66, PAC/13.

58. Enke to Maxwell, 1/21/66, FAA 6174/92; Enke to McNamara, "General McKee's Phase III Financial Plan," 1/28/66, memo, PAC/13.

59. Osborne to McKee, 1/24/66, and attached McCone, "FAA-SST Financial Plan—1/17/66," 1/19/66, FAA 6174/96.

60. FAA, "Summary Status Report to the President's Advisory Committee on Supersonic Transport," 2/15/66, p. 13, PAC/13; FAA, "SST Phase III Financial Plan for Prototype Construction and Test," 2/14/66, PAC/13.

61. Parsons to SS-2 et al., "Contractor Comments—Financial Plan and Major Issues," 3/2/66, memo, FAA 1647/2; Parsons to Edwards, 3/2/66, FAA 1647/2.

62. "Public-Private Enterprise," *Aviation Week and Space Technology*, 12/6/65, p. 2.

63. McCone to McNamara, 2/16/66, PAC/13.

64. Enke to McNamara, "Important SST Matters Requiring Your Attention," 2/25/66, memo, PAC/36.

65. Enke and Walgreen, "Financial Plan: Cost Sharing and Recoupment Formula," 3/3/66, PAC/13; "Financial Plan and Proposed Recoupment Formula beyond Phase III," circa 3/3/66, PAC/36; Steadman to McNamara 3/3/66, memo, PAC/36.

66. Maxwell to SST Division Chiefs, Contracts and Economics, "Meeting of President's Advisory Committee for Supersonic Transport," 2/26/66, memo, FAA 6174/96; Steadman to McCone, 3/1/66, PAC/13.

67. Maxwell to the record, "Meeting of President's Advisory Committee for Supersonic Transport Members," 3/10/66, memo, FAA 6174/96; McNamara to McKee, 3/12/66, and attached "Agreements Reached at Informal March 9 Meeting of Several Members of the PAC-SST," PAC/13; McNamara to Schultze, 3/12/66, PAC/13.

68. Dean to files, "Discussion with Prestemon on SST Developments," 3/15/66, memo, FAA 6174/14.

69. For continued criticisms by Enke immediately before the May 6 PAC meeting, see: Enke to McNamara, "PAC/SST May 6 Meeting," 5/3/66, memo, and attached "check list," PAC/4; Enke to McNamara, "Draft Letter to PAC-SST Members and Attendees," 4/25/66, memo, and attached letter, PAC/4; Enke to McNamara, 4/29/66, memo, PAC/4; DOD, "Comments on 'Financial Issues,' " n.d. (for PAC 5/6/66 meeting), PAC/4; DOD, "Candidate Agenda Items," n.d. (for PAC 5/6/66 meeting), PAC/4; DOD, "Proposed Agenda—May 6 Meeting of the President's Advisory Committee on Supersonic Transport," PAC/4; Enke to McNamara, 5/6/66, memo, PAC/4.

70. Enke to McNamara, "The Draft SST RFP to Be Printed April 25 for Further Comments by Manufacturers and Airlines," 4/19/66, memo, PAC/4; Enke, "Comments on FAA Digest of Manufacturers' Responses," 4/26/66, PAC/4; Enke to McNamara, "Preparations for May 6 PAC/SST Meeting," 4/29/66, memo, PAC/4; DOD, "Critical Assessment of SST Program," 4/29/66, PAC/4.

71. For the Bureau of the Budget's review of the financing issue, see: Parsons to Henry Rowen, 3/18/66, and attached "Index of Enclosures," FAA 1647/1; various drafts by FAA, BOB, DOD, CAB, and Treasury for final BOB financial issues paper, circa 3/24/66–4/1/66, FAA 6174/96; Parsons to Rowen, 4/1/66, and attached statements by and/or summaries of responses by Boeing, G.E., Lockheed, and Pratt & Whitney, FAA 1647/1; Enke, "Comments on Financial Issues Paper by BOB," 4/28/66, PAC/4; Schultze to McNamara, 4/26/66, and attached "Summary of Major Financial Issues," 4/25/66, "Recoupment Issues," 4/25/66, and "Allocations," n.d., PAC/4.

72. For the FAA views on financing by the May 6 PAC meeting, see: McKee to McNamara, 4/25/66, PAC/4; "Agenda Item II" (McKee's statement from backup material for 5/6/66 PAC meeting), FAA 6174/96; "PAC Position Papers," circa 4/66, FAA 6174/96; Learned to McKee, 5/4/66, FAA 6174/96; various FAA backup materials for 5/6/66 PAC meeting, FAA 6174/96. For the discussion and decisions on financing at the 5/6/66 PAC meeting, see: "Draft Proceedings of President's Advisory Committee on Supersonic Transport," 5/6/66, pp. 1–2, 30–71, PAC/32.

73. Osborne, "Notes on PAC/SST Problems," 6/27/66, PAC/4.

74. For FAA's delight with Osborne's paper, see: Maxwell to McKee, "Mr. Osborne's Paper," 6/14/66, memo, FAA 6174/95; Osborne to McNamara, "Supersonic Transport Documents," 6/30/66, PAC/4; McKee to McNamara 6/30/66, PAC/4.

75. Learned to McKee, 7/5/66, PAC/13; Learned to McKee, 7/9/66, FAA 1647/6.

76. Schultze to Steadman, 6/27/66, memo, and attached BOB to PAC, "Recoupment of the Government's SST Investment: Royalties *vs.* Pooling," 6/27/66, memo, PAC/13; "Summary Conclusions," "Detailed Evaluation," a set of comparative charts, "Analysis of Financial Outcomes," and "Addendum for Investment Recovery in Model Contract for Phase III," PAC/13; Steadman to Maxwell, 6/29/66, and attached revised page 2 of BOB, "Summary Conclusions," FAA 1647/16.

77. DOD, "Comments on Schultze's Paper on Royalties & Pooling," n.d., PAC/4; DOD, "Comments on the Pooling Arrangements Developed under BOB Sponsorship," n.d. PAC/4; DOD, "Comments on FAA Paper, 'Suggested Royalty Recoupment Clauses,' " n.d., PAC/4; Enke, "Comments on 'Notes on PAC/SST Problems' by Mr. Osborne," 6/30/66, PAC/4; J. W. Traenkle to Enke, "Comments on Mr. Stanley de J. Osborne's Letter of June 28, 1966, as Related to Pooling versus Royalty," 6/30/66, draft memo, PAC/4; Enke to Steadman, "More Comments on the Osborne Paper," 7/5/66, memo, and attached Walgreen, "Comments on Osborne Paper by Rastatter," 7/1/66, and Walgreen, "Comments on Osborne's Notes on PAC/SST Problems," 7/1/66, PAC/4.

78. Defense Department briefing paper on agenda items for 7/9/66 PAC meeting, PAC/4; "Draft Proceedings of President's Advisory Committee on Supersonic Transport," 7/9/66, pp. 50–77, PAC/32.

79. For the discussions involving Traenkle, Learned, and Maxwell, see: Learned to Maxwell, 7/26/66, FAA 6174/95; Traenkle to Enke, 7/26/66, memo, PAC/14; Traenkle to Enke, "Meeting between General Maxwell and Mr. Traenkle," 7/29/66, memo, PAC/14; Enke to McNamara, "SST Developments (Economic) since July PAC Meeting," 8/11/66, memo, PAC/14.

80. D. J. Haughton to Maxwell, 7/29/66, FAA 6174/7; Traenkle to F. L. Train, 8/5/66, FAA 1647/7; Traenkle to Enke, "J. Traenkle Visit to Discuss Recoupment with Boeing," 8/25/66, memo, FAA 1647/7; E. E. Hood to Maxwell, 8/1/66, FAA 1647/7.

81. For the final four draft contracts and for the decline in intensity of the whole financing debate before the 10/10/66 PAC meeting, see: Haldi to Steadman, 9/28/66, route slip, and attached "SST: PAC Policy Issues, Pooling, Royalty," 9/66, PAC/14; Walgreen, "Comments on Teaching Aid," 9/29/66, PAC/14; McKee and Schultze to PAC, "Recoupment Issues," 9/23/66, memo, attached "Abstract of Royalty and Pooling Investment Recovery Agreements," 9/23/66, and "Appendix: Draft Investment Recovery Agreements" (with two complete drafts, pooling and royalty, for the nonterminating or profit-sharing approach and two sets of additional provisions, pooling and royalty, for the terminating or nonprofit-sharing approach), 8/13/66, PAC/14.

82. John Haldi, "Points to Consider on a Key Issue for PAC Decision: Profit Sharing," draft paper, PAC/14; FAA, "Recoupment Statement," "Profit Sharing," and "Interest Recommendation," 10/5/66 (FAA briefing materials for PAC 10/6/66 meeting), FAA 6174/95; Traenkle to Steadman, "Comments on Two Issues Where PAC Guidance Is Requested," 9/23/66, memo, PAC/4; Traenkle to Steadman, "Comments on Pooling and Royalty Plans as of 9/23/66," 9/22/66, memo, PAC/14; Walgreen, "Comments on Pooling and Royalty Recoupment Proposals," 9/26/66, PAC/14; "Draft Proceedings of the President's Advisory Committee on Supersonic Transport," 10/6/66, pp. 43–66, PAC/32.

Chapter 10

1. Joseph Califano to McNamara, "Supersonic Transport Advisory Committee," 11/6/64, memo, PAC/10.

2. "Draft Proceedings of the President's Advisory Committee on Supersonic Transport," 5/21/65, pp. 42–50, PAC/32.

3. Raymond L. Bisplinghoff to NASA Administrator, "Sonic Boom Research Program," 6/21/65, memo, and attached "NASA Proposals for a Sonic-Boom Research Program," 6/18/65, FAA 6174/95; "Summary of Preliminary Program Recommendations," 8/3/65, PAC/12.

4. John M. Steadman to McNamara, "SST Sonic Boom Studies," 8/10/65, memo, PAC/12; McNamara to Donald F. Hornig, 8/12/65, PAC/12.

5. Bain to McKee, "Daily Highlight Report No. 9," 7/16/65, memo, FAA 6174/95.

6. McKee to Hornig, 9/1/65, LBJ Oversize.

7. B. N. Lockett to the file, "Telephone Conversation—September 3, 1965—Dr. Golovin/Mr. Lockett," 9/3/65, memo, FAA 6174/95; "Notes for *Record*, Discussion with Golovin," 10/6/65, NAS.

8. FAA briefing material for the 10/9/65 PAC meeting on "Agenda Item No. 3a," FAA 6174/96; "Draft Proceedings of the President's Advisory Committee on Supersonic Transport," 10/9/65, pp. 76–79, PAC/32; Hornig to McNamara, 10/9/65, FAA 6174/96; Maxwell to McKee, "Daily Highlights Report," 10/11/65, memo, FAA 6174/95.

9. Maxwell to McKee, 10/20/65, memo, FAA 6174/96; McKee to Hornig, 10/20/65, FAA 6174/96.

10. For the founding and initial membership of the OST Coordinating Committee, see: Hornig to McKee, 10/22/65, NAS; Hornig to Dunning, 10/22/65, NAS; Hornig to the President, 12/13/65, memo, LBJ EX CA.

11. "National Sonic Boom Program Working Group Meeting," 10/25–28/65, FAA Building, Washington, D.C., NAS.

12. "Draft Proceedings of the President's Advisory Committee on Supersonic Transport," 11/6/65, pp. 26, 34–37, 77, PAC/32.

13. "Memo for the President: Third Interim Report of the President's Advisory Committee on Supersonic Transport," 11/15/65, PAC/32.

14. For the pessimism at the White House, see: W. Marvin Watson to Califano, 9/25/65, memo, and attached statement by McKee and James E. Webb, LBJ Oversize; Califano to Hornig, 9/27/65, memo, LBJ Oversize; Hornig to Califano, 9/28/65, memo and attached list and "Presentation to OST: Ad Hoc Jet Aircraft NASA Panel," LBJ Oversize; Califano to Watson and Bill Moyers, 9/30/65, memo, LBJ Oversize; Moyers to Watson, 9/25/65 and 9/27/65, memos, LBJ Oversize; Moyers to Califano, 10/11/65, memo, LBJ Oversize.

15. For the controversy over the community overflights, see: NAS Committee on SST-Sonic Boom, "Notes on Second Meeting of OST Coordinating Committee," 11/22/74, memo, NAS; Golovin to Hornig, "Proceedings of the Second Meeting, November 22, 1965, of the OST Coordinating Committee for Sonic Boom Studies," 11/23/74, memo, NAS; Hewitt T. Wheless to Deputies et al., "Air Force Responsibility in National Supersonic Transport Sonic Boom Program," 12/7/65, memo, NAS; SST Economic Task

Force, "The Economics of a U.S. Supersonic Transport (Report of Findings to Date)," 8/20/65, pp. 27–29, PAC/25; Hutchinson to members and consultants, Coordinating Committee on Sonic Boom, 11/17/65, NAS; Hornig to McNamara, 11/24/65, NAS.

16. For the SRI plan, see: Park to members, NAS Committee on Sonic Boom-SST, "December 10, 1965 Meeting of Office of Science and Technology Coordinating Committee at SRI," 12/22/65, memo, NAS; "National Sonic Boom Evaluation Program: Report of the Office of Science and Technology," 12/10/65, PAC/36; Golovin to Hornig, "Conclusions of the Third Meeting (December 10, 1965) of the OST Coordinating Committee for Sonic Boom Studies," 12/24/65, memo, NAS; Maxwell, "Supersonic Transport Development: Daily Highlights," 12/14/65, FAA 6174/95.

17. To McKee, circa 10/12/65, draft memo, FAA 6174/95.

18. Maxwell to Golovin, 2/9/66, NAS

19. For discussions on the SRI plan, see: SRI, "Definition Study of National Sonic Boom Program," presentation of 1/15/66, Washington, D.C., PAC/36; SRI, "Definition Study of National Sonic Boom Program," presentation of 2/11/66, Washington, D.C., PAC/36; Golovin to Hornig, "Summary Proceedings of the OST Sonic Boom Studies Coordinating Committee Meeting of February 11, 1966," 2/15/66, memo, FAA 6174/92.

20. Golovin to Maxwell, 2/16/66, FAA 6174/96.

21. For Enke's views, see: Enke to McNamara, "Dr. Hornig's Recommendations to You That the Sonic Boom Program be Initiated Now by the USAF," 2/20/66, draft memo, PAC/36; Lockett to Maxwell, 3/4/66, transmittal slip, and attached Charles R. Foster to Brigadier General A. J. Evans, "Summary Proceedings of First Program Review Meeting of 25 February 1966," 2/28/66, memo, FAA 6174/92; Enke to Steadman, "Delay in Selecting a Contractor for Psycho-Sociological Surveys as Part of the Community Overflight Program," 3/14/66, memo, PAC/36; Steadman to McNamara, 3/3/66, memo, PAC/36.

22. Enke to McNamara, "Important SST Matters Requiring Your Attention," 2/25/66, memo, PAC/36.

23. Steadman to Harold Brown, 2/21/66, memo, and attachment, PAC/36; Brown to Steadman, 3/2/66, PAC/36; Steadman to McNamara, 3/5/6, memo, PAC/36; Steadman to Brown, 3/8/66, memo, PAC/36; Maxwell to McKee, 3/9/66, memo, FAA 1647/2.

24. Maxwell to the record, "Meeting of President's Advisory Committee for Supersonic Transport Members," 3/10/66, memo, FAA 6174/96; "Agreements Reached at Informal March 9 Meeting of Several Members of PAC-SST," circa 3/12/66, PAC/13.

25. Hornig to Brown, 3/14/66, draft letter, LBJ EX CA; Hornig to Watson, 3/15/66, memo, LBJ EX CA; Hornig to the President, 3/15/66, memo, LBJ EX CA; Watson to the President, 3/15/66, memo, LBJ EX CA.

26. Vierling, "Supersonic Transport Development: Daily Highlights," 3/15/66, FAA 6174/95; Dunning to Seitz, 3/23/66, FAA 6174/92.

27. Golovin to Hornig, "Summary, Proceedings of the OST Sonic Boom Studies Coordinating Committee Meeting of April 1, 1966," 4/6/66, memo, FAA 6174/92.

28. OST, "Status of the Sonic Boom Studies Program as of April 20, 1966," 4/20/66, PAC/4.

29. For Enke's warnings, see: Enke to Steadman, 4/1/66, memo, PAC/36; Captain Cha-

bon to Steadman, "Coordination of Air Force Proposed Press Release on Sonic Boom Overflight Program," 4/7/66, memo, PAC/36; Enke to Steadman, "Letter to Congressmen," 4/7/66, memo, PAC/36; Enke, "Suggested Changes in Letter to Congressmen," 4/7/66, PAC/36; Captain Chabon to Enke, 4/8/66, memo, and attached "Media Announcement to Be Made Shortly after Congressional Notification of Proposed Sonic Boom Overflights," PAC/36; Enke to Steadman and Colonel Taylor, 4/8/66, PAC/36; Captain Chabon to Enke, 4/27/66, memo, PAC/36.

30. Enke, "Comments on FAA Paper Regarding Role of PAC in Source Selection," 4/22/66, PAC/4.

31. OST, "Status of the Sonic Boom Studies Program as of April 20, 1966," 4/20/66, PAC/4.

32. Hornig to the President, "Supersonic Boom—Public Acceptability Tests," 4/23/66, memo (attached to Califano to the President, 4/27/66, memo), LBJ Oversize.

33. Hornig to Califano, 4/26/66, and attached Golovin to Hornig, "Relative Timing of Studies I and II in the Sonic Boom Program," 4/25/66, memo, LBJ Oversize.

34. Schultze to the President, "Sonic Boom Tests for the SST," 4/23/66, memo (attached to Califano to the President, 4/26/66, memo), LBJ Oversize.

35. McKee to Hornig, circa 4/66, draft letter, LBJ FAA Hist. Doc.; R. E. Parsons to Maxwell, "Expenditure Savings on Sonic Boom Overflights," 4/1/66, memo, FAA 6174/92.

36. McKee to Harold Brown, 4/5/66, draft letter, FAA 6174/92; McKee to Brown, 4/14/66, FAA Hist. Doc.

37. For McKee's move to delay these community overflights and for support for this action, see: McKee to the President, 4/20/66, draft letter, FAA 6174/95; FAA backup material used at 5/6/66 PAC meeting for "Agenda Item 5," FAA 6174/96; McKee to the President, "Community Overflight Sonic Boom Tests," 4/21/66, memo (attached to Califano to the President, 4/27/66, memo), LBJ Oversize; Bisplinghoff to McKee and Maxwell, "Sonic Boom," 5/11/66, memo, FAA 6174/92.

38. Califano to the President, 4/27/66, memo, LBJ Oversize; Richard J. Kent, Jr., *Safe, Separated, and Soaring* (Washington, D.C.: U.S. Government Printing Office, 1980), p. 144.

39. "[Draft] Sonic Boom Position Paper," 4/29/66, FAA 1647/6.

40. "Draft Proceedings of the President's Advisory Committee on Supersonic Transport," 5/6/66, pp. 11, 73–76, PAC/32.

41. Maxwell to Golovin, 5/24/66, FAA 1647/6.

42. *Aviation Daily*, 6/13/66, p. 357.

43. *New York Times*, 6/16/66, PAC/14.

44. For the protest over sonic boom damage claims in Oklahoma City and for the FAA's response to this development, see: "Supersonic Transport Development: Daily Highlights," 5/11/66, 5/26/66, FAA 6174/95; Lockett to Maxwell, "Oklahoma City Sonic Boom Claims Data," 5/12/66, memo, and attached "OKC Sonic Boom Claims Data: 5/2/66 through 5/6/66," FAA 1647/6; J. K. Power to Chief, Technical Operations Division, "Sonic Boom Activities—5/9–5/13/66," 5/16/66, memo, FAA 1647/6; Maxwell to Senator A. S. "Mike" Monroney, 5/20/66, FAA 1647/6; Lockett to Maxwell, "E. E.

Gravelle et al., Lawsuit," 5/31/66, 6/6/66, memos, FAA 1647/6; "Flash Highlights Report: Operations and Training Branch," 6/13/66, FAA 1647/6; "Flash Highlights Report: Operations Environment Branch," 6/16/66, 6/23/66, 6/29/66, 7/15/66, FAA 1647/6; "Flash Highlights Report: Technical Operations Division," 6/27/66, FAA 1647/6; Johnny M. Sands, "Trip Report," 6/30/66, FAA 1647/6; Power to Chief, Technical Operations Division, "Oklahoma City Field Office," 7/19/66, memo, FAA 1647/6.

45. Bo Lundberg, "The Menace of the Sonic Boom to Society and Civil Aviation," circa 5/66, PAC/36.

46. Stewart L. Udall to Frederick Seitz, 6/27/66, PAC/14.

47. "Draft Proceedings of the President's Advisory Committee on Supersonic Transport," 7/9/66, pp. 16–18, PAC/32.

48. Charles A. Lindbergh to Stewart L. Udall, 7/18/66, PAC/14.

49. For the FAA film, see: Johnny M. Sands, "Trip Report," 5/13/66, FAA 1647/6; Lockett to Maxwell, "Sonic Boom Movie," 5/23/66, memo, FAA 1647/6; Kenneth Youel to Lockett, 6/2/66, FAA 1647/6; Dunning to Youel, 6/9/66, FAA 1647/6; William P. Briley to Jim Helliwell, 6/9/66, FAA 1647/6.

50. For the first phase of the Edwards sonic boom tests, see: "Supersonic Transport Development: Daily Highlights," 5/23/66, FAA 6174/95; "Flash Highlight Report: Technical Operations Division," 5/23/66, 6/19/66, FAA 1647/6; H. B. Alexander to McKee, 6/10/66, memo, FAA 1647/6; McKee, "Effect of B-70 Loss on SST Program," 6/10/66, FAA 1647/6; C. L. Blake to the record, "B-70 Accident Sequence," 6/19/66, memo, FAA 1647/6; McKee to Representative George H. Mahon, 6/14/66, FAA 1647/6; Maxwell to McKee, "Dr. Adams' Telephone Call re XB-70," 6/22/66, memo, FAA 1647/6; C. M. Plattner, "Sonic Boom Tests Seek Forecast Data," *Aviation Week and Space Technology*, 6/20/66, pp. 55–57; FAA, "Section C—Sonic Boom Investigations" of the "Summary Status Report to President's Advisory Committee on Supersonic Transport," 6/30/66, pp. 14–18, PAC/4; Lockett to Assistant for Plans and Programs, "Bi-Monthly Status Report to the PAC, Your Memo Dated 6/23/66," 6/24/66, memo, and attached "Section C—Sonic Boom," FAA 1647/6.

51. Lockett to Maxwell, "Proposed Study of Sonic Boom Reports," 6/30/66, memo, FAA 1647/6.

52. T. H. Higgins to Chief, Operations Environment Branch, "Boeing SST (B2707) Sonic Boom," 6/15/66, memo, FAA 1647/6.

53. DOD, "Comment on Major SST Program Events (Including Economic Studies)" (for PAC meeting of 7/9/66), PAC/14; Maxwell to McKee, "Meeting of OST Sonic Boom Coordinating Committee on July 8, 1966," 7/12/66, memo, FAA 1647/6.

54. "Draft Proceedings of the President's Advisory Committee on Supersonic Transport," 7/9/66, pp. 31–36, PAC/32.

55. Golovin to Hornig, "Sonic Booms and the SST Program," 8/16/66, memo, FAA 1647/7.

56. Maxwell to McKee, "Discussion with Dr. Hornig," 7/14/66, memo, FAA 1647/6.

57. For the discussion on aircraft engine noise measurements, see: Boeing Co., "Aircraft Noise Reduction Status Report," DG-23482, 9/68; U.S. Department of Transportation, *Second Federal Noise Abatement Plan: FY 1970–71* (Washington, D.C.: GPO, 1971);

Wyle Laboratories Research Staff, *Noise Primer for the Supersonic Transport* (El Segundo, Calif.: Wyle Laboratories, 1971). The various examples of PNdB levels are from ibid.

58. National Sonic Boom Evaluation Office, *Sonic Boom Experiments at Edwards Air Force Base, Interim Report,* no. NSBEO-1-67, Alexandria (Virginia: Defense Documentation Center, 7/28/67).

59. Personal interviews.

60. Karl Kryter, "Sonic Booms from Supersonic Transport," *Science*, 1/24/69, p. 363.

61. For the FAA's attempt to control the SRI Edwards report, see: Lockett to the files, "Stanford Research Institute Coordination," 8/3/66, memo, FAA 1647/7; Colonel Charles R. Foster to Lockett, 8/17/66, FAA 1647/7; Lockett to Chief, Operations Environment Branch, "Sonic Boom Paper for Submission to Source Selection Advisory Staff (SSAS)," 8/8/66, memo, FAA 1647/7.

62. Maxwell, "Comments on August 16 [1966] Letter—Golovin to Hornig," n.d., memo, FAA 1647/7.

63. R. L. Bisplinghoff and Lockett to Maxwell, "Comments on Golovin Paper of August 16, 1966," 8/19/66, memo, FAA 1647/7; Albert J. Evans to Golovin, 9/12/66, FAA 1647/7; Dunning to Seitz, 9/13/66, NAS; Ibid., 9/14/66, NAS.

64. Maxwell to Golovin, 9/15/66, FAA 6174/95.

65. For the inability of Hornig and McKee to submit a joint statement on the sonic boom, see: Hornig to McKee, 8/17/66, 8/26/66, and attached McKee to McNamara, 8/26/66, draft letter, FAA 1647/7; McKee to McNamara, 9/1/66, PAC/14; Hornig to McNamara, 9/21/66, LBJ Oversize.

66. Joe Califano to the President, 9/17/66, memo, LBJ Oversize.

67. Enke to McNamara, "SST Developments (Economic) since July PAC Meeting," 8/11/66, memo, and attached Enke et al., "Costs and Benefits of Two SST Manufacturer Prototype and Manufacturing Programs," PAC/14.

68. Stewart L. Udall to McNamara, circa early 9/66, handwritten letter, PAC/14; Steadman to McNamara, 9/2/66, memo, PAC/14; McNamara to Udall, 9/3/66, PAC/14.

69. Steadman to McNamara, 9/26/66, memo, and attached "Proposed Agenda [for PAC meeting on 10/6/66]," PAC/14; Charles L. Schultze to McNamara, 10/1/66, PAC/15.

70. "Draft Proceedings of the President's Advisory Committee on Supersonic Transport," 10/6/66, pp. 12–14, 19–43, PAC/32; "FAA Position Regarding SST Sonic Boom Problem," "Sonic Boom," and "Program Status Report" (from FAA briefing material for the PAC meeting on 10/6/66), FAA 6174/95.

71. Lockett to Chief, Operations Environment Branch, "SSAS Task Assignment Concerning Sonic Boom," 10/31/66, FAA 1647/95.

72. John P. Taylor to [NAS] Committee Members, Consultants and Interested Staff Personnel, "Informal Notes on OST Coordinating Committee Meeting, October 27, 1966," and attachments, NAS; Hornig to McNamara, 11/1/66, and attached "Report of the OST Coordinating Committee on Sonic Boom Studies," 11/1/66, PAC/4.

73. Maxwell to Hornig, 11/10/66, FAA 6174/92.

74. "Draft Proceedings of the President's Advisory Committee on Supersonic Transport," 11/17/66, p. 2, PAC/32.

75. Ibid., 12/7/66, pp. 3–4, 49–52, PAC/32.

76. DOD, "The Effect on SST Redesign of Ignoring Sonic Boom Constraints" (paper for PAC meeting on 12/7/66), PAC/15; "Draft Proceedings of the President's Advisory Committee on Supersonic Transport," 12/7/66, pp. 3–10, PAC/32.

77. FAA, "SST Growth" (paper for PAC meeting on 12/10/66), PAC/15.

78. FAA, "Status of Legislation" (paper for PAC meeting on 12/10/66), PAC/15.

79. "Draft Proceedings of the President's Advisory Committee on Supersonic Transport," 12/10/66, pp. 13–20, 28–36, 50, PAC/32.

80. "Memorandum for the President, Fourth Interim Report of the President's Advisory Committee on Supersonic Transport," 12/22/66, PAC/32.

81. For warnings to the president on noise and sonic booms, see: McKee to Califano, 1/10/66, and attached McKee to Donald W. Douglas, Jr., 1/7/66, McKee to William M. Allen, 1/7/66, and McKee to Daniel J. Haughton, 1/7/66, LBJ Oversize; Califano to Bill Moyers, 1/9/66, memo, and attached Hornig to Califano, 1/7/66, memo, and "Summary Report of the October 29, 1965, Meeting by the Ad Hoc Panel on Jet Aircraft Noise," LBJ Oversize; Bill Moyers to Califano, 1/11/66, note, LBJ Oversize; McKee to Califano, 5/11/66, memo, and "Proposed White House Press Release," LBJ Oversize.

82. Calilfano to the President, 12/26/66, memo, LBJ Oversize.

83. Golovin to Hornig, "Summary, Proceedings of the Twelfth Meeting of the OST Coordinating Committee on Sonic Boom Studies, January 16, 1967, and Agenda Items for the Thirteenth Meeting, April 4, 1967," 3/27/67, memo, p. 2, FAA 1647/9.

84. Golovin to Karl Kryter, 2/23/67, FAA 1647/9; Kryter, "A Review of: 'Menace of the Sonic Boom to Society and Civil Aviation,' by Dr. Bo Lundberg," 3/17/67, draft, PAC/36.

85. Maxwell to Golovin, 4/11/67, FAA 1647/9.

86. For continued FAA attempts to modify the SRI draft report on the Edwards sonic boom tests, see: Maxwell to Golovin, 5/2/67, and attached written comments, FAA 6174/92; Golovin to Hornig, "Summary, Proceedings of the Fourteenth Meeting of the OST Coordinating Committee on Sonic Boom Studies, May 17, 1967, and Agenda Items for the Fifteenth Meeting, July 26, 1967," 6/26/67, memo, FAA 1647/11; S. J. Converse to the files, "Report on May 17 Meeting of OSTSBCC," 5/25/67, memo, FAA 1647/10; McKee to Hornig, 7/25/67, FAA 6174/95.

87. Albert J. Evans to Golovin, 4/12/67, FAA 6174/92.

88. Maxwell to Dunning, 4/24/67, NAS; Dunning to Hornig, circa early 5/67, FAA 6174/92.

89. Hornig to Dunning, 7/27/67, FAA 1647/11.

90. National Sonic Boom Evaluation Office, *Sonic Boom Experiments at Edwards Air Force Base*.

91. For the development and planning of SR-71 sonic boom tests, see: J. W. Howell to the files, "Meeting with Mr. Harper of NASA on Monday (January 23, 1967)," 1/26/67, memo, FAA 1647/8; Lockett, "Flash Highlight Report—Technical Operations Division," 2/28/67, FAA 1647/9; Maxwell to McKee, "SR-71 Flights," 3/17/67, memo, FAA 1647/9; Converse to the files, "Record of the Eighth Meeting of the National Sonic Boom

Evaluation Office Program Review Group," 3/24/67, memo, FAA 1647/9; Lockett to Maxwell, "SR-71 Flights," 3/23/67, memo, FAA 1647/9; McKee to Dunning, 5/8/67, FAA 1647/10; McKee to Colonel Donald W. Paffel, 5/9/67, FAA 1647/10; Kryter to OST Coordinating Committee on Sonic Boom Studies, "Proposed Cities for Sonic Boom (SR-71)—Environmental Noise Studies," 5/12/67, memo, FAA 1647/10; Hoover to Maxwell, "Amendment of NASA Contract to Include Sonic Boom Human Response Studies," 9/18/67, memo, FAA 1647/12; Golovin to Hornig, "Summary, Proceedings of the Fourteenth Meeting of the OST Coordinating Committee on Sonic Boom Studies, May 17, 1967 and Agenda Items for the Fifteenth Meeting, July 26, 1967," 6/26/67, memo, FAA 1647/11; Golovin to Maxwell, "Funding Requirements for Sonic Boom Studies," 5/18/67, memo, FAA 6174/95; Converse to the files, "Report on May 17 Meeting of OSTSBCC [OST Sonic Boom Coordinating Committee]," 5/25/67, memo, FAA 1647/10.

92. For FAA public relations policy and FAA communication with the air force during the SR-71 tests, see: Maxwell to McKee, 6/16/67, note, FAA 1647/10; Maxwell to Director, Office of Noise Abatement, "FAA Support of SR-71 Sonic Boom Research Program," 8/22/67, memo, FAA 1647/11; D. D. Thomas to Regional Directors et al, "FAA Support of SR-71 Sonic Boom Research Program," 8/23/67, memo, FAA 1647/11; Maxwell to McKee, "SR-71 Overflight Program," 7/7/67, 7/21/67, 7/22/67, memos, FAA 1647/11; Howell to the file, "Meeting with Colonel Foster Regarding Weekly Report on SR-71 Overflight Program—7/25/67," 7/25/67, memo, FAA 1647/11; Lockett to Maxwell, "Summary of SR-71 Flights," 7/27/67, memo, FAA 1647/11; "SR-71 Status Report—Flights: 1–28 July; Complaints: 1 July–1 August," FAA 1647/11; Colonel Harold E. Collins to Maxwell, 9/5/67, and attached "SR-71 Status Report: 1 July–24 August," FAA 1647/11; Collins to Maxwell, 9/7/67, and attached "SR-71 Status Report: 1 July–31 August," FAA 1647/11; Colonel Robert C. Harrison to Maxwell, 9/8/67, and attached "SR-71 Status Report: 1 July–7 September," FAA 1647/11; Howell to Maxwell, "SR-71 Status Report," 9/11/67, FAA 1647/11.

93. Tracor, *Public Reactions to Sonic Booms,* NASA CR-1665 (Washington, D.C.: NASA, September 1970).

94. For the transfer of sonic boom testing and evaluation responsibility from OST to the Department of Transportation, see: memo of telephone call to Maxwell, 4/20/67, and attached Hoover to Associate Administrator for Programs, "Transfer of Sonic Boom Functions to Noise Abatement Staff," 4/18/67, memo, FAA 6174/92; Maxwell to PD-1, "Identification of Sonic Boom Functions Responsibility," 4/18/67, memo, FAA 6174/92; Howell to Chief, Technical Operations Division, "Transfer of Sonic Boom Functions to Noise Abatement Staff," 4/19/67, memo, FAA 1647/10; Converse to SS-200, "Meeting with PD-5 re Sonic Boom and Noise Programs," 6/23/67, memo, FAA 1647/10; James E. Densmore to the Secretary of Transportation, Alan S. Boyd, "Transfer of Responsibilities for Aviation Noise and Sonic Boom Activities from the Office of Science and Technology to the Department of Transportation," 8/2/67, memo, NAS; Boyd to Hornig, 8/3/67, NAS; Boyd to McNamara, 8/12/67, FAA 6174/95; Lt. General Joseph R. Holzaips to the record, "Air Force Support of Sonic Boom Research," 8/22/67, memo, NAS; Maxwell to the FAA Administrator, "Transfer of OST Sonic Boom and Noise Responsibilities to the Department of Transportation," 8/25/67, memo, FAA 1647/11; Hornig to Maxwell, 8/29/67, FAA 1647/11; Hornig to Boyd, 8/25/67, FAA 1647/11; Densmore to Members, Observers, and Participants, OST Coordinating Committee, "16th Meeting of Coordinating Committee on Sonic Boom Studies," 9/21/67, memo, NAS; Charles R.

Foster to Executive Secretariat et al., "Responding to Noise, SST and Sonic Boom inquiries," 10/3/67, memo, FAA 1647/12; Boyd to McKee, "Aircraft Noise Abatement," 12/12/67, memo, FAA 1647/13; Thomas to Boyd, "Aircraft Noise Abatement," 12/21/67, memo, FAA 1647/13; Hornig to Boyd, 12/15/67, FAA 1647/13; Hornig to Boyd, 3/22/68, FAA 6174/92.

95. For the increasing public reporting and protest and for the FAA's concern over the SST's potential sonic boom hazards, see: Dean Rusk to American Embassy, Bern, Switzerland, "Letter Dated July 8, 1966, from the International Association against Noise," 8/5/66, DOS airgram, FAA 1647/7. The anti-SST attacks by Elizabeth Borish and Harvard University biology professor John T. Edsall, who wrote separate letters to the editor of the *New York Times* in August 1966, marked the real beginning of public opposition to the SST. Borish and Edsall soon became active in the first anti-sonic-boom protest group. Edsall called the disturbance from the "carpet" of the sonic boom "the gravest objection to the SST." See *New York Times*, 8/31/66, p. 42, FAA 6174/95. For the FAA's early concern with such adverse publicity, see: W. F. Boone to Dr. Mueller et al., 9/1/66, memo, FAA 6174/95. For airlines' doubts about the SST because of the sonic boom, see: Neil Gleeson (Aer Lingus) to Maxwell, 10/28/66, FAA 1647/8. Gleeson said that the sonic boom "represents an unresolved problem of a very serious nature" and that his airline was "not satisfied" that the manufacturers' stated sonic boom overpressures would be acceptable to the public. At about the same time, the aviation director of the Environmental Sciences Services Administration publicly warned that the sonic boom might be the SST's most difficult problem. See *Aviation Daily*, 10/31/66, p. 360, PAC/15. On the same day the *New York Times* reported that the SST "for all practical purposes" had become an over-water airplane because of the sonic boom. See *New York Times*, 10/31/66, pp. 1, 71. For FAA concern about the international repercussions of banning the SST because of the sonic boom, for PAC concern over the apparent lack of interest in the sonic boom problem by the British and French, and for PAC concern over a possible lack of authority by the FAA to ban SSTs because of the sonic boom, see: Maxwell to Giller et al., 12/7/66, and attached statement, FAA 1647/8; FAA, "Regulatory Authority for Certification and Operation of Foreign Manufactured Aircraft: Noise and Sonic Boom" (paper for PAC meeting of 12/7/66), FAA 1647/1; "Draft Proceedings of the President's Advisory Committee on Supersonic Transport," 12/7/66, pp. 28–29, PAC/32.

Chapter 11

1. B. J. Vierling to SST Staff, "Source Selection Procedures Team," 1/12/66, memo, FAA 1647/2.

2. For the interim design assessment in late 1965, see: "Draft Proceedings of the President's Advisory Committee on Supersonic Transport," 5/6/66, pp. 2–3, PAC/32; FAA briefing material for the PAC meeting of 5/6/66 on "Agenda Item 1," FAA 6174/96; "Assessment of Supersonic Transport Airframe Designs by Joint Government Technical Team Commences Today," 11/15/66, FAA news release 65-110, LBJ FAA Hist. Doc.; Maxwell, "Supersonic Transport Development: Daily Highlights," 12/13/66, FAA 6174/95; Maxwell to Steadman, 1/5/66, memo, FAA 6174/96; FAA, "Summary Status Report to PAC," 2/15/66, pp. 2–3, PAC/13; McKee to McNamara, 1/5/66, FAA 6174/96; I. H. Hoover to McKee, "White House Report," 1/10/66, memo, FAA 6174/95; Charles L.

Blake to McKee, "SST Technical Assessment, December 17, 1965," 12/20/65, memo, FAA 1647/2; *New York Times*, 3/5/66, p. 54.

3. For Boeing's SST organizational changes, its design changes, and its comments on private 747 financing and for the FAA's reactions to Boeing's actions, see: "Draft Proceedings of the President's Advisory Committee," 5/6/66, pp. 2–4; FAA briefing materials for the PAC meeting of 5/6/66 on "Agenda Item I"; Maxwell, "Supersonic Transport Development: Daily Highlights," 1/20/66, 1/28/66, 2/9/66, 2/21/66, 2/23/66, 4/1/66, 4/20/66, 5/6/66, FAA 6174/95; I. H. Hoover to McKee, "White House Report," 2/7/66, memo, FAA 6174/95; "Summary Status Report to PAC," 2/15/66, pp. 2, 14–15; Maxwell to McKee, "North American Assistance to Boeing," 1/12/66, memo, FAA 6174/95; McKee's handwritten notation of 2/10/66 on Maxwell, "Supersonic Transport: Daily Highlights," 2/9/66, FAA 6174/95; Vierling, "Supersonic Transport Development: Daily Highlights," 1/5/66, 4/5/66, FAA 6174/95; Maxwell to McKee, "Boeing Progress," 1/20/66, memo, FAA 6174/95; "Program Status," circa 2/28/66, p. 3, FAA 6174/95; Hoover to McKee, "White House Report," 3/7/66, FAA 6174/95; *New York Times*, 4/15/66, p. 60; William V. Vitale, "Meeting with Boeing Company," 3/10/66, FAA 6174/95; "Meeting with Mr. T. Wilson, the Boeing Company, Apr. 14, 1966—4:00 P.M.," n.d., FAA 1647/4; Allen to McKee, 4/29/66, FAA 6174/95; Maxwell to the record, "Mr. McCone's Visit to the Boeing Company—April 28, 1966," 5/22/66, memo, and attached "Notes on Mr. McCone's Visit to the Boeing Company," FAA 6174/91; Maynard Pennell to Maxwell, 5/4/66, FAA 6174/95.

4. For the evolution of Lockheed's design during the first half of 1966 and for the FAA's opinion of the Lockheed model, see: "Draft Proceedings of the President's Advisory Committee," 5/6/66, pp. 3–7; FAA briefing materials for the PAC meeting of 5/6/66 on "Agenda Item I"; FAA, "Summary Status Report to PAC," 2/15/66, pp. 2–3; *New York Times*, 4/14/66, p. 26; Maxwell to McKee, "North American Assistance to Boeing," 1/12/66; Maxwell, "Supersonic Transport Development: Daily Highlights," 1/20/66, 2/2/66, 2/10/66, 2/16/66, 2/21/66, 2/24/66, 2/29/66, 3/24/66, 4/1/66, 4/19/66, FAA 6174/95; "Program Status," circa 2/28/66; Hoover to McKee, "White House Report," 3/7/66; Vierling, "Supersonic Transport Development: Daily Highlights," 3/15/66, FAA 6174/95; "Lockheed Boosts Lift/Drag Ratio of SST," *Aviation Week and Space Technology*, 3/28/66, p. 40; R. R. Heppe to Charles C. Blake, 2/4/66, FAA 1647/4; F. E. Ellis, Jr., to Configuration Manager, "Lockheed Model and Performance Specification Comments," 2/10/66, memo, FAA 1647/4; William V. Vitale, on meeting with Lockheed, circa 3/8/66, memo, FAA 6174/95.

5. For the evolution of the engine designs and for the FAA's assessment of the engine work, see: "Draft Proceedings of the President's Advisory Committee," 5/6/66, pp. 4–7; FAA briefing materials for the PAC meeting of 5/6/66 on "Agenda Item I"; "Summary Status Report to PAC," 2/15/66, p. 4; Maxwell, "Supersonic Transport Development: Daily Highlights," 1/20/65, 3/15/66, 4/1/66, 4/15/66, FAA 6174/95; "Program Status," circa 2/28/66; Hoover to McKee, "White House Report," 3/21/66, memo, FAA 6174/95; McKee to McNamara, 4/1/66, LBJ FAA Hist. Doc.

6. For the organization and procedures for the evaluation that were being considered by the FAA during the first half of 1966, see: Maxwell to McKee, "SST Evaluation and Source Selection," 1/27/66, memo, FAA 1647/1; "Supersonic Transport Development Program (Phase III): An Approach to the SST Evaluation and Source Selection," 2/7/66, FAA 1647/1; B. N. Lockett to Assistant for System Integration et al., "Evaluation

Criteria Needed for Operational Assessment of Both Airframe and Engine Proposals," 2/25/66, memo, FAA 1647/2; D. D. Davis to Division Chiefs et al., "Source Selection Planning," 3/23/66, memo, FAA 1647/2; Davis to Blake et al. "Source Selection Planning," 5/5/66, memo, FAA 1647/2; Davis to Skaggs et al., "Source Selection Planning," 5/20/66, 5/26/66, memos, FAA 1647/2.

7. For Enke's criticism of FAA management of the competition, for his urging more PAC involvement, and for opposing FAA views, see: Maxwell to Enke, 4/18/66, PAC/4; Enke to McNamara, "The Draft SST RFP to Be Printed April 25 for Further Comments by Manufacturers and Airlines," 4/19/66, memo, PAC/4; McKee to McNamara, 4/25/66, PAC/4; FAA, "Role of the PAC in the Source Selection Process," FAA 6174/96; DOD, "Comments on 'Financial Issues' Paper," PAC/4; Enke, "Comments on FAA Paper Regarding Role of PAC in Source Selection," PAC/4; Enke, "Critical Assessment of SST Program," 4/29/66, PAC/4; Enke to McNamara, 5/6/66, memo, PAC/4; FAA backup material for PAC meetings of 5/6/66 on "Agenda Item II—General McKee's Statement," FAA 6174/96.

8. "Draft Proceedings of the President's Advisory Committee," 5/6/66, pp. 76–79.

9. For the establishment and structure of the Airline SST Committee, see: "Report of Meeting of General Maxwell and Mr. Vierling with Mr. Mentzer, Chairman of the Airlines SST Committee," 12/8/65, FAA 6174/95; W. C. Mentzer to F. C. Wiser et al., 1/10/66, and attached "Minutes of the Second Airlines SST Committee Meeting," 1/7/66, FAA 1647/1; Mentzer to Maxwell, 2/4/66, FAA 1647/1; Mentzer to Wiser et al., 2/3/66, and attached "Formation of and Instructions for Airline SST Specialist Teams," 2/2/66, FAA 1647/1; Maxwell to McNamara, "Supersonic Transport Status Report," 2/15/66, memo, PAC/13; B. N. Lockett to Vierling, "Office of Supersonic Transport Development Airline Coordinating Committee," 3/10/66, memo, FAA 1647/1; Mentzer to Maxwell, 3/22/66, FAA 1647/1; Maxwell to Mentzer, 3/26/66, 3/29/66, FAA 1647/1.

10. "Draft Proceedings of President's Advisory Committee," 5/6/66, pp. 7–8.

11. "Airlines Plan for Both Large Jets, SST," *Aviation Week and Space Technology*, 3/7/77, pp. 180–183.

12. For Pan American's delay in participating, see: John G. Borger to Vierling, 5/2/66, FAA 1647/6; David E. Galas, "Flash Highlights Report," 5/5/66, FAA 1647/6; McKee, "Memorandum for Record," 5/5/66, FAA 6174/95.

13. For the airlines' critiques that stressed the potential harm of government involvement, see: Maxwell to McKee, "Airline Letter re: REP," 6/3/66, memo, FAA 6174/95; Robert F. Six to McKee, 5/26/66, FAA 1647/6; Tillinghast to Maxwell, 6/1/66, FAA 1647/6; Donald W. Nyrop to Maxwell, 6/1/66, FAA 1647/6; W. T. Seawell to Maxwell, 5/27/66, FAA 1647/6; G. E. Keck to Maxwell, 5/27/66, FAA 1647/6; S. B. Kauffman to Maxwell, 5/27/66, FAA 1647/6.

14. Maxwell to PAC, "Supersonic Transport Status Report," 6/30/66, memo, PAC/4; Maxwell to Mentzer, 6/30/66, FAA 1647/6; McKee to Trippe, 6/30/66, FAA 1647/6.

15. "Draft Proceedings of the President's Advisory Committee on Supersonic Transport," 7/9/66, pp. 1–22, PAC/32; Laurence S. Kuter to Maxwell, 8/30/66, FAA 1647/7; *New York Times*, 9/22/66, p. 90.

16. For the airline specialist teams' reviews and for the FAA's praise of the airline teams' work, see: Chief, Propulsion Branch, to SS-140 et al., "Summary of Meeting with Airline SST Propulsion Specialist Team on July 16, 1966," 8/16/66, memo, FAA 1647/1;

R. S. Stahr to R. W. Pinnes, 10/28/66, and attached "Minutes of SST Propulsion System Specialist Team Meeting," 10/7/66, Chicago, FAA 1647/1; Pinnes to Maxwell, "Airline SST Propulsion System Specialist Team, Comments on Phase III Proposals," 11/4/66, memo, FAA 1647/1; R. A. Peterson to Chief, Propulsion Branch, "Comments on Airline SST Ground Support Committee Recommendations for Boeing and Lockheed," 9/6/66, memo, and attachments, FAA 1647/1; Vierling to Mentzer, 9/13/66, FAA 1647/1; Pinnes to Stahr, 11/7/66, FAA 1647/1.

17. For the airline evaluations, see: Chavkin to Airline Evaluation Team Leaders, "Airline Evaluation Instructions," 10/26/66, memo, FAA 1647/7; Hoover to McKee, "White House Report Items," 11/31/66, memo, FAA 6174/95; George A. Spater (American Airlines) to McKee, 10/28/66, FAA 1647/7; letters to Maxwell from: F. D. Hall (Eastern Airlines)—10/31/66, M. B. Fannon (Eastern Airlines)—11/11/66, L. B. Maytag, Jr. (National Airlines)—10/31/66, Donald W. Nyrop (Northwest Airlines)—11/9/66, R. F. Six (Continental Airlines)—10/28/66, David C. Garrett, Jr. (Delta Airlines)—10/27/66, Laurence S. Kuter (Pan American)—10/31/66, J. T. Dyment (Air Canada)—11/10/66, FAA 1647/8; Qantas Empire Airways, "Extract from U.S. SST Evaluation Report: Summary of Recommendation," FAA 1647/8; M. A. Cristadoro, Jr., to Maxwell, "Review of Pan American Airlines Proposal Evaluation," 11/10/66, memo, FAA 1647/8; Kuter to Maxwell, 12/9/66, telex, FAA 1647/8; Kuter to Maxwell, 12/9/66, FAA 1647/8; "Draft Proceedings of the President's Advisory Committee on Supersonic Transport," 11/17/66, p. 34, PAC/32; "Draft Proceedings of the President's Advisory Committee on Supersonic Transport," 12/7/66, pp. 34–35, PAC/32.

18. For the evolution of Boeing's design, see: T. A. Wilson to Maxwell, 5/17/66, FAA 1647/6; James R. Ashlock, "Boeing Moves Engine in SST Redesign," *Aviation Week and Space Technology*, 6/20/66, pp. 34–35; C. C. Blake to Maxwell, "Boeing Alternate Prototype Proposal," 6/15/66, memo, FAA 1647/6; FAA, "Summary Status Report to the PAC," 6/30/66, pp. 1–6, PAC/32; "Draft Proceedings of the President's Advisory Committee on Supersonic Transport," 7/9/66, pp. 23–34, PAC/32; FAA, "Summary Status Report to PAC," 8/31/66, PAC/24; Vierling to the file, "Telephone Conversation, 7/13/66, with Mr. Maynard Pennell," 7/13/66, memo, FAA 1647/4; Blake to Maxwell, "Telephone Conversation with Bob Withington on Boeing Configuration," 7/14/66, memo, FAA 1647/6; Blake to Maxwell, "Highlights from Boeing Visit," 8/3/66, memo, FAA 1647/6; I. H. Hoover to McKee, "White House Report," 7/11/66, memo, FAA 6174/95; *New York Times*, 9/30/66, p. 49, 10/9/66, sec. X, p. 1.

19. FAA, "Summary Status Report to PAC," 6/30/66, pp. 3–8, PAC/32; "Draft Proceedings of the President's Advisory Committee," 7/9/66, pp. 24–25.

20. For the evolution of the engine design and for the FAA's and airframe manufacturers' views, see: "Draft Proceedings of the President's Advisory Committee," 7/9/66, pp. 24–25; "Summary Status Report to the PAC," 8/31/66, pp. 6–7; "Draft Proceedings of the President's Advisory Committee on Supersonic Transport," 10/6/66, pp. 5–6, PAC/32; Pennell to Maxwell, 11/11/66, 11/23/66, FAA 1647/8; Blake to Maxwell, "Boeing Production Airplane Performance Objectives Based upon General Electric Improved Engine," 11/30/66, memo, FAA 1647/9; Pennell to Maxwell, 12/12/66, FAA 1647/9.

21. For Boeing's subdued promotional style, see: Pennell to Maxwell, 11/3/66, FAA 1647/8; Vierling to the record, "Boeing Presentation—November 14—9 AM," 11/8/66, memo, FAA 1647/4; "Boeing Modifies SST Design," *Aviation Week and Space Technology*, 12/26/66, p. 31.

22. For Lockheed's promotional activities, see: FAA, "Summary Status Report to PAC," 6/30/66, pp. 3–8, PAC/32; "Draft Proceedings of the President's Advisory Committee," 7/9/66, pp. 24–25; UPI dispatch no. 119, 7/15/66, FAA 1647/4; *New York Times*, 8/10/66, pp. 21–22; Lockett to Contracts Manager, "Proposed Article for Lockheed Horizons: Supersonic Transport Design Evaluation," 8/22/66, memo, FAA 1647/7; G. D. Waring to J. W. Locraft, 8/26/66, and attached proposed release, FAA 1647/7; Locraft to the records, "Lockheed Press Release SST2799, 8/26/66, 'Lockheed Supersonic Transport Proposal Submitted Today—FAA Evaluation Begins,' " 9/6/66, memo, FAA 1647/7; C. L. Johnson to McKee, 8/15/66, FAA 6174/95; W. M. Magruder to Lockett, 8/13/66, FAA 1647/7; C. L. Johnson to Maxwell, 9/19/66, FAA 1647/7; Lockett, "Flash Highlights: Technical Operations Division," 9/26/66, FAA 1647/7; Robert A. Bailey to Maxwell, 8/19/66, FAA 6174/95; *New York Times*, 12/2/66, p. 29; D. J. Haughton to Maxwell, 9/6/66, FAA 1647/7; D. J. Haughton to McKee, 9/6/66, FAA 1647/7; C. L. Johnson to Maxwell, 10/24/66, FAA 1647/4.

23. For the structure of the design competition evaluation, see: Maxwell, "Supersonic Transport Development: Daily Highlights," 3/29/66, FAA 6174/95; Lockett to Maxwell, "NASA Support of the SST Source Selection Evaluation," 6/30/66, memo, FAA 1647/6; McKee to James E. Webb, 7/26/66, and attachment, FAA 1647/7; Webb to McKee, 8/5/66, FAA 1647/7; Maxwell to McKee, "Status of Personnel for Source Selection Evaluation," 6/17/66, memo, FAA 1647/2; McKee to McNamara, 7/18/66, FAA 1647/6; Goranson and Pinnes to Chairman, Source Selection Procedures Team, "Revised Scoring Technique for Phase II-C Evaluation," 5/16/66, memo, FAA 1647/2; FAA, "[Draft] Supersonic Transport Economic Model Ground Rules," 5/1/66, FAA 1647/4; F. E. Ellis to (Skaggs) Chief, Economics Staff, "Supersonic Transport Economic Model Ground Rules," 6/23/66, memo, FAA 1647/6; Vierling to Members, Source Selection Planning Team, "Source Selection Task Assignments," 5/24/66, memo, FAA 1647/2; "Role of the Foreign Airlines in the Source Selection Process" (from FAA backup material for PAC meeting of 5/6/66), FAA 6174/96; Maxwell to Robert R. Margrave, 5/27/66, FAA 1647/6; Maxwell to G. Stroh, 7/18/66, FAA 1647/2; Maxwell to Charles S. Murphy, 6/17/66, FAA 1647/6; FAA, "Summary Status Report to PAC," 6/30/66, pp. 43–44, PAC/32.

24. "Draft Proceedings of the President's Advisory Committee," 7/9/66, pp. 38–39.

25. For the organization of the design competition evaluation, see: FAA, "Summary Status Report to PAC," 8/31/66, pp. 37–40, PAC/24; John D. Calhoun to Colizink, "1966 Source Selection Participants," 9/18/68, memo, FAA 1647/2; McKee to the record, "Summary of Comments Made by Administrator, FAA, to the Source Evaluation Group on SST," 9/6/66, memo, FAA 6174/95; Hoover to McKee, "Items for White House Report," 9/12/66, 9/26/66, memos, FAA 6174/95; "Supersonic Transport Development Program Evaluation and Source Selection Team Personnel List," 9/13/66, FAA 1647/2; Hoover to R. A. Bailey, 9/18/66, FAA 1647/7; Maxwell to McKee, "Daily Diary—Evaluation/Source Selection Activities," 9/6/66, 9/7/66, 9/14/66, 9/15–16/66, 10/8/66, 10/10/66, memos, FAA 6174/95; Locraft to Chief, Administrative Staff, "Request from SS-30 for Conference Loans," 9/16/66, memo, and attachment, FAA 1647/7; Robert C. Robinson to SS-305 et al., "Phase III Contract Negotiations," 9/19/66, memo, FAA 1647/7; Locraft to SS-10 et al., "Phase III Negotiation," 9/19/66, memo, and attachment, FAA 1647/7; Locraft to SS-10 et al., "Review of Work Statements Prior to Negotiation of Phase III," 9/20/66, memo, FAA 1647/7; Locraft to Mr. Reuben Stanley, "Review of Statements of Work Prior to Negotiations of Phase III," 10/5/66, memo, FAA 1647/7; Locraft to SS-3,

"Problems of Negotiation Scheduling," 10/11/66, memo, FAA 1647/7; Maxwell to Pennell, 10/19/66, telex, FAA 1647/7; Blake to SS-1 et al., "Lockheed and Boeing Phase II-C Progress Review," 10/24/66, memo, FAA 1647/7; *New York Times*, 11/1/66, p. 25; *Aviation Daily*, 11/1/66, p. 1; Hoover to McKee, "White House Report Item," 10/24/66, memo, FAA 6174/95; Maxwell to Col. Maurice A. Cristadoro, 8/17/66, FAA 1647/7; Cristadoro to SS-100 et al., "SSAS Planning," 8/9/66, memo, FAA 1647/7; Hoover to Cristadoro, "SSAS Planning Meeting—August 10, 1966," 8/11/66, memo, FAA 1647/7; Hoover, "SSAS Activities—September 7, 1966," 9/8/66, memo, FAA 1647/7; Cristadoro to Blake et al., "Methodology Briefing for Administrator, 15 September 1966," 9/9/66, memo, FAA 1647/7; Cristadoro to Hoover et al., "Study Activity for SSAS Data and Information Base," 10/20/66, memo, and attachment, FAA 1647/7; G. T. Haugen to Chairman, Source Selection Advisory Staff, "SSAS Analysis Topics," 10/11/66, memo, and attachments, FAA 1647/7; Haugen to SSAS, "Initial List of Programming Options," 9/29/66, and attachment, FAA 1647/7; Cristadoro to Maxwell, "SSAS Analytical Activity," 10/31/66, memo, FAA 1647/8; Cristadoro to SSAS Task Leaders et al., "Dry Run SSAS Presentations to General Maxwell," 11/9/66, memo, FAA 1647/8.

26. For outside advice on the design competition evaluation, see: Vierling, "Supersonic Transport Development: Daily Highlights," 8/23/66, FAA 6174/95; Lockett to Maxwell, "Progress Report on Cornell's Study of the Variable-Sweep versus Delta SST," 7/29/66, memo, FAA 1647/7; Maxwell to McKee, "Scientific and Technical Review of Supersonic Transport Evaluation Results," 9/9/66, memo, FAA 6174/95; R. L. Bisplinghoff to McKee and Maxwell, "SST Advisory Board," 11/18/66, memo, FAA 6174/95.

27. For the establishment and views of the source selection council, see: Maxwell to the record, "Draft Agency Order," 9/2/66, memo, FAA 6174/95; "Supersonic Transport Source Selection Council," 9/2/66, FAA Order, FAA 6174/95; "SST Program Briefing for General McKee and Source Selection Council," 11/21–23/66, FAA 6174/95; source selection council to McKee, "Source Selection Council Report," 12/9/66, memo, FAA 6174/14.

28. For the FAA's management, control procedures, and managerial style during the evaluation, see: "Draft Proceedings of the President's Advisory Committee on Supersonic Transport," 10/6/66, pp. 7–11, PAC/32; Maxwell to Gordon Thiel (GE), 9/16/66, telex, FAA 1647/7; *New York Times*, 9/24/66, p. 50; Maxwell to John H. Thorn (CIA), 10/7/66, FAA 1647/7; Maxwell to all SST, SSAS and SSEG personnel, "Security of Evaluation Results," 10/7/66, memo, FAA 1647/2; Hoover to Cristadoro, "Daily Source Selection Activities Report," 9/7/66, memo, FAA 6175/95; Cristadoro to Blake et al., "Source Selection Advisory Staff (SSAS) Staff Meetings," 9/9/66, memo, FAA 1647/7.

29. For the evaluation findings comparing the Boeing-General Electric and the Lockheed-Pratt & Whitney designs and for the debate between developing one or two prototype models, see: "SST Growth," circa 12/7–15/66, FAA briefing paper, FAA 1647/1; "SST Program Options," FAA background paper for the PAC meeting of 12/7/66, PAC/4; FAA briefing charts for the PAC meeting of 12/7/66, FAA 1647/1; "Draft Proceedings of the President's Advisory Committee," 12/7/66, pp. 1–25.

30. For the debate between the FAA and the Defense Department over the decision on how to proceed, see: source selection council to McKee, "Source Selection Council Report," 12/9/66; Steadman to McNamara, 12/5/66, memo, PAC/4; "Program Options and Readiness to Proceed," circa 12/5/66, DOD briefing paper, PAC/4.

31. For the debate at the December 7 PAC meeting and for the airline demands, see: "Draft Proceedings of the President's Advisory Committee," 12/7/66; "Comparison of Airline Inputs and Assumptions with FAA Validated Data," FAA background paper for PAC meeting of 12/7/66, PAC/4: "Comments on Comparison of Airline Inputs and Assumptions with FAA Validated Data," Defense Department staff background paper for PAC meeting of 12/7/66, PAC/4; FAA briefing charts for PAC meeting of 12/7/66, FAA 6174/96.

32. "Draft Proceedings of the President's Advisory Committee on Supersonic Transport," 12/10/66, pp. 35–63, PAC/32.

33. McKee and Boyd to McNamara, "Draft Fourth Interim Report of the President's Advisory Committee on Supersonic Transport," 12/12/66, memo, FAA 6174/95; McKee to McNamara, 12/16/66, 12/19/66, FAA 6174/95.

34. "Memorandum for the President, Fourth Interim Report of the President's Advisory Committee on Supersonic Transport," 12/22/66, PAC/32.

35. Califano to the President, 12/26/66, memo, LBJ Oversize.

36. Goodrich to McKee, "Interpretation of Section 2, Fourth Interim Report of PAC," 12/28/66, memo, FAA 6174/95; McKee to Schultze, 12/30/66, and attachments, FAA 6174/95.

37. Maxwell to McKee, "SST Fiscal Year 1968 Financing," 12/19/66, memo, FAA 6174/95; McKee to McNamara, "SST Fiscal Year 1968 Financing," 12/19/66, memo, LBJ Oversize.

38. Califano to the President, 12/26/66, memo, p. 6, LBJ Oversize; Ramsey Clark to Califano, 12/28/66, memo, LBJ Oversize.

39. McKee to Schultze, 12/30/66 [this letter discusses McKee's December 29 letter to the president], FAA 6174/95.

40. *New York Times*, 1/1/67, pp. 1, 38, 42; *Aviation Daily*, 1/5/67, p. 19.

41. For reactions at Lockheed and Boeing, see: D. J. Haughton to All Members of Supervision, "Effects of the SST Loss," 1/2/67, memo, FAA 6174/95; statement by D. J. Haughton, 12/31/66, FAA 6174/95; *New York Times*, 1/1/67, p. 28.

42. *Washington Post*, 12/29/66. Enke a few months later published an academic summary of his views on the SST, see: Stephen Enke, "Government-Industry Development of a Commercial Supersonic Transport," *American Economic Review*, 57 (May, 1967) pp. 71–75.

Chapter 12

1. "The Supersonic Decision," *Aviation Week and Space Technology*, 1/9/67, p. 11.

2. Elwood Quesada to McKee, 1/13/67, FAA 1647/8.

3. Humphrey quote from "Let's Get on with the SST," *Machinist*, 2/16/67, NAS Cmte newsletter, no. 3, 2/24/67, p. 1.

4. Halaby quote from "The Supersonic Transport," *Congressional Record*, 2/16/67, p. A673, NAS Cmte newsletter, no. 3, 2/24/67, p. 2.

5. Marion Sadler to McKee, 1/4/67, FAA 1647/8; *New York Times*, 1/12/76, p. 60.

6. B. J. Vierling to Associate General Council-Legislative Division, "Notes on Staff

Report Policy Planning for Aeronautical Research and Development—Issues for Further Congressional Consideration," 1/6/67, memo, FAA 1647/8.

7. Boyd quote from *Aviation Daily*, 2/14/67, p. 261, NAS Cmte newsletter no. 2, 2/21/67, p. 1.

8. "Summary of Statements against the SST," circa 1/31/67, LBJ EX CA.

9. McNamara quote from *Aviation Daily*, 2/23/67, p. 311, NAS Cmte newsletter, no. 4, circa 2/24–3/9/67, p. 2.

10. Maxwell to McKee, 1/4/67, transmittal slip, FAA 6174/95; McKee to Schultze, 1/4/67, draft letter, FAA 6174/95; McKee to Schultze, 1/6/67, FAA 6174/95.

11. Califano to McNamara, 1/4/67, memo, LBJ Oversize.

12. McKee to McNamara, 1/12/67, and attached "SST Manufacturers' Cost Sharing Issues," FAA 1647/8.

13. Maxwell to the record, "Conference on SST Financing, 13 January 1967," 1/13/67, memo, and "Conference on SST Financing, 16 January, 1967," 1/17/67, memo, FAA 6174/95.

14. McKee to Schultze, 1/21/67, draft letter, FAA 6174/95; McKee to Schultze, 1/21/67, and attached excerpts of 10/9/66 PAC proceedings, FAA 6174/95.

15. Maxwell to McKee, "Airline Investment in SST Development Program," 1/16/67, memo, FAA 6174/95; McKee to the President, "SST: Airline Risk Payments," circa 1/20/67, draft memo, FAA 6174/95; Skaggs to Maxwell, "Airline Financial Participation," 1/26/67, memo, FAA 1647/8; Maxwell to McKee, "Pan Am Suggestion on Airline Participation in Phase III Financing," 1/28/67, memo, FAA 6174/95.

16. For pressure for airline risk participation, see: Maxwell to the record, "Conference on SST Financing—16 January, 1967," 1/17/67, memo, FAA 6174/95.

17. Schultze to the President, "SST," 1/31/67, memo, LBJ EX CA.

18. McKee to the President, 2/8/67, LBJ EX CA.

19. McKee to Representative George H. Mahon, 2/1/67, FAA 6174/95.

20. For Pan American's views, see: Lawrence S. Kuter to Maxwell, 12/9/66, FAA 6174/95; Borger to Vice-President, Technical Staff, "Relative Operating Costs of SST's," 11/9/66, 1/4/67, memos, FAA 6174/95; Kuter to McKee, circa 1/10/67, handwritten note, FAA 6174/95.

21. Maxwell to McKee, "General Kuter's Handwritten Note," 1/11/67, memo, FAA 6174/95; McKee to McNamara, 1/12/67, FAA 6174/95.

22. Skaggs to Maxwell, "Pan Am Financing Proposal," 1/16/67, memo, and attached "Evaluation of an Airline Financing Proposal," 1/16/67, FAA 6174/95; Maxwell to McKee, "Pan Am Suggestion on Airline Participation in Phase III Financing," 1/28/67, memo, FAA 6174/95; Maxwell to the record, "Conference on SST Financing—16 January, 1967."

23. Maxwell to McKee, "Airline Payments on SST," 2/7/67, memo, FAA 6174/95; McKee to the President, 2/7/67, LBJ EX CA.

24. For the status of airline SST delivery positions, see: Boyd to the President, 2/11/67, LBJ FAA Hist. Doc.; Tillinghast to Boyd, 2/14/67, FAA 1647/9; Kuter to McKee, 2/10/67, LBJ FAA Hist. Doc.; F. D. Hall (Eastern Airlines) to Maxwell, 2/10/67, telex, FAA

1647/9; Vierling to the record, "Telephone Conversations with Mr. Redman of Airlift International, Inc.," 2/10/67, memo, FAA 1647/9; Vierling to the record, "Telephone Conversations with Mr. Brian A. Cooke of World Airways, Inc.," 2/10/67, memo, FAA 1647/9.

25. "President Readies SST Message," *Aviation Week and Space Technology*, 2/20/67, pp. 26–27; McKee to Symington, 2/13/67, and attachments, FAA 6174/95; McCone to McKee, 2/21/67, FAA 1647/9.

26. Maxwell to McKee, "Negotiations with the Boeing Company on Airline Financial Participation," 2/23/67, memo, FAA 6174/95.

27. McKee to Allen, 2/23/67, and attachment, FAA 6174/95.

28. For Boeing's "capitulation" and for its new SST sales guidelines for American-flag carriers, see: Allen to McKee, 2/24/67, telex, LBJ FAA Hist. Doc.; Maxwell to Boyd, "Airline Letters re Risk Contribution," 2/28/67, memo, FAA 1647/9; Boyd to Trippe, 2/28/67, LBJ FAA Hist. Doc; Boyd to Charles H. Dolson (Delta Airlines), 2/28/67, FAA 1647/9.

29. For Boeing's negotiations and agreements with the airlines, see: "Airline Participation Agreements and Policy," 3/20/67, LBJ FAA Hist. Doc.; Maxwell to Boyd, "Airline Participation in the SST Financing—Status Report," 3/27/67, memo, FAA 1647/9; Maxwell to Boyd, "Airline Contribution for SST Development," 3/7/67, memo, FAA 1647/9; Maxwell to McKee, "Airline Agreements with Boeing," 3/20/67, memo, FAA 1647/95; Maxwell to Boyd, "Airline Contributions for SST Development," 3/6/67, memo, and attached to Smith et al., 3/6/67, telex, FAA 1647/9; Skaggs to Maxwell, "Investment Tax Credit," 3/6/67, memo, FAA 1647/9; Hans A. Klagsbrunn to Maxwell, 3/17/67, FAA 1647/9; Rummel to Boyd, 3/7/67, FAA 1647/9; Rummel to Boyd, 3/7/67, FAA 1647/9; Tillinghast to the Boeing Company, 3/7/67, FAA 1647/9; Rummel to Maxwell, 4/5/67, FAA 1647/9; "Minutes of April 6, 1967, FAA Airline Meeting," FAA 1647/9; Maxwell to Klagsbrunn et al., 4/7/67, FAA 1647/9.

30. For the determination of the general FAA policy about delivery positions and airline contributions, see: Parsons to Deputy Director, Office of Budget, "Apportionment Request," 3/1/67, memo, and attached "Financing the SST Prototype Phase," 3/1/67, FAA 1647/9; Chatkin, "Re Foreign Carriers Making Risk Contribution for SST Delivery Positions," 2/10/67, record of telephone call, FAA 1647/9; Vierling to Maxwell, "Foreign Airline Participation in SST Risk Contribution," 2/17/67, memo, FAA 1647/9; Vierling to the record, "Telephone Conversation with Mr. Kenneth Rutledge of Air Canada Airlines," 2/15/67, memo, FAA 1647/9; Maxwell to C. H. Jackson (BOAC), 2/23/67, FAA 1647/9; "Airline Participation Agreements and Policy," 3/20/67, LBJ FAA Hist. Doc.; Maxwell to Boyd, "Airline Participation in the SST Financing—Status Report," 3/27/67; Maxwell to McKee, "Foreign Carrier Participation in SST Financing," 3/20/67, memo, FAA 1647/9; McKee to Maxwell, 3/21/67, handwritten note, FAA 1647/9; Maxwell to D. H. Hardesty (Northwest Airlines) et al., 5/15/67, FAA 1647/10; Vierling to McKee, "Revised Allocation Policy," 5/11/67, memo, FAA 6147/10; Tillinghast to Maxwell, 5/22/67, FAA 1647/10; "Release Schedule of Airline Delivery Position Policy," 5/22/67, FAA 6174/95; Cecil MacKay to Maxwell, 5/26/67, route slip, FAA 6174/95; Department of State to Athens et al., "Civil Aviation—U.S. Policy on SST Delivery Position Reservations," 5/26/67, DOS airgram, FAA 1647/10.

31. "New Policy Governing SST Delivery Priorities Announced by FAA," 6/5/67, FAA news release no. 67-52, LBJ FAA Hist. Doc.

32. McKee to Marvin Watson, 2/11/67, handwritten note, LBJ EX CA; from the President, 2/11/67, 8:00 P.M., typed message, LBJ EX CA.

33. Califano to the President, 3/1/67, memo, and attached Boyd to the President, "Supersonic Transport," memo, "SST Tally—1/2/67, U.S. Airline Contributions to SST," draft presidential message, and draft White House press release, 3/1/67, LBJ Oversize.

34. Califano to the President, 3/13/67, and attached Boyd to the President, "Supersonic Transport Program," 3/13/67, memo, LBJ Oversize; President to Boyd, 3/13/67, draft letter, and attached "SST Tally—3/13/67," LBJ Oversize.

35. "SST Tally—3/20/67," LBJ Oversize; "SST Headcount as of March 30, 1967," LBJ EX Misc.

36. For the activities of the pro-SST lobbying effort, see: undated message beginning "On March 1, our tally . . . ," FAA 6174/95; "SST Tally—3/20/67"; Walt Rostow to the President, 3/16/67, LBJ National Security Files; Boyd to the President, "SST," 4/3/67, memo, LBJ EX CA; Boyd to the President, "SST," 4/5/67, memo, LBJ Oversize; Boyd to Califano, "SST," 3/28/67, memo, LBJ Oversize; Boyd to the President, "Conversation with Senator George D. Aiken (Vt.) on the SST," 3/29/67, memo, LBJ Oversize; Robert E. Kintner to the President, 3/24/67, memo, LBJ EX CA; Boyd to the President, "SST," 3/23/67, memo, LBJ Oversize; C. R. Smith to McKee, 3/31/67, and attached C. R. Smith to Senator E. L. Bartlett, 3/31/67, FAA 6174/95; Smith to Rostow, 4/3/67, LBJ Gen. Misc.

37. Mike Manatos to Watson, 4/14/67, memo, LBJ EX CA.

38. Maxwell to McKee, "Delay of Phase III Go-Ahead-Adverse Effects," 4/10/67, memo, FAA 1647/9.

39. Califano to the President, 4/20/67, memo, LBJ Oversize.

40. Ibid., 4/21/67, and attached Schultze to the President, "SST," 4/21/67, memo, the President to Boyd, Schultze to the President, the President to the House Speaker, and "Statement by the President on the Supersonic Transport," draft letters and draft statement, LBJ Oversize.

41. Boyd to the President, "SST," 4/28/67, memo, LBJ Oversize; Califano to the President, 4/28/67, memo, and the President to Califano, handwritten notation, LBJ Oversize.

42. "Statement by the President on the Supersonic Transport," 4/29/67, White House release, LBJ Oversize; White House news conference transcript, 4/29/67, LBJ National Security Files; the President to Boyd, 5/1/67, FAA 6174/95; Vierling to McKee, "Item for White House Report," 5/4/67, memo, FAA 6174/95.

43. *New York Times*, 4/30/67, pp. 1, 85.

44. Tillinghast to the President, 5/3/67, telegram, LBJ Oversize; Tillinghast to McKee, 5/3/67, 5/19/67, FAA 1647/10.

45. McKee to George Keck et al., 5/15/67, individual letters, FAA 1647/10.

46. Harold D. Watkins, "SST Program Responsibility Shifts," *Aviation Week and Space Technology*, 5/8/67, pp. 28–29.

47. Maxwell, "Remarks before the Aviation/Space Writers Association," 5/16/67, Las Vegas, Nevada, LBJ FAA Hist. Doc.

48. Symington to McKee, 4/13/67, FAA 1647/4.

49. McKee to Edmund P. Learned, 3/14/67, FAA 1647/9.

Chapter 13

1. For the FAA's perceptions of problems at Boeing in March 1967, see: Parsons to Assistant to the Director, "Proposed Boeing Brochure," 3/1/67, memo, FAA 1647/9; Richard Glassler, "Trip Report," 3/2/67, FAA 1647/9; Carroll to Howell, 3/8/67, memo, FAA 1647/9.

2. Pennell to Maxwell, 2/9/67, 1647/9.

3. "Withington to Head Boeing SST Effort," *Aviation Week and Space Technology*, 3/20/67, p. 38.

4. C. L. Blake, "Flash Highlights: Engineering Division," 7/18/67, FAA 1647/11.

5. "Boeing to Subcontract 69% of SST," *Aviation Week and Space Technology*, 1/9/67, p. 28.

6. Maxwell to McKee, "Martin Co. Participation in SST Program," 9/5/67, memo, FAA 6175/95; Charles A. Fuchs "Trip Report," 9/11–14/67, FAA 1647/12.

7. For the Boeing SST's design difficulties that emerged in the 1966 evaluation, see: Lockett to Skaggs, "B2707-100 Weights," 1/13/67, memo, FAA 1647/8; R. G. O'Leone, "Boeing Focuses on Range-Payload Goal," *Aviation Week and Space Technology*, 1/9/67, pp. 26–27; Blake to the record, "B2707-100 Performance," 1/31/67, memo, FAA 1647/8.

8. For continued doubts in March and April 1967 about Boeing's SST design, see: Wesley M. Maulden to Maxwell, 3/17/67, FAA 1647/9; Parson to Maxwell, 3/23/67, memo, FAA 1647/9; H. W. Withington to Maxwell, 4/17/67, FAA 1647/9; Vierling to Executive Secretary, "White House Report," 4/20/67, memo, FAA 1647/10; R. E. Parsons, "White House Report," 4/20/67, FAA 1647/10; Maxwell, "Trip Report—Boeing Program Review Meeting Action Items," 4/26/67, FAA 1647/10; Robert C. Robinson to SS-3 et al., "Boeing Program Review Meeting Action Items," 4/24/67, memo, and attachment, FAA 1647/10; R. W. Pinnes to Maxwell, "Comments on SST Program Review Meeting," 4/24/67, memo, FAA 1647/10.

9. Colonel James H. Voyles, Jr., "Trip Report," 5/11–12/67, FAA 1647/10.

10. "Trip Report—Summary: The Boeing Company," 5/10–11/67, visited by Messrs. Bartholomew, Drant, Goranson, Rosenbaum, and Tamburello, FAA 1647/10; T. H. Drant, "Trip Report: The Boeing Company," 5/10–11/67, FAA 1647/10.

11. For Boeing's efforts at designing the B2707-200, see: Charles J. V. Murphy, "Boeing's Ordeal with the SST," *Fortune*, 10/68, pp. 129–132; Blake, "Flash Highlights: Engineering Division," 7/18/67, 7/20/67, FAA 1647/11; D. O. Waller, "Analysis and Control Division—Configuration Management Branch: Trip Report," 7/18–20/67, FAA 1647/11; Withington to Maxwell, 8/10/67, FAA 1647/11.

12. For the design problems (especially increasing weight) with the B2707-200, see: Withington (signed by Maulden) to Maxwell, 8/4/67, FAA 1647/11; Maxwell to Withington, 8/10/67, FAA 1647/11; Blake, "Flash Highlights: Engineering Division," 8/10/67, FAA 1647/11; Lockett to Blake, "Recommended SST Prototype Weight Savings," 8/7/67, memo, FAA 1647/11.

13. Blake to Maxwell, "Acceptance of the B-2707-200 Design," 8/18/67, memo, FAA 1647/11.

14. F. E. Ellis, Jr., "Trip Report," 9/5–7/67, FAA 1647/11.

15. For pessimistic reports on the B2707-200, see: Drant to Blake, "Status of Airframe Structure," 9/14/67, memo, FAA 1647/12; C. R. Ritter and John Shoates, "Trip Report," 9/6–15/67, FAA 1647/12; Vierling to McKee, "Topics of Discussion with Mr. Bill Allen, President of Boeing," 9/29/67, memo, FAA 6174/95.

16. Murphy, "Boeing's Ordeal."

17. R. L. Bisplinghoff to Maxwell, "Boeing SST Review, November 14, 1967," 11/16/67, memo, FAA 1647/12.

18. "Boeing Reports U.S. SST Program Progress," 11/28/67, Boeing news release, no. S-9485, FAA 6174/95; Richard G. O'Leone, "Final SST Prototype Design Established," *Aviation Week and Space Technology*, 12/4/67, pp. 34–35.

19. For the FAA's seeemingly positive posture on Boeing's redesign efforts at the end of 1967, see: McKee to Boyd, "Proposed Design Changes," 12/4/67, memo, FAA 1647/13; McKee to Mahon et al., 12/4/67, FAA 6174/95; Boyd to the President, "Proposed Design Changes to the SST," 12/11/67, memo, LBJ EX CA; Harley O. Staggers to McKee, 12/11/67, FAA 6175/95; McKee to Staggers, 12/20/67, FAA 6174/95.

20. R. Fabian Goranson to Maxwell, "Changes Affecting SST Performance," 12/19/67, memo, and attachments, FAA 6174/13; Blake to Bisplinghoff, 12/21/67, FAA 6174/13.

21. Blake, "Flash Highlights: Engineering Division," 1/3/68, FAA 905/1.

22. Allen to McKee, 1/2/68, LBJ FAA Hist. Doc.

23. McKee to Senator John Stennis, 1/26/68, FAA 6174/95; McKee to Representative Edward P. Boland, 1/26/68, LBJ Hist. Doc.; Boyd to the President, "SST Program Schedule," 1/24/68, memo, LBJ EX CA.

24. For Boeing's decision to ask for a year-long delay and for the FAA's decision to grant this request, see: Murphy, "Boeing's Ordeal"; Boyd to the President, 2/14/68, memo, LBJ FAA Hist. Doc.; Hutchinson to McNamara, "SST Program Delay," 2/14/68, memo, LBJ FAA Hist. Doc.

25. Bisplinghoff to Maxwell, "Boeing Visit 15 February 1968," 2/19/68, memo, FAA 905/1.

26. McKee to Boyd, "SST Configuration Improvement Plan," 3/22/68, memo, FAA 905/2.

27. Califano to the President, 2/19/68, memo, LBJ EX CA.

28. Hutchinson to the President, 2/20/68, memo, FAA 6174/95; President to Watson, handwritten note (on Hutchinson to the President, 2/20/68, memo), LBJ EX CA.

29. *New York Times*, 2/21/69, p. 1; *Seattle Post-Intelligencer*, 2/21/68.

30. For the effect of the *New York Times* report, see: Watson to the President, 2/21/68, memo, LBJ EX CA; Harry C. McPherson to Jim Jones, 2/21/68, telex, LBJ EX CA; Hutchinson to the President, 2/21/68, memo, LBJ EX CA.

31. Boeing news release, 2/22/68, FAA 6175/95.

32. Allen to McKee, 2/22/68, FAA 6174/95.

33. "Sample of Long and Short Letters Sent to Congressmen—2/22/68," attached Hutchinson to Magnuson, 2/22/68, Hutchinson to Senator Carl Hayden, 2/22/68, FAA 6174/95; Maxwell to Senator Robert A. Taft, Jr., 2/22/68, FAA 905/2; Hutchinson to Russell W. Hale (National Aeronautics and Space Council), 2/22/68, FAA 6174/95;

Hutchinson to McNamara, "SST Program Delay," 2/22/68, memo, FAA 6174/95; Steadman to McKee, 2/26/68, FAA 6174/95; Maxwell to Sam Hughes (Bureau of the Budget), "Supersonic Transport Program," 2/22/68, memo, FAA 6174/95.

34. *Aviation Daily*, 3/8/68, p. 37, NAS Cmte Newsletter, no. 33, 3/15/68, p. 3.

35. Contract Amendment (Request No. WA-SS-68-54, Contract No. FA-SS-67-3, Modification No. 15, effective date: 3/29/68), signed on 4/25/68 by Boeing and the FAA, FAA 6174/95; Maxwell to Withington, 4/2/68, FAA 6174/95; McKee to Boyd, "Contract Modification for SST Design Improvement Program," 4/15/68, memo, FAA 905/2.

36. Allen to McKee, 4/9/68, LBJ FAA Hist. Doc.

37. McKee to Allen, 4/16/68, FAA 905/2.

38. Maxwell to Hale, 4/30/68, FAA 6174/95; Maxwell to Lehan, "Contract Amendment for the SST Design Improvement Program," 4/26/68, memo, FAA 905/2.

39. For the beginning of the year-long redesign effort by Boeing, see: Withington to Maxwell, 3/15/68, FAA 6174/95; R. W. Pinnes, "Trip Report: SST Propulsion Technical Review—The Boeing Company, Seattle, Washington, April 8–10, 1968," 4/23/68, FAA 905/2; Blake, "Record of Telephone Call from Maulden," 4/8/68, FAA 905/2.

40. For Boeing's work on the SCAT-15F, see: C. R. Ritter to the record, "SCAT-15F Operational Empty Weight," 4/10/68, memo, FAA 905/2; Blake to Maxwell, "SCAT-15F Operational Empty Weight," 4/11/68, memo, FAA 905/2; Irons to Blake, "Estimated DOC for SCAT 15F," 4/12/68, memo, FAA 905/2; Withington to Blake, 5/3/68, FAA 6174/92.

41. For the status of the Boeing design effort in May and June, 1968, see: Withington to Maxwell, 5/23/68, FAA 6174/1; Blake, "Trip Report: The Boeing Company—June 6–7, 1968," 6/17/68, FAA 6174/1; R. F. Goranson to the record, "Notes on Boeing Configuration Review, June 6–7, 1968," 6/20/68, memo, FAA 6174/1; R. H. Praut, "Trip Report Summary: The Boeing Company—June 6–7, 1968," FAA 6174/1; R. L. Krick to Blake, "Comments on June 6–7, 1968 Boeing Review," 6/13/68, memo, FAA 6174/1; Ellis to Blake, "Boeing Review—June 6–7, 1968, SS-100 Memorandum of June 12, 1968," 6/19/68, memo, FAA 6174/1; V. Tamburello and R. N. Bell, "Trip Report: The Boeing Company and NASA Aims—June 18–20, 1968," FAA 6174/1; Ellis, "Flash Highlights: Engineering Division," 7/3/68, FAA 6174/1.

42. For the status of the Boeing redesign effort in August and September 1968, see: Blake, "Flash Highlights: Engineering Division," 8/5/68, FAA 6174/1; Holtby to distribution list, "SST Internal Evaluation Data," 8/6/68, memo, FAA 6174/92; Ellis, Tamburello, and Bell, "Trip Report: The Boeing Company—August 13–14, 1968," 8/22/68, FAA 6174/1; F. P. Blakeslee (Boeing) to F. Goranson, 9/23/68, FAA 6174/1; Vierling to Acting FAA Administrator, "White House Report," 9/19/68, memo, FAA 6174/1; *New York Times*, 10/22/68, p. 93; *Aviation Daily*, 10/22/68, p. 71, NAS Cmte, Newsletter, no. 48, 10/17–31/68, p. 2.

43. *Aviation Daily*, 4/26/68, p. 288, NAS Cmte Newsletter, no. 37, 5/20/68, p. 4; *New York Times*, 5/1/68, p. 94; *Tacoma News Tribune*, 5/2/68, p. D-12, NAS Cmte Newsletter, no. 37, 5/20/68, p. 5; "NASA's SCAT-15F Model: A New SST Contender," *Aerospace Technology*, May 20, 1968, p. 26.

44. *Washington Post*, 7/9/68, p. 1, NAS Cmte Newsletter, no. 41, 7/1/68, p. 5; Harold D. Watkins, "Boeing Approaches Critical SST Decision," *Aviation Week and Space Tech-*

nology, 7/22/68, pp. 31–32; *Washington Post*, 9/4/68, p. A5, NAS Cmte Newsletter, no. 46, 9/13–30/68, p. 2; *New York Times*, 9/14/68, pp. 1, 7, 9/19/68, p. 93.

45. For the debate on whether the FAA's contract with Boeing permitted the design change from a swing-wing to a fixed-wing model, see: Locraft to Chief, ATC and Communications Contracts Branch, "The Boeing Company's Impending Change to Fixed Wing from a Variable Sweep Wing Configuration," 7/17/68, memo, FAA 6174/1; Locraft to Chief, Advisory Services Division, "Redesign Effort—Contract FA-SS-67-3, The Boeing Company," 8/30/78, memo, FAA 6174/2; Elmer M. Staats (U.S. Comptroller General) to Representative Sidney Yates, 6/21/68, FAA 6174/1; Representative Harley O. Staggers to Boyd, 9/20/68, FAA 6174/95; John E. Robson to Staggers, 10/25/68, FAA 6174/95; Maxwell to Thomas, 10/24/68, memo, FAA 6174/95.

46. For the findings of the interagency evaluation, see: "Briefing for Airline SST Committee Participating in Evaluation of the Proposed Boeing SST B-2707-300 Configuration," 2/6/69, "Validation Summaries: Airplane Technical; System Integration; and Manufacturing/Economics," 1/24/69, "Assessment Summaries: Airplane Technical," 2/5/69, "System Integration," 1/31/69, and "Manufacturing/Economics," 2/5/69, FAA 6174/97.

47. Vierling et al. to Maxwell, "SST Program Integration Board Report," 2/13/69, memo, FAA 6174/95.

48. For the formation and activities of Bisplinghoff's group, see: Maxwell to Dr. Arthur Raymond, 9/4/68, FAA 6174/2; Maxwell to Dr. Ernest E. Sechler, 9/4/68, FAA 6174/2; Blake to Bisplinghoff, 7/26/68, FAA 6174/1; Calhoun to F. G. Coffey (Boeing), 10/7/68, FAA 6174/2; Holtby to Bisplinghoff, 10/25/68, FAA 6174/2; Calhoun to Bisplinghoff, 1/29/69, FAA 6174/3; Calhoun to Bisplinghoff, Raymond, and Sechler, 11/4/68, FAA 6174/7; Calhoun to Program Integration Board, "February 6 Is Date for Briefing to Airlines and Dr. Bisplinghoff's Committee," 11/4/68, memo, FAA 6174/2; Calhoun to SS-100 et al., "Dr. Bisplinghoff's Committee to Meet February 5, 6, & 7, 1969," 1/7/69, memo, FAA 6174/3.

49. For the findings of Bisplinghoff's group, see: Calhoun to Chairman, PIB, "Plans and Requirements for Dr. Bisplinghoff's Team during the Evaluation," 9/17/68, memo, FAA 6174/2; Calhoun, "Report of Dr. Bisplinghoff's Committee Meeting [of 10/1/68]," FAA 6174/2; "Agenda for Dr. Bisplinghoff's Committee [for 10/28–29/68]," FAA 6174/97; Calhoun to PIB et al., "Dr. Bisplinghoff to Visit Here Monday, December 2, 1968," 11/27/68, memo, FAA 6174/97; Raymond, Sechler, and Bisplinghoff to Maxwell, "SST Technical Evaluation Committee Report," 2/7/69, memo, FAA 6174/95; Bisplinghoff to Thomas and Maxwell, "Personal Views on the B-2707-300 Airplane and on the SST Program," 2/7/69, memo, FAA 6/79/95; Sechler to Thomas and Maxwell, "Personal Comments Generated from a Review of the SST B-2707-300 as a Member of the Dr. R. Bisplinghoff Committee—Primarily Devoted to the Area of Structures and Materials," 2/7/69, memo, FAA 6174/95; Raymond to Thomas and Maxwell, "Personal Views on Current Boeing SST Design Following Review by Bisplinghoff Group, February 5–7, 1969," 2/7/69, memo, FAA 6174/95; Calhoun to Bisplinghoff, 2/11/69, FAA 6174/3.

50. Maxwell to Airlines Holding Delivery Positions for the American SST, "Preliminary Reports of Government Technical Review of Boeing SST B2707-300," 1/28/69, memo, FAA 6174/3; Mentzer to Airline SST Committee, 1/30/69, FAA 6174/3.

51. "Briefing for Airline SST Committee Participating in Evaluation of the Proposed Boeing SST B-2707-300 Configuration," 2/6/69, FAA 6174/97.

52. Mentzer to Maxwell, 2/10/69, FAA 6174/3.

53. R. V. Carleton (Braniff) to Maxwell, 2/14/69, FAA 6174/4; Harding L. Lawrence (Braniff) to Maxwell, 2/17/69, FAA 6174/4; G. E. Keck (United) to Thomas, 2/18/69, FAA 6174/4; C. H. Dolson (Delta) to Thomas, 2/20/69, FAA 6174/4.

54. Harold D. Watkins, "Airlines Appear to Be Diverging in Degrees of Support for SST," *Aviation Week and Space Technology*, 2/17/69, p. 25.

55. Maxwell to Thomas, "Results of Evaluation of the Boeing Company's Proposed Redesign of the SST (B-2707-300)," 2/14/69, memo, FAA 6174/95.

Chapter 14

1. Raymond B. Maloy to Deputy Administrator, "History of the Supersonic Transport Program—International," 11/22/26, memo, FAA Historian's Files; American Embassy, Paris, to Secretary of State, Washington, D.C., 6/1/64, DOS cable, FAA 1647/12.

2. Maloy to Deputy Administrator, "History of the Supersonic Transport Program—International," 11/22/66; Califano to McNamara, "Supersonic Transport Advisory Committee," 11/6/64, memo, pp. 7–10, PAC/10.

3. D. J. Haughton to Halaby, 1/1/65, and attached, "Lockheed Comments on FAA Concorde Paper of October 1, 1964," FAA 6174/95.

4. For British-American discussions on the SST at the beginning of 1965, see: Maloy to Deputy Administrator, "History of the Supersonic Transport Program—International," 11/22/66; Bain to Halaby, "U.S.–Concorde Information," 2/6/65, memo, and attached charts and tables, FAA 6174/95; Lockett to Halaby, "Revised Comparative Data," 2/10/65, memo, and attached tables, FAA 6174/95.

5. "Tripartite Meeting between Ministers Jacquet, Jenkins and Mr. Halaby" (minutes of the meeting of 2/16/65), LBJ FAA Hist. Doc.

6. "Draft Proceedings of the President's Advisory Committee on Supersonic Transport," 3/30/65, pp. 1–7, 77–79, PAC/10.

7. For the anti-Concorde views of Enke and his group, see: Enke, "The SST Venture—Issues and Questions," SST 1965 Task Force Internal Paper, no. 1, 4/12/65, pp. 24–28, PAC/41; Enke, "The Profitability and Place of Supersonic in Air Transport (an Interim Statement)," SST 1965 Task Force Internal Paper, no. 13, 4/29/65, revised 5/6/65, pp. 12–14, PAC/41; D. J. Edwards, "Some Aspects of SST Development," SST 1965 Task Force Internal Paper, no. 9, 4/29/65, pp. 3–6, PAC/14; DOD, "Memorandum for the President's Advisory Committee on Supersonic Transport," 5/1/65, memo, pp. 13–14, PAC/11. For the TSR-2 cancellation, see: Richard J. Kent, Jr., *Safe, Separated, and Soaring* (Washington, D.C.: U.S. Government Printing Office, 1980), p. 140.

8. "Draft Proceedings of the President's Advisory Committee on Supersonic Transport," 5/5/65, pp. 9–19, PAC/32.

9. Bain to Halaby, "Summary Information—Paris Air Show," 6/22/65, memo, FAA 6174/96; Bain to McKee, "Concorde Certification Data," 8/17/65, memo, FAA 6174/95; Califano to McNamara, "SST Briefing for Congressional Leadership," 6/30/65, memo, and attached, "SST Briefing Outline," 6/30/65, p. 4, FAA 6174/96.

10. SST Economics Task Force, "The Economics of a U.S. Supersonic Transport (Report of Findings to Date)," 8/20/65, pp. 30–33, PAC/25.

11. FAA briefing material on "Agenda Item No. 4" for the 10/9/65 PAC meeting, FAA 6174/96.

12. "Draft Proceedings of the President's Advisory Committee on Supersonic Transport," 10/9/65, pp. 3–7, 45–47, PAC/32.

13. Herbert J. Coleman, "Concorde Hews to Schedule for First Flight in March 1968," *Aviation Week and Space Technology*, 10/25/65, pp. 130–133.

14. FAA, "Concorde Intelligence Summary," 10/29/65, FAA 6174/96; Maxwell to Steadman, 10/29/65, FAA 6174/96.

15. Maxwell to Maloy, "Status of Concorde," 11/1/65, memo, FAA 6174/95; Maloy to McKee, "Concorde Evaluation," 11/2/65, memo, FAA 6174/96.

16. "Draft Proceedings of the President's Advisory Committee on Supersonic Transport," 11/6/65, pp. 4, 44, 51–53, PAC/32.

17. "Memorandum to the President, Third Interim Report of the President's Advisory Committee on Supersonic Transport," and "Annex," 11/15/65, PAC/3.

18. Enke, "Continued Interest of French and British Officials in an Agreement with the U.S. to Rationalize SST Competition," 1/24/65, PAC/13.

19. Maloy to Deputy Administrator, "Visit of Dr. Stephen Enke, Deputy Assistant Secretary, Department of Defense, to Paris, January 10–11, 1966," 2/2/66, memo, and attached Maloy, circa early 1/66, cable, and article from 1/2/66 issue of the *Economist*, "Washington's Supersonic Offer," FAA 6174/96.

20. Maxwell to Steadman, 2/2/66, FAA 6174/96.

21. For favorable reviews of the Concorde, see: Lockett to Maxwell, "Trans World Airlines Analysis of the Concorde," 1/10/66, memo, FAA 1647/3; Maxwell, "Supersonic Transport Development: Daily Highlights," 1/27/66, FAA 6174/95.

22. Maxwell, "Supersonic Transport Development: Daily Highlights," 1/21/66, 2/7/66, FAA 6174/95.

23. FAA, "Summary Status Report to the President's Advisory Committee on Supersonic Transport," 2/15/66, p. 9, PAC/13.

24. For PAC views and decisions in March 1966, see: Maxwell to the record, "Meeting of President's Advisory Committee for Supersonic Transport Members," 3/10/66, memo, FAA 6174/96; McNamara to McKee, and attached, "Agreements Reached at Informal March 9 Meeting of Several Members of the PAC-SST," PAC/13. For Enke's views in March, see: Enke to McNamara, "Rationalizing SST Competition: Some British Interest," 2/25/66, memo, PAC/13; Enke to McNamara, "Important Matters Requiring Voter Attention," 2/25/66, memo, PAC/36; Enke to McNamara, 3/3/66, memo, PAC/13.

25. FAA, "The Effect of a Longer Interval between the Concorde and the U.S. Supersonic Transport," and "Concorde Impact on SST Schedules" (reports to the PAC for the 7/9/66 PAC meeting), PAC/4.

26. For the March 1966 CIA study, see: Enke to McNamara, "Preparations for May 6 PAC/SST Meeting," 4/29/66, memo, PAC/1; FAA, "Summary Status Report to President's Advisory Committee on Supersonic Transport," 6/30/66, p. 39, PAC/32.

27. "Draft Proceedings of the President's Advisory Committee on Supersonic Transport," 5/6/66, pp. 9–12, 16, 19–28, 76, PAC/32.

28. For various reports on the Concorde received by the FAA, see: B. J. Vierling to Maloy, "Concorde Program," 5/17/66, memo, FAA 1647/3; Lockett to Vierling, "Accuracy of Sud/BAC Discussion with Aviation Space Writers, May 23, 1966," 6/6/66, memo, FAA 1647/6; "Concorde Aircraft," 6/7/66, DOD Intelligence Information Report, no. 1-836-0258-66, FAA 1647/6; Maloy to Vierling, 6/24/66, telex, FAA 1647/6; "Concorde SST," 6/27/66, DOD Intelligence Information Report, no. 1-832-0135-66, FAA 1647/7; Maloy to FAA, Washington, D.C., 6/29/66, telex, FAA 1647/6; Maloy to Maxwell, 6/29/66, telex, FAA 1647/6; C. L. Blake to J. W. Howell, "Concorde Olympus 593 Engine Ratings," 7/20/66, memo, FAA 1647/6; Maloy to Maxwell, "Concorde SST," 7/7/66, memo, FAA 1647/6; American Embassy, London, to Secretary of State, "Civair," 9/9/66, DOS cable, FAA 1647/7; Clifton F. von Kann to Maxwell, "Visit to British Aircraft Corporation Facilities at Bristol, England," 9/14/66, memo, FAA 1647/3; Maxwell to von Kann, 9/22/66, 9/30/66, FAA 1647/3. For Trippe's comments, see: "Draft Proceedings of the President's Advisory Committee on Supersonic Transport," 7/9/66, pp. 7–9, 19–20, PAC/32.

29. FAA, "Summary Status Report to President's Advisory Committee on Supersonic Transport," 6/30/66, pp. 30–40, 8/31/66, PAC/32; Steadman to McNamara, 9/2/66, memo, PAC/14.

30. Robert G. Prestemon to Allen H. Skaggs, 6/23/66, memo, FAA 1647/6.

31. DOD, "Comments on the FAA Paper on the PAC Assignment of the Effect of a Longer Interval between the Concorde and the SST" (briefing paper for 7/9/66 PAC meeting), PAC/4.

32. W. C. Mentzer to Maxwell, 11/16/66, FAA 1647/8; N. R. Parmet to Maxwell, 11/17/66, and attachments, FAA 1647/8.

33. McKee to Califano, 10/19/66, and attached Raymond L. Bisplinghoff to McKee and Maxwell, "Concorde Airplane," 10/14/66, memo, LBJ Oversize.

34. "Draft Proceedings of President's Advisory Committee on Supersonic Transport," 10/6/66, pp. 15, 27–29, PAC/32.

35. Maxwell to McNamara, "Foreign Supersonic Transport Programs—Status Report," 11/7/66, memo, and attached "Foreign Supersonic Transport Programs Status Report," 10/27/66, PAC/4.

36. "Draft Proceedings of the President's Advisory Committee on Supersonic Transport," 12/7/66, pp. 27–33, 49–62, 74–75, PAC/32; Donald F. Chamberlain, CIA Director of Scientific Intelligence, to Office of Supersonic Transport Development, FAA, "Concorde Product Negotiations," 12/8/66, memo, PAC/15; E. Drexel Godfrey, Jr., CIA Director of Current Intelligence, to Office of Supersonic Transport Development, FAA, "FAA Request for Political Evaluation of the Concorde Project as Related to British Entry in the EEC," 12/9/66, memo, PAC/15; "Draft Proceedings of the President's Advisory Committee on Supersonic Transport," 12/10/66, pp. 13–14, 17–21, 28–36, 49–50, 54–55, PAC/32.

37. "Memorandum to the President: Fourth Interim Report of the President's Advisory Committee on Supersonic Transport," 12/22/66, PAC/32.

38. For the more passive and less useful intelligence on the Concorde after 1966, see: Lockett to Assistant for System Integration et al., "SST Intelligence," 2/15/67, memo, FAA 1647/9; Charles O. Cary to Maxwell, "Foreign SST Intelligence," 3/68, memo,

FAA 905/2; John J. Carroll to Frank Sheridan (CIA), 8/2/68, FAA 6174/1; Maxwell to Cary, "Classified Foreign SST Developments Report Dated April 2, 1968," 4/12/68, memo, FAA 1647/12.

39. Maloy to Deputy Administrator, "Concorde Status Report," 6/28/67, memo, FAA 1647/10; Maloy to Maxwell, "Hamilton/Forestier Visit, Monday, July 10, 1967," 7/3/67, memo, FAA 1647/11.

40. B. A. Schriever, B. U. Marschner, and B. H. Goethert, "Concorde Supersonic Transport: Report of Visit to Construction Plants and Test Facilities," 7/28/67, FAA 6174/95; Maloy to Assistant Administrator for International Aviation Affairs, IA-1, "Concorde Weight/Power Problem," 7/29/67, memo, FAA 1647/11; Maxwell to McKee, "Concorde Information," 11/1/67, memo, FAA 6174/95.

41. For delays in the Concorde program in 1967 and 1968, see: Maloy to Deputy Administrator, "Concorde Status Report," 1/4/68, memo, FAA 905/1; FAA translation of "Six Reasons for the Concorde's Six Months' Delay," *Paris-Match,* 3/30/68, FAA 6174/1; Maloy to Deputy Administrator, "Visit of Mr. Jean Forestier to the United States in April," 4/1/68, memo, FAA 905/2; American Embassy, Paris, to Department of State, "CIVAIR-Russian SST and Concorde," 5/16/68, DOS airgram, FAA 6174/1; American Embassy, Paris to Department of State, "CONCORDE Aircraft—Pacing Items," 4/25/68, DOS airgram, FAA 6174/1; Maloy to file, "Conversation with Forestier, on May 9, 1968," 5/10/68, memo, FAA 6174/1; Maloy to file, "Conversation with Paul Besson of Air France, May 9, 1968, re: SST and the Air Bus," 5/10/68, memo, FAA 6174/1; Department of State to the American Embassy, Paris, "Strike Settlement of Sud-Aviation," 6/24/68, DOS airgram, FAA 6174/1; Vierling to Maxwell, 9/17/68, memo, FAA 6174/2. For continued delays in the Concorde program in September and October 1968, see: Cary to Maxwell, 9/10/68, route slip, FAA 6174/2; American Embassy, Paris, to Secretary of State, "Concorde Project," 10/8/68, DOS cable, FAA 6174/2; Moore to FAA, "Concorde Project," 10/25/68, cable, FAA 6174/2; American Embassy, Paris, to Secretary of State, "Concorde Project," 9/16/68, and 10/24/68, DOS cables, FAA 6174/2.

42. Vierling to Director, Information Services, "Weekly Highlight Report," 12/5/67, memo, FAA 6174/95; Paul Frankfurt to Maxwell, "Concorde Roll Out," 12/8/67, memo, FAA 1647/12.

43. Secretary of State to the American Embassy, New Delhi, "CIVAIR-Air India Concorde Options," 3/26/68, DOS cable, FAA 905/2; U.S. Embassy, New Delhi, to the Secretary of State, "CIVAIR-Air India Concorde Options," 4/4/68, DOS cable, FAA 905/2; Lockett to the files, "Concorde Options," 4/9/68, memo, FAA 905/2.

44. Maxwell to Representative Philip J. Philbin, 6/27/68, FAA 6174/1; Maxwell to Director of Budget, "International Transportation Memorandum," 9/12/68, memo, and attached "Comments/Suggested Revisions: Draft Program Memorandum III—International Transportation (Supersonic Transport Program)," FAA 6174/2.

45. American Embassy, Paris, to Department of State, "Concorde First Taxi Trails," 9/19/68, DOS airgram, FAA 6174/2.

46. American Embassy, London, to Department of State, "Roll Out of Concorde 002 at Filton," 9/14/68, DOS airgram, FAA 6174/2.

47. Various Associated Press dispatches on the Concorde, 3/2/69, FAA 6174/4.

48. D. D. Thomas to Boyd, "French-British Concorde," 2/24/69, memo, FAA 6174/4.

49. American Embassy, Paris, to Department of State, "Report of Selection of U.S. Navigation System for Concorde," 3/27/69, DOS airgram, FAA 6174/4; Paul A. Volcker to James Beggs, 3/18/69, FAA 6174/95.

50. John H. Shaffer to John A. Volpe, "Concorde," 4/4/69, memo, FAA 6174/95; American Embassy, London, to Secretary of State, 4/3/69, DOS cable, FAA 6174/4; "Bristol—Add Concorde (8)," 4/9/69, news telex, FAA 6174/4.

51. Volpe to the President, "Status Report—Supersonic Transports," 8/15/69, memo, FAA 6174/95; "Concorde 001 Flies Supersonically," *Aviation Week and Space Technology*, 10/6/69, p. 23.

52. Francis E. Rundell to SS-1, "Telephone Conversation with Earl Schaffer—Senator Percy's Office," 2/16/70, memo, FAA 6174/93.

53. U. Alexis Johnson to Senator William Proxmire, 5/7/70, FAA 6174/93.

54. Maloy to file, "Conversation with Forestier, on May 9, 1968," 5/10/68, memo, FAA 6174/1.

55. Maxwell to Thomas, "Supersonic Flight over Britain and France," 12/5/68, memo, FAA 6174/92.

56. U. Alexis Johnson to Senator William Proxmire, 5/7/70.

57. Cary to Acting Administrator, "International Implications of U.S. Decision to Proceed with or Cancel Supersonic Transport Program," 2/5/69, memo, FAA 6174/4.

58. Robert Hotz, "An-22 SST Model Climax Soviet Display," *Aviation Week and Space Technology*, 6/21/65, pp. 28–31.

59. Bain to Halaby, "Summary Information—Paris Air Show," 6/22/65, memo, pp. 4–5, FAA 6174/95.

60. Califano to McNamara, "SST Briefing for Congressional Leadership," 6/30/65, memo, and attached "SST Briefing Outline," 6/30/65, p. 5, FAA 6174/96.

61. Donald F. Chamberlain, CIA Director of Scientific Intelligence, to Maxwell, "Status Report on the Soviet Supersonic Transport TU-144," 8/19/66, memo, included in FAA, "Summary Status Report to President's Advisory Committee on Supersonic Transport," 8/31/66, pp. 32–35, PAC/4.

62. Chamberlain to Maxwell, "TU-144 SST Bi-Monthly Status Report, October 1966," 10/27/66, memo, included in FAA, "Foreign Supersonic Transport Programs Status Report," 10/27/66, PAC/4.

63. Annex of "PAC to the President: Fourth Interim Report of the President's Advisory Committee on Supersonic Transport," 12/22/66, p. 7, PAC/32; Califano to the President, 12/26/66, memo, p. 3, LBJ Oversize.

64. Associated Press dispatch, 1/18/67, FAA 1647/8.

65. American Embassy, Paris, to Secretary of State, "CIVAIR—Russia SST" (received by SST Office on 4/16/68), DOS cable, FAA 905/2; American Embassy, Budapest, to American Embassy, Paris, "CIVAIR—Russian SST" (received by SST Office on 4/16/68), DOS cable, FAA 905/2.

66. American Embassy, Paris, to Department of State, "CIVAIR—Russian SST and Concorde," 5/17/68, DOS airgram, FAA 6174/1.

67. Lockett to Chief, Analysis and Control Division, "Status Report for the PAC," 7/5/

68, memo, and attached "Section F: Foreign SST Programs," draft of 7/3/68, p. 3, FAA 6174/1; American Embassy, Moscow, to Department of State, "A View of Aeroflot's Future," 10/1/68, DOS airgram, FAA 6174/2.

68. Associated Press dispatch, no. 68, 12/31/68, FAA 6174/3.

69. For airline and industry views of the TU-144, see: R. W. Rummel (TWA) to Volpe, 6/16/69, FAA 6174/6; James M. Beggs to Rummel, 7/3/69, FAA 6174/6; Rummel to Secor Browne, 8/12/69, and attached N. R. Parmet to Airline SST Committee, "Review of Russian TU-144 SST," 8/5/69, memo, and "TU-144 Inspection Notes," 8/5/69, FAA 6174/95.

70. Ralph N. Read to Acting Director, SST Development, "Tokyo–London Flight Times," 12/24/69, memo, FAA 6174/5.

Chapter 15

1. For Gibson's article and for the FAA's reaction to it, see: John E. Gibson, "The Case Against the SST," *Harper's* (7/66):76–93; Maxwell to John Fischer, 5/19/66, 6/10/66, FAA 1647/6; Maxwell to Gibson, 5/20/66, FAA 1647/6; Fischer to Maxwell, 5/25/66, 7/14/66, FAA 1647/6; Rosemary Wolfe to Maxwell, 6/14/66, FAA 1647/6; J. M. Chavkin to Division Chiefs and Chief, Economics Staff, "'The Case Against the SST,' by John E. Gibson," 8/8/66, memo, FAA 1647/1; FAA assignments as noted on copy of the Gibson article, FAA 1647/1; FAA, "Comments on the Case Against the SST: Harper's Magazine," n.d., FAA 1647/1; David E. Galas to Branch Chiefs, "'The Case Against the SST,' by John E. Gibson," 8/19/66, memo, FAA 1647/7; L. Trenary to Chief Engineering Division, "Systems Branch Comments on 'The Case Against the SST,'" 8/12/66, memo, FAA 1647/1; R. W. Pinnes to Chief, Engineering Branch, "'The Case Against the SST,' by John E. Gibson," 8/25/66, memo, FAA 1647/7; C. L. Blake to Chief, Analysis and Control Division, "Comments on the Gibson Report," 9/16/66, memo, and attachments, FAA 1647/1.

2. *Wall Street Journal*, 7/28/66, p. 14.

3. Ibid., 2/9/67, p. 10.

4. Robert A. Bailey to B. J. Vierling, 8/9/66, FAA 1647/7.

5. For the article by Hohenemser and for its impact, see: "Supersonic Transport Development: Daily Highlights," 6/22/66, FAA 6174/95; Maxwell to Bisplinghoff, "Article by Professor Hohenemser," 6/22/66, memo, FAA 1647/6; Kurt H. Hohenemser, "The Supersonic Transport," *Bulletin of the Atomic Scientists* (12/66):8–12; Maxwell to Reid, 2/13/67, FAA 1647/9.

6. For FAA press monitoring and approaches to the press, see: R. E. Parsons to McKee, "Comments on *Wall Street Journal* Article," 4/17/66, memo, and attached *Wall Street Journal*, 5/16/66, pp. 1, 12, FAA 1647/1; F. G. Coffey (Boeing Co.) to Maxwell, 1/25/67, telex, FAA 1647/8; Maxwell to McKee, "News Clips Attached," 7/25/67, memo, FAA 6174/95; "Ambassy," London to Department of State, "Civil Aviation: House of Commons Debate on Sonic Boom," 2/9/67, DOS airgram, FAA 1647/95; Vierling to Deputy Assistant Administrator for International Aviation Affairs, "*New York Times* article by John L. Hess," 8/3/67, memo, FAA 1647/11; Maxwell to the editor of the *Wall Street Journal*, 7/30/66, FAA 1647/7.

7. For the FAA's contact with the *Washington Evening Star*, see: Allen H. Skaggs to

Acting Chief, Analysis and Control Division, "Sunday's Star Editorial, August 14, 1966," 8/15/66, memo, FAA 1647/7; James H. Polve to the record, "Rebuttal to Evening Star Editorial," 8/16/66, memo, FAA 1647/7; *Washington Evening Star*, 11/15/66, p. A-5; Maxwell to McKee, "STAR Article—February 7," 2/8/67, memo, FAA 1647/9; McKee to Newbold Noyes, 5/2/67, FAA 1647/10; Maxwell to William Hines, 11/17/67, FAA 1647/12.

8. "SST Progress Not Supersonic," *Washington Daily News*, 3/28/67, FAA 1647/9.

9. Converse to Chief, Technical Operations Division, "NET Film 'Noise: The New Pollutant,'" 3/15/67, memo, FAA 1647/9; Lockett to Maxwell, "NET Film 'Noise: The New Pollutant,'" 4/6/67, memo, FAA 1647/9.

10. For the FAA's communication with Senator Moss, see: Senator Frank E. Moss to McKee, 1/12/67, FAA 1647/9; McKee to Moss, 2/10/67, FAA 1647/9; Moss to McKee, 4/10/67, and attached editorial entitled "Sonic Boom No 'Boom' to Us," from the *Richfield [Utah] Reaper*, FAA 1647/9.

11. For the FAA's communication with Senator Symington, see: *New York Times*, 5/2/67, p. 46; Symington to McKee, 5/2/67, 5/12/67, FAA 6174/95; McKee to Symington, 5/10/67, FAA 1647/10.

12. For the FAA's communication with other senators, see: Maxwell to Senator Edward M. Kennedy, 5/25/67, FAA 1647/1; Maxwell to Senator Clifford Case, 10/27/67, FAA 1647/3; Maxwell to Edmund Learned, 8/22/67, FAA 1647/11.

13. For pro-SST pieces, see: *Aerospace News*, 5/6/66, press release, FAA 1647/3; Maxwell to Edgar E. Ulsamer (Air Force Association), 7/11/66, FAA 1647/6; Gerald B. Grinstein to Ulsamer, 2/10/67, FAA 6174/95.

14. John Mecklin, "The $4 Billion Machine That Reshapes Geography," *Fortune* (2/67):112–117; Maxwell to Mecklin, 2/9/67, FAA 1647/9; Skaggs to Eleanor Carruth, 2/8/67, FAA 1647/9.

15. S. G. Tipton to McKee, 4/4/67, and attached "The SST: An Essential Investment in the U.S. Economy," FAA 1647/10.

16. Richard Slawsky to the President, 4/25/67, LBJ EX CA; S. Royer Wolin (Pan American) to George Christian, 5/3/67, and attached *Miami Herald*, 5/2/67, p. 67, LBJ EX CA.

17. F. E. Ellis to Stuart M. Levin (*Space/Aeronautics*), 6/7/67, FAA 1647/10.

18. James H. Straubel (Air Force Association) to Maxwell, 7/13/67, FAA 6174/95; Maxwell to Straubel, 7/17/67, FAA 6174/95.

19. Blake to Chief, Special Projects Branch, "Comments on Luncheon Talking Paper with *New York Times* Staff," 11/28/67, memo, FAA 1647/12.

20. Richard Hellman, "SST: Not All Smooth Flying," *Challenge* (7–8/67):34–37; Lambert P. Irons to Assistant to the Director, "Comments on the *Supersonic Transport—Not All Smooth Flying*, by Richard Hellman, July–August, 1967, issue of *Challenge Magazine*," 10/17/67, memo, FAA 6174/92.

21. For the Evans-Novak columns and the FAA's response, see: *Washington Post*, 8/10/67, p. A17, 8/31/67, p. A21; FAA to Senator John Stennis, n.d., draft letter, FAA 6174/95; FAA to Representative Edward P. Boland, 9/15/67, draft letter, FAA 6174/95; McKee to Boyd, 12/6/67, and attached article from the *Washington Post*, 12/20/75, FAA 6174/95.

22. For other negative pieces on the SST, see: Irons to Assistant to the Director, "Com-

ments on Article by George W. Hilton in *Business Horizon*, Volume 10, #2, *Federal Participation in the SST Program*," 9/20/67, memo, FAA 1647/12; Vierling to Director, Office of Information Services, "Weekly Highlights Report for the Secretary of Transportation," 9/6/67, memo, FAA 1647/11; John D. Calhoun to Mr. Stimpson, "Anti-SST Comment by Mr. G. Ray Woody, National Airlines," 10/25/67, memo, FAA 1647/12; *Washington Post*, 12/20/67; McKee to Boyd, 12/20/67, memo, FAA 6174/95.

23. Juan T. Trippe to McKee, 2/9/67, FAA 1647/9; McKee to Trippe, 2/20/67, FAA 1647/9.

24. Charles C. Tillinghast, Jr., to McKee, 6/29/67, FAA 6174/95.

25. Withington to Maxwell, 5/31/67, FAA 1647/10; T. A. Wilson to Maxwell, 9/28/67, and attached Vermont Royster (editor of *Wall Street Journal*) to Lowell Mickelwait (Boeing), 9/26/67, FAA 6174/95.

26. For the organization and activities of the SST Office's special projects staff, see: Brian Tennan to Col. J. H. Voyles, Jr., 12/4/67, FAA 1647/13; Paul Frankfurt to Nathan Goldstein (*New York Times*), 11/2/67, FAA 1647/12; Frankfurt to Carl Cleveland (Boeing), 11/2/67, FAA 1647/12; Frankfurt to Alfred Kildow (American Institute of Aeronautics and Astronautics), 11/2/67, FAA 1647/12; Maxwell to SST Divisions and Staffs, "Public Appearances by SST Office Personnel," 12/1/67, memo, FAA 1647/13.

27. Eugene S. Kropf to Frankfurt, 12/4/67, FAA 1647/13.

28. For public relations coordination between the FAA and the contractors, see: Voyles to John H. Newland (Boeing), 12/6/67, FAA 1647/13; Newland to Voyles, "(1) Heading for SST Public Information Guide Pages; (2) Progress on Motion Picture Script," 12/5/67, memo, FAA 1647/12; Voyles to Newland, 12/8/67, FAA 1647/13; M. Philip Copp to Newland, 12/27/67, FAA 905/1; David I. McGinnis (General Electric), to Voyles, 12/7/67, FAA 1647/13; Voyles to McGinnis, 12/8/67, FAA 1647/13.

29. Maxwell to Representative Thomas P. O'Neill, Jr., 4/5/67, FAA 1647/9.

30. Maxwell to Ansel Adams, 2/1/67, FAA 1647/9.

31. Charles G. Warnick to McKee, "Requests for Reports Regarding Sonic Boom Tests," 4/19/67, memo, FAA 1647/10.

32. For Lundberg's anti-SST work and its impact on the FAA, see: Bo Lundberg to Senator Warren Magnuson, 12/9/66, FAA 1647/9; McKee to Boyd, "Letter from Bo Lundberg Regarding Effects of Sonic Boom and People at Sea," 3/2/67, FAA 1647/9; McKee to Magnuson, 4/6/67, FAA 1647/1; Donald G. Agger to Boyd, "Proposed Trip by Dr. Lundberg," 5/19/67, FAA 1647/10; Maxwell to Assistant Administrator for Congressional Liaison, "Letter from William G. Wells, Jr.—Office of Congressman Karth," 2/13/67, FAA 1647/9; Maxwell to T. A. Wilson, 8/24/67, and attached Blake to Assistant to the Director, "Background of Bo Lundberg," 8/24/67, memo, and Blake to Assistant to the Director, "Comments on the Uneconomic, Unwanted SST, by Bo Lundberg," 8/24/67, memo, FAA 1647/11.

33. For Shurcliff's background, for the early anti-SST activities by Shurcliff and for the handling of his complaints about the NASA sonic boom report, see: Tracy Kidder, "Tinkering with Sunshine," *Atlantic Monthly* (10/77):70–83; *New York Times*, 12/14/80, sec. II, p. 10; William A. Shurcliff, "A Brief History of the Citizens League Against the Sonic Boom," 1/25/67, CL; Shurcliff to FAA, 10/7/66, FAA 1647/7; O'Neill to McKee, 1/19/67, and attached Shurcliff to O'Neill, 1/14/67, FAA 1647/8; Shurcliff to FAA, 1/11/

67, FAA 1647/8; Proxmire to FAA Congressional Liaison, 1/18/67, inquiry, and attached Shurcliff to Proxmire, 1/10/67, FAA 1647/8; Shurcliff to Boyd, 1/15/67, FAA 1647/9; Paul A. Sitton to Shurcliff, 2/3/67, FAA 1647/9; Maxwell to Shurcliff, 1/20/67, FAA 1647/8; Maxwell to O'Neill, 2/6/67, 2/25/67, FAA 1647/9; Maxwell to Proxmire, 2/14/67, FAA 1647/9.

34. For Shurcliff's style and fund-raising methods and for the beginning of the Citizens League, see: personal interviews; Shurcliff, "Brief History"; Citizens League advertisement, *New York Times*, 6/16/67, p. 22; Joel Primack and Frank Von Hippel, *Advice and Dissent* (New York, N.Y.: Basic Books, 1974), pp. 15–16; Shurcliff to Theodore Berland, 6/27/72, CL.

35. For the anti-SST campaign by the Citizens League between March and August 1967, see: *Christian Science Monitor*, 3/18/67; Senator Edward Brooke to FAA Congressional Liaison, inquiry, FAA 1647/9; Shurcliff to Senator Kennedy, 3/27/67, FAA 1647/9; Maxwell to Senator Brooke, 4/18/67, FAA 1647/9; Maxwell to Senator Kennedy, 4/18/67, FAA 1647/9; Shurcliff to LBJ, 4/21/67, FAA 1647/10; Shurcliff to Dean Rusk, 7/12/67, FAA 1647/11; Shurcliff to Maxwell, 4/22/67, FAA 1647/10; Shurcliff to FAA, Office of Information, 3/29/67, FAA 1647/10; Shurcliff to FAA Office of Information, 4/24/67, and attached Citizens League Fact Sheet, no. 11, FAA 1647/10; Congressman Reid to FAA Congressional Liaison, 5/18/67, and attached Citizens League advertisement, FAA 1647/10; Senator Robert Kennedy to FAA Congressional Liaison, 5/24/67, and attached Shurcliff to Senator Robert F. Kennedy, 5/26/67, FAA 1647/10; Maxwell to Senator Robert F. Kennedy, 6/26/67, FAA 1647/10; Shurcliff to Senator Edward Kennedy, 5/18/67, FAA 1647/10; Shurcliff to Maxwell, 6/3/67, FAA 1647/10; Maxwell to Senator Edward Kennedy, 6/9/67, FAA 1647/10; Hoover to Shurcliff, 5/8/67, FAA 1647/10; Shurcliff to Niles, 5/19/67, FAA 1647/10; Patrick Cody to Shurcliff, 6/7/67, FAA 1647/10; Maxwell to Shurcliff, 6/13/67, FAA 1647/10; Shurcliff to Maxwell, 6/30/67, FAA 1647/11; Maxwell to Shurcliff, 7/10/67, FAA 1647/11; Shurcliff to LBJ, 7/17/67, FAA 1647/11; Shurcliff to Senator Magnuson, 8/5/67, FAA 1647/11

36. For the anti-SST activities of the Citizens League between August and December 1967, see: Citizens League, "Rising Tide of Protest against the SST and Its Inevitable Sonic Boom," 8/23/67, FAA 6174/92; Citizens League Fact Sheet, no. 11a, 7/67, CL; Maxwell to Representative Albert H. Quie, 9/28/67, FAA 1647/12; Shurcliff to McKee, 10/30/67, FAA 1647/12; Susan Aronson (Citizens League) to Charles J. Collis (Republic Aviation), 10/24/67, FAA 6174/92; Shurcliff to Maxwell, 10/30/67, 11/6/67, 11/10/67, 11/20/67, FAA 1647/12; Charles R. Lowell (Citizens League) to Vierling, 11/3/67, FAA 1647/12; Shurcliff to Representative Hastings Keith, 11/6/67. FAA 1647/12; Shurcliff to Vierling, 11/11/67. FAA 1647/12; Maxwell to Representative Charles E. Bennett, 12/4/67, FAA 1647/13; Maxwell to Representative Hastings Keith, 12/5/67, FAA 1647/13; Irons to Special Assistant for Supersonic Transport Development, "Answer to Dr. Shurcliff's Letter Dated November 6, 1967," 11/22/67, memo, FAA 1647/12; Shurcliff to the President, 12/6/67, LBJ Gen. CA; Maxwell to Acting Director Noise Abatement, "Letter to Dr. Shurcliff," 12/13/67, memo, FAA 6174/92; Hoover to Shurcliff, 12/29/67, FAA 905/1; Shurcliff to Vierling, 12/22/67, FAA 1647/13; Vierling to Shurcliff, 11/15/67, FAA 1647/12; Maxwell to Shurcliff, 11/16/67, 11/21/67, FAA 1647/12; Maxwell to Senator Clifford Case, 1/8/68, FAA 905/1; Maxwell to Representative Hastings Keith, 1/8/68, FAA 905/1.

37. For examples of the anti-SST protest letters by individuals and their impact during the latter half of 1967, see: Elizabeth P. Borish to Boyd, 7/12/67, FAA 1647/11; Maxwell

to Senator Percy, 7/24/67, FAA 1647/11; Maxwell to Julian Loebenstein, 7/29/67, FAA 1647/11; Maxwell to Borish, 8/16/67, FAA 1647/11.

38. Senator A. S. "Mike" Monroney to McKee, 5/1/67, FAA 1647/10.

39. *Chicago American*, 8/4/67, p. 10.

40. Dr. Henry F. Allen to Boyd, 9/12/67, FAA 6174/92.

41. Maxwell to Mayor W. Don MacGillivray of Santa Barbara, California, 9/21/67, FAA 1647/12.

42. For FAA responses to the new protest in 1967, see: Vierling to Halaby, 5/1/67, FAA 1647/1; Maxwell to McKee, "Letter to Congressman Boland," 7/12/67, FAA 1647/11; McKee to Congressman Edward Boland, 7/17/67, FAA 1647/11; Maxwell to Julian Loebenstein, 7/29/67, FAA 1647/11; Charles Williams to Congressman Ray Madden, 8/3/67, FAA 1647/11.

43. Alexander Garvis to Shurcliff, 8/9/67, FAA 1647/11.

44. Maxwell to Senator Clifford Case, 1/8/68, FAA 905/1.

45. For FAA-industry coordination in public relations, see: Blake to Voyles, "Comments on Boeing Draft Brochure," 1/22/68, memo, FAA 905/1; Voyles to Maxwell, "The U.S. SST Program Report," 3/26/68, memo, FAA 905/2; Voyles to Daniel B. Priest (ATA), 3/26/68, FAA 905/2; Voyles to Glen Bayless (AIA), circa 3/26/68, FAA 905/2; Lockett to Special Assistant for Supersonic Transport Development, "Comments on Proposed Boeing SST Press Information Kit," 7/3/68, memo, FAA 6174/1; Stephen Fishe to Gregg Reynolds (Boeing), 7/30/68, FAA 6174/1; Reynolds to Col. J. D. Calhoun, 8/14/68, FAA 6174/1; J. W. Locraft to Reynolds, 11/12/68, FAA 6174/3; Reynolds to Col. Donald Robb, 2/4/68, FAA 6174/3; Robb to Col. William McGinty (Boeing), 2/20/68, FAA 6174/4; Walter C. Swan (Boeing) to Calhoun, 3/6/69, FAA 6174/4; McGinty to Robb, 3/12/69, FAA 6174/4; Parsons to Frankfurt, "Boeing Brochures; SS-4 Memo of 3/13/69," 3/18/69, memo, FAA 6174/4; Robb to McGinty, 3/19/69, FAA 6174/4; Parsons to Voyles, "Boeing Brochure—The United States Supersonic Transport, April 1969," 4/10/69, memo, FAA 6174/4; McGinty to Robb, 8/1/69, FAA 6174/6; G. T. Haugan to Voyles, "Boeing Press Kit Info," 8/8/69, memo, FAA 6174/6.

46. For FAA-Boeing communication about media issues, see: Reynolds to Calhoun (FAA) and W. A. Schoneberger (GE), 7/5/68, memo, FAA 6174/1; Calhoun to C. W. Duffy (Boeing), 7/26/69, FAA 6174/1; Calhoun to Duffy, 9/26/68, FAA 6174/2; W. E. Clothier (Boeing) to Robb, 8/4/68, FAA 6174/6.

47. Reynolds to Calhoun, 7/31/68, FAA 6174/1.

48. Ibid., 6/28/68, FAA 6174/1.

49. For Boeing's SST advertising, see: Newland to Voyles, 1/2/68, FAA 6174/1; Robb to McGinty, 2/4/69, FAA 6174/3; Robb to McGinty, 4/15/69, FAA 6174/11; Reynolds to Voyles, 5/29/68, FAA 6174/1.

50. Reynolds to Robb, 1/29/69, FAA 6174/3.

51. Reynolds to Jack Mitchell (Boeing), 6/15/68, FAA 905/2.

52. For the distribution of Boeing's SST brochures, see: Reynolds to Voyles, 5/6/68, 6/14/68, FAA 6174/1; C. W. Duffy (Boeing) to Calhoun, 6/7/68, FAA 6174/1; Calhoun to Duffy, 6/21/68, FAA 6174/1.

53. For the FAA's monitoring of Boeing's public relations effort, see: Clothier to Frank-

furt, 6/27/69, and attached press release, FAA 6174/6; Voyles to Reynolds, 2/8/68, FAA 905/1; Voyles to Duffy, 3/26/68, 4/8/68, FAA 905/2; Duffy to Calhoun et al., "SST Program Information Ground Rules," 5/20/68, memo, FAA 6174/1; Reynolds to Calhoun, 9/11/68, FAA 6174/2; Maxwell to T. A. Wilson, 2/25/69, FAA 6174/4; Frankfurt to Calhoun, "Boeing Press Briefing," 12/6/68, memo, FAA 6174/3; Frankfurt to SST Division Chiefs, "Public Release of New SST Configuration," 10/9/68, memo, FAA 905/2.

54. For the FAA-General Electric coordination in public relations and General Electric's media work, see: David A. McGinnis (GE) to Voyles, 4/23/68, FAA 905/2; Vierling to Deputy Director, Aeronautical Center, " 'Washington to London' Pamphlet," 5/28/68, memo, FAA 6174/1; D. D. Leshner (GE) to Frankfurt, 7/11/68, FAA 6174/1; Robb to SS-1 et al., "GE Publicity Material," 1/16/69, memo, FAA 6174/3; W. K. Linnercoth to W. J. Wallace (GE), 3/11/69, FAA 6174/4; Wallace to Ray A. Milagrane, 8/20/69, FAA 6174/6; Locraft to Wallace, 2/3/69, FAA 6174/3; W. A. Schoneberger (GE), 3/26/69, FAA 6174/4; Schoneberger to Duffy and Voyles, 4/2/68, FAA 905/2; Leshner to Frankfurt, 4/17/68, FAA 6174/4.

55. Voyles to Maxwell, "SST Coverage—Network TV," 4/2/68, memo, FAA 905/2.

56. For the three-way dialogue among Boeing, General Electric, and the FAA about SST public relations matters, see: Schoneberger to Voyles, 1/30/68, FAA 905/1; Reynolds to Schoneberger, 4/1/68, FAA 905/2; Schoneberger to Voyles, 4/11/68, FAA 905/2; Schoneberger to Calhoun, Duffy, and Reynolds, 6/20/68, FAA 6174/1; Frankfurt, "Trip Report—October 17–18, 1968, General Electric Company, Evendale, Ohio," FAA 6174/2.

57. Hugh C. Newton to Voyles, 2/5/68, FAA 905/1; ibid., 5/3/68, FAA 6174/1.

58. Eliot Janeway column, "McKee Says SST No 'Fly-by-Night' Project," 2/7/68, *Chicago Tribune Press Service;* Voyles to Maxwell, "Radio Tape," 4/26/68, memo, FAA 905/2.

59. Frankfurt to Maxwell, "Huntley-Brinkley Report," 3/25/68, memo, FAA 905/2; Voyles to Maxwell, "Huntley/Brinkley Show," 4/25/68, memo, FAA 905/2; Frankfurt to Maxwell, "NBC-TV" 6/21/68, memo, FAA 6174/1.

60. Reynolds to Calhoun, 9/18/68, FAA 6174/2; A. Scofield (Boeing) to J. Demeter (FAA), 1/20/69, telex, FAA 6174/3.

61. Col. James H. Polve to Lloyd E. Cooney, 2/12/69, FAA 6174/4.

62. Maxwell to Representative Dan Kuy Kendall, 4/1/69, FAA 6174/4.

63. *Washington Post,* 1/29/68, p. A15; draft letter to "Chairman," circa 1/29/68, FAA 6174/95.

64. Lambert P. Irons to Special Assistant for SST Information, "Answers to Questions by Air Travel," 3/5/68, memo, FAA 905/2; Blake to IS-5, "Comments on Questions from 'Air Travel,' " 3/5/68, memo, FAA 905/2.

65. For the *Aerospace Technology* issue on the SST, see: Voyles to Blake, "Special Report—Technology of the SST," 3/14/68, memo, FAA 905/2; R. F. Goranson to Voyles, "Contributors to Aerospace Technology Magazine Article," 5/17/68, memo, FAA 6174/1; Maxwell to Michael Getler, 5/20/68, FAA 6174/1; Maxwell to Wayne W. Parrish, 5/20/68, FAA 6174/1.

66. McKee to Mrs. Fredrica S. Friedman, 4/5/68, FAA 905/2.

67. Alexander Garvis to Charles Murphy, 7/23/68, FAA 6174/1; Garvis to the record, "Request for SST/Sonic Boom Information," 8/8/68, memo, FAA 6174/1.

68. For other examples of FAA assistance to the media for SST coverage, see: James K. Thompson to Chief, Technical Operations Division, "Weekly Activities Report," 8/16/68, memo, FAA 6174/1; Paul Hersh (*Science and Technology*) to Blake, 2/15/69, FAA 6174/4; "Comments on Paul Hersh's SST Draft," n.d., FAA 6174/4; Blake to Hersh, 3/20/69, FAA 6174/4; Calhoun to Bob Hotz (*Aviation Week and Space Technology*), 4/8/69, FAA 6174/4; Hal Watkins to C. F. Mass, "Request for Information," 4/9/69, record of telephone call, FAA 6174/4.

69. Charles Gerras (*Prevention*) to Maxwell, 5/7/68, and attached letter from Maxwell to *Prevention*, FAA 6174/92; Maxwell to Gerras, 5/9/68, FAA 6174/92; Maxwell to Gerras, 10/15/68, FAA 6174/2.

70. McKee to Mayor John V. Lindsay, 5/10/68, and attached article from the *New York Times*, 5/10/68, FAA 6174/92.

71. *Newsweek*, 5/27/68, p. 20; Maxwell to Osborn Elliot (*Newsweek*), 5/24/68, FAA 6174/1.

72. Maxwell to Melvin L. Selzer, 5/31/68, FAA 6174/1.

73. Stephen Shepard, "The Supersonic Boom," *Atlantic Monthly* (8/68):10–14; Roy M. Green (*Atlantic Monthly*) to Maxwell, 7/15/68, FAA 6174/1; Maxwell to Green, 7/29/68, FAA 6174/1.

74. "Breakfast with General Maxwell," 1/25/68, FAA 905/1; "Press Conference Attendees," 1/25/68, FAA 905/1.

75. Frankfurt to Maxwell, "Public Appearances," 2/22/68, memo, FAA 905/2; Frankfurt to Maxwell, "Public Relations Activities in New York City," 3/1/68, memo, FAA 905/2; Reynolds to Voyles, 3/6/68, FAA 905/2.

76. For Wings Club speech, see: Frankfurt to Maxwell, "PR Activities in New York," 3/22/68, memo, FAA 905/2; Maxwell to Bob Considine, 4/4/68, FAA 905/2; Maxwell to McKee, "Bob Considine Articles," 4/10/68, memo, FAA 905/2.

77. Calhoun to the record, "Background Information on the University of Portland—Dining-In," 3/14/68, memo, FAA 905/2.

78. Voyles to Maxwell, "Luncheon Meeting National Association of Manufacturers," 3/29/68, memo, FAA 905/2; Voyles to Maxwell, "National Association of Manufacturers Luncheon Speech—15 May," 4/19/68, memo, FAA 905/2; W. P. Gallander (NAM) to Maxwell, 5/27/68, FAA 6174/1.

79. R. H. Dickerson to Maxwell, 6/3/68, FAA 6174/1.

80. R. W. Smart (Air Force Association) to Maxwell, 9/5/68, FAA 6174/2; Charles G. Warnick to OP-1 et al., "Seminars," 9/10/68, memo, FAA 6174/2; Frankfurt to Messrs. Blake, Lockett, and Parsons, "Director's Speech to the AFA," 9/13/68, memo, FAA 905/2; Edward B. Giller to Maxwell, 9/9/68, FAA 6174/2; Smart to Maxwell, 9/25/68, FAA 6174/2.

81. For Maxwell's speeches on the SST in 1969, see: "Firm Speaking Engagements for General Maxwell," 1/31/69, FAA 6174/3; Robb to Maxwell, "Speech to Santa Barbara Channel City Club," 1/23/69, memo, FAA 6174/3; Henry Handler (Armed Forces Management Association) to Maxwell, 9/6/68, FAA 6174/2; E. A. Allard (Boeing Manage-

ment Association) to Maxwell, 3/25/69, FAA 6174/4; Maxwell to Clifton (Air Transport Association), 3/25/69, FAA 6174/4; H. Neil Carson (Council on Transportation Law of the Federal Bar Association) to Maxwell, 4/18/69, FAA 6174/4.

82. Calhoun to SST Division Chiefs, "Speaking Engagements," 9/30/68, memo, FAA 6174/4.

83. For the 1968 speaking appearances of officials from the SST Office, see: Calhoun to Lockett and Irons, "Trip to Dartmouth, March 7, 1968," 3/6/68, memo, 905/2; G. Douglas Doil to Voyles, "Public Appearances and Speeches, of SS-210 Personnel," 3/29/68, memo, FAA 905/2; Robert D. Izer (American Society of Mechanical Engineers) to Maxwell, 9/5/68, FAA 6174/2; Blake to Sheldon Goldstein (ASME Student Branch of NY City College), 9/25/68, FAA 6174/2; D. M. Sherwood (Management Club, North Hollywood, California) to Blake, 9/13/68, FAA 6174/2; Leonard T. Glasser (Tullahoma Post, Society of American Military Engineers) to James H. Polve, 7/23/68, FAA 6174/1; Polve to Glasser, 7/29/68, FAA 6174/1.

84. For the AIAA meeting and for Maxwell's role in organizing its SST portion, see: Maxwell to Mentzer, 4/30/68, FAA 905/2; Maxwell to Cary, "Request for Assistance in Getting Soviet Speaker," 4/19/68, memo, FAA 905/2; Maxwell to Associate Administrator for Plans, "Schedule for AIAA Meeting," 5/24/68, memo, FAA 6174/1; Maxwell to Cary, "Soviet Speaker for AIAA," 6/5/68, memo, FAA 6174/1; Calhoun to Civil Aviation Officer, "Inviting British or French to Speak on the Concorde at AIAA," 5/14/68, memo, FAA 6174/1; Cary to Gifford D. Malone (Department of State), "Soviet Participation in AIAA Meeting," 6/6/68, memo, FAA 6174/1; Rusk to Amembassy, Moscow, "Invitation to Tupolev to Participate in AIAA Meeting, October 21–25, 1968," 6/13/68, DOS airgram, FAA 6174/1; Calhoun to Mr. Bakke, "AIAA Meeting, October 1968, Philadelphia," 8/5/68, memo, FAA 6174/1; Calhoun to Mentzer, 8/9/68, FAA 6174/1; Maxwell to General Forestier (Concorde), 5/24/68, FAA 6174/1; Maxwell to J. A. Hamilton (Concorde), 5/24/68, FAA 6174/1; Calhoun to the record, "Concorde Papers for AIAA Meeting," 6/21/68, FAA 6174/1; Maxwell to Mentzer, 7/1/68, FAA 6174/1; Vierling to M. J. de Lagarde (SUD-Aviation), 8/9/68, FAA 6174/1; Calhoun to Mrs. Rosalind Rosen (AIAA), 8/9/68, FAA 6174/1; J. de Lagarde to Maxwell, 8/2/68, FAA 6174/1; J. Giusta (SUD-Aviation) to Mentzer, 8/28/68, FAA 6174/2; Mentzer to Calhoun, 9/13/68, FAA 6174/2; Calhoun to Rosen, 9/20/68, 9/25/68, FAA 6174/2; J. de Lagarde to Mentzer, 9/10/68, FAA 6174/2; Maxwell to Hamilton, 10/28/68, FAA 6174/2; Maxwell to Forestier, 10/28/68, FAA 6174/2; Maxwell to Session Chairmen and Authors, "Instructions and Suggestions for Supersonic Transport Sessions: AIAA Annual Meeting October 21 and 22, 1968,—Philadelphia," 6/12/68, memo, and attachments, FAA 6174/97; Maxwell to George Henderson et al., all dated 6/13/68, individual letters, FAA 6174/1; Blake to SS-4, "General Maxwell's Address to AIAA June 18—Possible Approach," 6/15/68, memo, FAA 6174/1; Calhoun to Mentzer, 8/5/68, FAA 6174/1; Calhoun to J. P. Taylor, 8/5/68, FAA 6174/1; Calhoun to Henderson, 8/5/68, FAA 6174/1; C. W. Duffy (Boeing) to Mentzer, 8/6/68, FAA 6174/1; G. T. Haugan to Special Assistant for SST Development, "Review of Paper to Be Presented at the 5th Annual Meeting of the AIAA," 8/23/68, memo, FAA 6174/1; S. J. Converse to Chief, Technical Operations Division, "Public Appearances by SST Office Personnel," 9/13/68, memo, FAA 6174/2; Calhoun to Locraft et al., "Papers Presented at the AIAA 5th Annual Meeting, Philadelphia, Pa., October 21–23, 1968," 10/25/68, memo, FAA 6174/2; James K. Thompson, "AIAA 5th Annual Meeting and Technical Display, Philadelphia, Pa.," 11/1/68, trip report, FAA 6174/2; R. Cargill Hall (Jet Propulsion Laboratory historian) to Maxwell, 9/4/68, FAA 6174/2; Alfred G.

Kildow (AIAA) to Thomas G. Foxworth et al., "Convention Television," 10/3/68, memo, FAA 6174/2.

85. For reactions to the AIAA meeting, see: Maxwell to James J. Harford (AIAA), 11/4/68, FAA 6174/2; Maxwell to Mentzer, 10/28/68, FAA 6174/2; Maxwell to Dunning, 10/28/68, FAA 6174/2; Maxwell to Henderson, 10/28/68, FAA 6174/2; Henderson to Thompson, 10/24/68, FAA 6174/2; Henderson to Maxwell, 10/24/68, FAA 6174/2; G. L. Hanson (General Dynamics) to Haugan, 10/28/68, FAA 6174/2.

86. James R. Kerr (President of Avco Corporation and Chairman of the Explorers Club, National Dinner Committee) to Maxwell, 7/18/68, FAA 6174/1.

87. Walter A. Wood to Maxwell, 9/26/68, FAA 6174/2.

88. George C. Martin to Maxwell, 3/13/69, FAA 6174/4.

89. Maxwell to George McTigue, 4/2/69, FAA 6174/4.

90. For the support of various organizations, see: Stuart G. Tipton (Air Transport Association) to the President, 3/28/69, FAA 6174/4; Shaffer to Volpe, "Air Transport Association Letter," 4/7/69, memo, FAA 6174/4; David A. Tate (Air Force Association's Northwest Evergreen Chapter) to the President, 4/8/69, FAA 6174/4; Robert E. Lehnherr (Air Force Association's Greater Seattle Chapter) to the President, 4/9/69, FAA 6174/4; Volpe to Tipton, 4/15/69, FAA 6174/4; A. A. West (Air Force Association—Virginia) to the President, 8/19/69, FAA 6174/6; Vierling to West, 9/8/69, FAA 6174/6; P. L. Siemiller (International Association of Machinists and Aerospace Workers) to the President, circa 4/69, telegram, FAA 6174/4; Maxwell to Shaffer, "Telegram from President of International Association of Machinists and Aerospace Workers," 4/16/69, memo, FAA 6174/4; Shaffer to Volpe, "Telegram from President of International Association of Machinists and Aerospace Workers," 4/16/69, memo, FAA 6174/4; Volpe to Siemiller, 4/22/69, FAA 6174/4.

91. For the SST Office's attempts to influence the public at large, see: Lockett to Chief, Special Projects Staff, "Comments Concerning Updated Slide Library and Associated Narrative," 8/5/69, memo, FAA 6174/6; Vierling to A. R. Roesch, 7/10/69, FAA 6174/6; Willis F. Briley (Thomas Craven Film Corp.) to Locraft, 6/27/68, FAA 6174/1; Voyles to Special Assistant for Supersonic Transport Development, "Final Clearance for SST Article in FAA Pamphlet; IS-5 Route Slip of 2/23/68," 2/23/68, memo, FAA 905/2; Lockett to Voyles, "Facts about the SST," 2/28/68, memo, FAA 905/2; Irons to Voyles, "The Facts about the SST," 2/28/68, memo, FAA 905/2; Blake to Voyles, "The Facts about the SST," 2/28/68, memo, FAA 905/2; Voyles to Maxwell, "Questions and Answers on the SST," 4/2/68, memo, FAA 905/2; Voyles, "Trip Report," 4/4/68, FAA 905/2; Mervin K. Strickler, Jr., to Maxwell et al., "Aviation Noise and Sonic Boom Educational Proposal," 6/6/68, memo, FAA 6174/1; Maxwell to Strickler, "Aviation Noise and Sonic Boom Educational Proposal," 6/14/68, memo, FAA 6174/1; Robert F. O'Neill to SS-4, "Suggested Follow-up to an Application for Membership in the SST Club," 1/30/69, memo, FAA 6174/3.

92. Representative Bob Mathias to FAA Congressional Liaison, 8/2/68, and attached newspaper articles, FAA 6174/1; Hoover to Deputy [SST] Director, "Noise Damage Lawsuit," 8/2/68, memo, FAA 6174/1.

93. Maxwell to C. R. Devine (Reader's Digest Association), 10/31/68, FAA 6174/2; Myles Callum (*Good Housekeeping*) to Boyd, 11/23/68, and attached article, 11/28/68, FAA 6174/3; *Sacramento Bee*, 12/7/68.

94. C. D. Alexander (Secretary of the California Senate) to the President, and attached, *Senate Joint Resolution*, no. 7, 1/30/69, FAA 6174/6.

95. Representative Sidney R. Yates to Everett Hutchinson, 3/12/68, FAA 905/2; Boyd to Yates, 3/25/68, FAA 6174/95.

96. For examples of individual protest letters to the Department of Transportation, the FAA, the White House, and Congress and for the effect on the SST Office, see: Louis L. Jaffe to Boyd, 3/19/68, FAA 905/2; Maxwell to Jaffe, 3/27/68, FAA 905/2; Maxwell to H. S. Greenawalt, 3/21/68, FAA 905/2; Maxwell to Representative John J. Rhodes, 3/25/70, FAA 905/2; McKee to Representative Richard L. Ottinger, 5/10/68, FAA 6174/1; Maxwell to Mrs. Charles E. Humberger, 6/16/68, FAA 6174/1; Maxwell to Mr. & Mrs. Richard Wagner, 6/19/68, FAA 6174/1; Maxwell to Representative Hastings Keith, 8/5/68, FAA 6174/1; Blake to George B. Gould (Administrative Assistant to Representative Charles H. Wilson), 8/8/68, and attachments, FAA 6174/1; Maxwell to Representative Edward P. Boland, 8/28/68, FAA 6174/12; Maxwell to Mrs. Henry Herbert Marshall, 9/9/68, FAA 6174/2; Julian Loebenstein to Boyd, 9/8/68, FAA 6174/2; Maxwell to Senator Henry M. Jackson, 9/12/68, FAA 6174/2; Maxwell to Senator George Murphy, 9/19/68, FAA 6174/2; Maxwell to Representative Thomas B. Curtis, 12/16/68, FAA 6174/3; Maxwell to Clarence Dahlke, 12/16/68, FAA 6174/3; Maxwell to Poul Anderson, 2/11/69, FAA 6174/3; Maxwell to Dr. J. H. Meier, 2/13/68, FAA 6174/3; Maxwell to Representative Joseph K. Karth, 2/18/69, FAA 6174/3; Maxwell to H. T. Fritz, 2/26/69, FAA 6174/3; Representative Lester L. Wolff to Beggs, 3/5/69, FAA 6174/97; Beggs to Wolff, 3/17/69, FAA 6174/97; Arlie J. Nixon to Volpe, 3/18/69, FAA 6174/4; Bert Vorchheimer to Beggs, 3/17/69, FAA 6174/4; Howard V. Plough to Volpe, 4/18/69, FAA 6174/4. For the exchange between Senator Smith and Maxwell, see: Senator Margaret Chase Smith to Maxwell, 5/17/68, FAA 6174/1. For the audiologist's comments, see: Patricia Ashley to the President, 3/17/69, telegram, FAA 6174/4.

97. For examples of pro-SST correspondence, see: Representative Clark MacGregor to Volpe, 3/21/69, telegram, FAA 6174/4; Maxwell to W. K. Smith, 3/25/69, FAA 6174/4; G. Y. Sails to the President, 4/7/69, FAA 6174/4; A. Pietromonaco to the President, circa 4/69, FAA 6174/4; Mrs. Joel B. Goldberg to Volpe, 4/18/69, FAA 6174/4.

98. For a statistical summary of SST-related correspondence, see: John D. Calhoun to Special Assistant to the Undersecretary, "Responses to Letters to the President on the SST Program," 4/4/69, memo, FAA 6174/4.

99. For the SST Office's defense, see: the letters by Maxwell cited in note 96 above.

100. For Shurcliff's anti-SST inquiries and for the FAA's response to them in 1968, see: Shurcliff to Vierling, 1/10/68, FAA 905/1; Proxmire to Maxwell, 2/15/68, FAA 6174/92; Shurcliff to Maxwell, 3/6/68, FAA 905/2; Maxwell to Shurcliff, 3/15/68, FAA 905/2; Edwin L. Weisl, Jr. (Assistant Attorney General, Civil Division) and John G. Caughlin (Chief of Torts Division) to Shurcliff, 4/1/68, FAA 905/2; S. J. Converse to Chief, Technical Operations Division, "Sonic Boom Litigation, Kane vs. Standard Accident Insurance Company," 5/3/68, memo, FAA 6174/1; Maxwell to Shurcliff, 9/16/68, FAA 6174/2; Shurcliff to Maxwell, 10/17/68, FAA 6174/3.

101. For the Shurcliff–FAA/Department of Transportation correspondence about Shurcliff's *SST and Sonic Boom Handbook*, see: Shurcliff to Calhoun, 11/22/68, FAA 6174/2; Shurcliff to Thomas, 11/22/68, 11/28/68, FAA 6174/3; Maxwell to Shurcliff, 11/25/68, FAA 6174/3; Shurcliff to Boyd, 11/28/68, 12/12/68, FAA 6174/3; Charles R. Foster to

Shurcliff, 12/9/68, FAA 6174/3; Thomas to Shurcliff, 12/18/68, FAA 6174/3; Shurcliff, 3/69 (accompanied 2/15/69 *SST and Sonic Boom Handbook*), FAA 6174/4.

102. For the detailed review of the handbook by the SST Office, see: Blake to Operations Environment Branch, "Comments on Draft of Handbook by William A. Shurcliff," 2/7/69, memo, FAA 6174/3; Converse to Blake, "Shurcliff's SST and Sonic Boom Handbook," 2/28/69, memo, FAA 6174/4.

103. Calhoun to Shurcliff, 12/5/68, 12/17/68, FAA 6174/3; Shurcliff to Calhoun, 12/9/68, FAA 6174/3; Shurcliff to FAA Office of SST Information, 12/11/68, FAA 6174/3.

104. Hoover to Shurcliff, 1/3/69, 1/27/69, FAA 6174/3; Hoover to Shurcliff, 4/4/69, FAA 6174/4.

105. For Shurcliff's anti-SST campaign in 1969, see: Maxwell to Shurcliff, 5/29/69, FAA 6174/6; Shurcliff to Vierling, 7/31/69, FAA 6174/6; Robb to Chief, Analysis and Control Division, "Comments on Shurcliff Fact Sheet," 7/16/69, FAA 6174/6. For the Citizens League's 6/69 release on SST economics, see: Shurcliff to Maxwell, 6/20/69, and attached "The Supersonic Transport Planes: Further Dimming of the Economic Prospects," for release on 4/4/69, FAA 6174/6.

106. For Shurcliff's attack on the notion of solely over-water SST routes, see: Shurcliff to Maxwell, 3/10/69, FAA 6174/4; Shurcliff to John H. Shaffer, 7/10/69, FAA 6174/6; Shurcliff to Proxmire, 9/1/69, FAA 6174/6.

107. For Shurcliff's contact with congressional SST opponents, see: Shurcliff to Proxmire, 1/10/68, 1/16/68, 3/11/68, FAA 6174/92; Maxwell to Representative Frank Horton, 3/14/68, FAA 905/2; Senator Clifford P. Case to Shurcliff, 11/12/68, FAA 6174/3; Shurcliff to Case, 10/21/68, FAA 6174/3; Maxwell to Case, 11/28/68, FAA 6174/3; Shurcliff to Senator Robert P. Griffin, 6/20/69, and attached Citizens League, "The Supersonic Planes: Further Dimming of the Economic Prospects," 7/4/69, FAA 6174/6; Shurcliff to Maxwell, 3/2/69, FAA 6174/4.

108. T. J. Flanagan (Pan American) to Shurcliff, 7/9/69, FAA 6174/6.

109. For the role of individual Citizens League members, see: Harold Blood to Volpe, 1/28/69, FAA 6174/4; J. Greene to the President, 1/29/69, FAA 6174/3; Shurcliff to Maxwell, 2/9/69, FAA 6174/3; Shurcliff to D. D. Thomas, 2/9/69, FAA 6174/4; Maxwell to Elizabeth P. Borish, 3/13/69, FAA 6174/4; Shurcliff to Volpe, 3/18/69, FAA 6174/4; John R. Allietta to Volpe, 4/18/69, FAA 6174/4; Allietta to Shaffer, 7/9/69, FAA 6174/6; Citizens League Newsletter 20, 1/22/69, FAA 6174/3; Citizens League, "Warning! Sonic Boom Hour of Decision Is at Hand," 3/11/69, circular, FAA 6174/4; Dr. Bernard D. Davis (Professor of Bacterial Physiology at Harvard Medical School) to Volpe, 3/24/69, FAA 6174/4.

110. Shurcliff to Maxwell, 6/20/79, FAA 6174/6.

111. Maxwell to Shurcliff, 7/9/69, FAA 6174/6.

112. For the anti-SST work of the Town-Village Aircraft Safety and Noise Abatement Committee and for the formation of NOISE in late 1969, see: Harrison D. Bergin (Director, Town-Village Aircraft Safety and Noise Abatement Committee) to Citizens League, 6/14/67, 6/15/67, CL; Bergin to Shurcliff, 7/18/67, 10/26/67, 1/21/69, 1/29/69, 2/5/69, 5/12/69, CL; Shurcliff to Bergin, 10/6/67, circa 10/67, 10/31/67, 4/29/69, CL; *Town of Hempstead News*, 10/9/68, 10/31/68, CL; Town-Village Committee, "The Jet Noise Problem," circa 1/69, announcement, CL; Town-Village Committee, copy of article from *American*

Aviation, 1/6/69, CL; Town-Village Committee, "Quotes," 1–12/69, 2–3/70, vol. 1, nos. 1–12, vol. 2, nos. 2–3, CL; Maxwell to Bergin, 4/11/69, FAA 6174/4; Clifford A. Deeds (Director of Town-Village Committee) to Shurcliff, 8/28/69, 9/22/69, 10/30/69, CL; Shurcliff to Deeds, 9/7/69 and 11/4/69, CL.

113. For the anti-SST work by the Citizens Committee for the Hudson Valley and by its director William Hoppen, see: William Hoppen to Shurcliff, 6/28/67, 7/12/67, 11/17/67, 4/27/68, circa 6–7/68, 9/17/68, CL; "Organizations and Chairmen or Heads of Groups," 6/9/67, luncheon list of Overseas Press Club, CL; Hoppen to "All Organizations," 8/67, announcement, and attached Citizens League Fact Sheet, no. 10A, 6/67, CL; Polly Eustis (Citizens League) to Hoppen, 11/27/67, CL; Walter S. Boardman to Representative Edward P. Boland, 10/16/67, CL; Hoppen to Shurcliff, 1/29/68, and attached mailing list, CL; Shurcliff to Hoppen, 2/6/68, 5/15/68, 7/6/68, 9/20/68, 12/8/68, CL; Bob Litch (Federation of Conservationists—U.S.) to Hoppen, circa 4/68, CL; Hoppen, 4/29/68, announcement, CL; "Conservation Bill of Rights," 11/30/64, 3/15/68, drafts and articles, CL; Hoppen to Senator Clifford P. Case, 5/27/68, CL; Hoppen to Mrs. James F. Graves, 6/8/68, CL; Graves to Hoppen, circa 6/68, CL; Hoppen to Shurcliff, 12/4/68, and attached notations and comments on Shurcliff, "SST and Sonic Boom Handbook," 11/15/68, CL; Hoppen to Shurcliff, 12/21/68, and attached list, CL.

114. For Angeluscheff's anti-SST statements and contact with Shurcliff, see: Zhivko D. Angeluscheff, "Sonic Boom—the Enemy of Civilization," presented to 5th International Congress for Noise Abatement, London, England, May 13–18, 1968, CL; Shurcliff to Angeluscheff, 11/3/68, 11/20/68, 2/21/68, CL; Angeluscheff to Shurcliff, 11/17/68, 2/25/68, CL; Angeluscheff to Bo Lundberg, 1/16/68, CL; Shurcliff, 2/21–23/68, form letter, and attached list, CL.

115. For the 1967 communication between Shurcliff and the Sierra Club and for the Sierra Club's position on the SST in 1967, see: Robin Way (Sierra Club) to Shurcliff, 5/25/67, 9/21/67, CL; Michael McCloskey (Sierra Club) to Mrs. M. E. Borish, 7/7/67, CL; Connie Flateboe (Sierra Club) to Shurcliff, 7/24/64, postcard, CL; David Brower (Sierra Club) to Shurcliff, 8/21/67, CL; Shurcliff to Brower, 8/26/67, CL; McCloskey to Shurcliff, 9/12/67, 9/25/67, CL; Shurcliff to McCloskey, 8/30/67, CL; Forsyth to Susan Aronson (Citizens League), 10/24/67, CL; Gary A. Soucie (Sierra Club) to Shurcliff, 9/19/67, CL; Shurcliff to Soucie, 9/24/67, CL; McCloskey to Senator Carl Hayden, 9/25/67, CL; George Marshall (Sierra Club) to Polly Eustis (Citizens League), 3/13/67, CL.

116. For the correspondence between William Shurcliff and Charles Shurcliff during the summer of 1968, see: William Shurcliff to Charles Shurcliff, 6/26/68, 7/7/68, 8/4/68, CL; Eustis to Charles Shurcliff, 6/27/68, CL; Shurcliff to Citizens League, 7/1/68, and attached list of activities, CL; Charles Shurcliff to William Shurcliff, 7/9/68, 7/29/68, CL; Charles Shurcliff to "Dear Sir," 7/25/68, CL; Charles Shurcliff to "All CLASB Directors and Personnel," 8/2/68, CL; Shurcliff to Eustis, 8/5/68, note, CL. For Sierra Club activity, see: Brower to members of Grand Canyon Task Force, 11/25/68, memo, and attached "Regional Disputes, Lobbying Shape Colorado River Bill," *Congressional Quarterly*, 11/1/68, pp. 3019–3031, CL; *Sierra Club Bulletin* (10/68):25.

117. For Baxter's anti-SST work and for the Baxter-Shurcliff correspondence, see: Lockett to Maxwell, 2/29/68, "Professor Baxter's Study of Legal Aspects of Aircraft Noise and Sonic Boom," 2/29/68, memo, FAA 905/2; Shurcliff to Professor William F. Baxter, 1/13/67, 4/12/68, 11/28/68, 12/28/68, 1/4/69, 4/26/69, CL; Baxter to Shurcliff, 9/22/67, 4/4/68, 4/17/68, 12/27/68, 5/5/69, CL; Baxter to Eustis, 10/13/67, CL; Eustis to Baxter,

4/18/68, CL; Baxter to Nancy K. Piore (Citizens League), 4/9/68, CL; William F. Baxter, "The SST: From Watts to Harlem in Two Hours," *Stanford Law Review* 21 (11/68):1–57; Charles J. Peters to Director, Office of Noise Abatement, and Maxwell, "Professor Baxter's report on Legal Aspects of Aircraft Noise and Sonic Boom," 3/5/68, memo, FAA 905/2; M. Cecil Mackey to Frank W. Lehan, 12/6/68, memo, FAA 6174/3; Maxwell to Shurcliff, 4/14/69, FAA 6174/4.

118. Rice Odell (Conservation Foundation) to Shurcliff, 1/18/68, CL; "Conservation Foundation Letter," 12/29/67, 8/30/68, 3/17/68, CL.

119. Mrs. Russell Brodine (Committee for Nuclear Information) to Shurcliff, 5/26/67, CL; Shurcliff to Brodine, 6/17/67, CL.

120. Herbert H. Mills (World Wildlife Fund) to Mrs. M. E. Borish, 6/9/67, CL; Piore to Davis, 10/3/67, CL.

121. Professor B. T. Feld (MIT) to Shurcliff, circa 12/67, and attached statement by Richard Garwin, "The Supersonic Transport Aircraft," 9/67, CL.

122. Caroline Weichlein (Society for the Psychological Study of Social Issues) to Elizabeth M. Champagne (Citizens League), 9/22/67, CL; Aronson to Weichlein, 10/23/67, CL; Weichlein to Aronson, 11/17/67, CL; Shurcliff to Weichlein, 1/15/68, and attached Shurcliff, "The SST—and the Unanswered Crucial Question," 1/14/68, CL.

123. Paul M. Tilden (*National Parks Magazine*) to Shurcliff, 5/17/68, CL; Shurcliff to Tilden, 5/22/68, CL.

124. René Dubos to W. H. Ferry, 1/24/68, CL; Barry Commoner to Ferry, 2/2/68, CL; Perry T. Rathbone (Director, Museum of Fine Arts, Boston) to Shurcliff, 8/13/68, CL.

125. Jill Sen (Stanford Conservation Group) to Citizens League, 8/18/69, CL; Shurcliff to Sen, 9/6/69, CL; Paul Kulkosky (Columbia University, Institute for the Study of Science in Human Affairs) to Citizens League, 7/8/69, CL; Shurcliff to Kulkosky, 7/13/69, CL; Paul Bezaire (National Anti-Pollution Association) to Shurcliff, 4/15/69, CL; Shurcliff to Bezaire, 4/27/69, CL.

126. For Shurcliff's failures to enlist the support of various individuals or groups, see; Stephen Enke to Shurcliff, 7/18/67, CL; Shurcliff to Enke, 11/15/67, CL; Luis W. Alvarez to Shurcliff, 7/20/67, CL; John Gardner to Shurcliff, 12/22/68, CL; Shurcliff to John Bailey (Democratic National Committee), 5/23/68, CL; Bailey to Shurcliff, 6/5/68, CL; Representative Hale Boggs to Shurcliff, 7/23/68, CL; Shurcliff to World Health Organization, 6/30/67, CL; Action on Smoking and Health to Shurcliff, 8/28/68, CL; Consumers Union to Shurcliff, 2/21/67, CL; Federation of American Scientists to Shurcliff, 4/25/67, CL; Charles F. Wurster (Environmental Defense Fund) to "Gentlemen," 12/23/68, CL; Shurcliff to Wurster, 1/5/69, CL; William D. Hurley (National Council on Noise Abatement) to J. Reece Roth, 12/6/68, and attached newsletter of 6/30/69, CL; Shurcliff to Hurley, 9/1/69, CL; Mrs. Bruce B. Benson (League of Women Voters) to Mrs. Frank J. Ingelfinger, 1/29/69, CL.

Chapter 16

1. Minutes of the eleventh meeting of the NAS Committee on SST-Sonic Boom, 10/26/65, and attachment, NAS.

2. Dunning to Seitz, 12/3/65, NAS.

3. Minutes of the fifteenth meeting of the NAS Committee on SST-Sonic Boom, 8/4/66, NAS.

4. Converse to Lockett, "Visit to Dean Dunning," 7/28/66, memo, FAA 1647/7.

5. Ibid.; Lockett to Maxwell, "Telephone Conversation with Dr. Dunning, August 25, 1966," 8/26/66, memo, FAA 1647/7.

6. Maxwell to Dunning, 4/24/67, FAA 1647/10.

7. Converse to the files, "Report on May 17 Meeting of OSTSBCC," 5/25/67, FAA 1647/10.

8. Raymond Bauer to Dunning, 4/24/67, NAS.

9. Dunning to Hornig, 7/26/67, FAA 6174/92; Hornig to Dunning, 7/27/67, FAA 1647/11; Dunning to Hornig, circa 8/67, NAS.

10. Lockett, "Trip Report," 9/23/66, FAA 1647/7.

11. Dunning to McKee, 2/3/67, and attached "Sonic Boom," FAA 1647/9.

12. John P. Taylor (NAS Committee executive secretary) to Lockett, 12/7/66, FAA 1647/8.

13. Lockett to Chief, Analysis and Control Division, "Project Manager Assignments," 4/7/67, memo, FAA 1647/10.

14. Vierling to McKee, 6/1/67, note, FAA 1647/10.

15. Maxwell to Dunning, 4/24/67, FAA 1647/10.

16. Dunning to Boyd, 5/10/67, FAA 6174/95.

17. Boyd to Dunning, 5/23/67, FAA 6174/95.

18. Maxwell to Dunning, 12/4/67, FAA 1647/13.

19. Taylor to Maxwell, 4/3/68, and attached Dunning to David Rose, 3/28/68, draft letter, FAA 905/2.

20. McKee to John V. Lindsay, 5/10/68, LBJ FAA Hist. Doc.

21. Minutes of the fourteenth meeting of the NAS Committee on SST-Sonic Boom, 5/26/66, NAS.

22. Minutes of the fifteenth meeting of the NAS Committee on SST-Sonic Boom.

23. Minutes of the fourteenth meeting of the NAS Committee on SST-Sonic Boom.

24. Lockett, *Trip Report*, 9/23/66, FAA 1647/7.

25. "Summary of Meeting—OST Sonic Boom Coordinating Committee, January 16, 1967," 1/18/67, memo, FAA 1647/8.

26. Minutes of the eighteenth meeting of the NAS Committee on SST-Sonic Boom, 2/20/67, NAS.

27. McKee to Senator Frank E. Moss, 2/10/67, FAA 1647/9; Maxwell to Senator Charles H. Percy, 7/24/67, FAA 1647/11.

28. Dunning to Shurcliff, 12/7/66, CL.

29. John T. Edsall to President Lyndon B. Johnson, 1/24/67, NAS; Hornig to Edsall, 2/2/67, NAS; Edsall to Hornig, 2/28/67, NAS.

30. Golovin to Dunning, 3/28/67, NAS.

31. Dunning to McKee, 4/24/67, FAA 1647/10.

32. Bauer to Dunning, 4/24/67, NAS.

33. Shurcliff to Dunning, 7/5/67, FAA 6174/92.

34. Dunning to McKee, 7/11/67, FAA 6174/92; McKee to Maxwell, 7/14/67, handwritten note, FAA 6174/92.

35. Shurcliff memo on telephone conversation with Dunning, 7/18/67, CL.

36. NAS Committee on SST-Sonic Boom, Subcommittee on Research, *Report on Generation and Propagation of Sonic Boom,* 10/67, NAS.

37. C. Richard Soderberg to Shurcliff, 12/16/67, FAA 1647/13.

38. NAS Committee on SST-Sonic Boom, Subcommittee on Physical Effects, *Report on Physical Effects of the Sonic Boom,* 2/68, NAS.

39. Maxwell to McKee, circa early 3/68, handwritten notes, FAA 6174/92.

40. NAS news release on *Report on Physical Effects of the Sonic Boom,* 3/5/68, FAA 6174/92.

41. NAS Committee on SST-Sonic Boom, *Report on Physical Effects,* pp. 2, 8, 12.

42. NAS news release on *Report on Physical Effects.*

43. *New York Times,* 3/5/68, p. 3, FAA 6174/92; *Washington Post,* 3/5/68, p. A-14, FAA 6174/92.

44. Shurcliff to Dunning, 3/18/68, CL.

45. Ibid., 5/29/68, CL.

46. Donald M. Weinroth to Taylor, 6/5/68, and attached Dunning to Shurcliff, draft letter, NAS.

47. Joseph H. Zettel to Taylor, 7/10/68, NAS.

48. Hallowell Davis to Taylor, 7/5/68, NAS.

49. Taylor to Dunning, "Subject: Recent Shurcliff Correspondence," 6/18/68, NAS.

50. Shurcliff to Members of the NAS Governing Board, 6/21/68, CL.

51. NAS Committee on SST-Sonic Boom, Subcommittee on Human Response, *Report on Human Response to the Sonic Boom,* 6/68, NAS.

52. NAS news release on *Report on Human Response to the Sonic Boom,* 6/25/68, FAA 6174/11.

53. NAS Committee on SST-Sonic Boom, Subcommittee on Research, *1968 Progress Report on National Research Program—Sonic Boom Generation and Propagation,* 6/68, NAS.

54. Shurcliff to Bauer, 7/2/68, CL.

55. Bauer to Shurcliff, 7/15/68, CL.

56. Shurcliff to Bauer, 7/25/68, CL.

57. Draft news release by the Citizens League Against the Sonic Boom for release on 12/20/68 (attached to Shurcliff to Seitz, 11/25/68), NAS.

58. C. S. French to Seitz, 8/12/68, CL.

59. Harvey Brooks to Dunning, 7/16/68, NAS.

60. Brooks to Shurcliff, 7/18/68, 8/5/68, CL.

61. Shurcliff to Brooks, 7/24/68, 8/7/68, CL.

62. John S. Coleman (NAS executive officer) to NAS members, 8/22/68, and attached Dunning to Seitz, 8/19/68 and "Statement of the Committee on SST-Sonic Boom," 8/19/68, NAS.

63. Dunning to Seitz, 8/23/68, NAS.

64. The history for 9/68 is recounted in the draft news release by the Citizens League for release on 12/20/68.

65. For the SST opposition's use of the modifying statement, see: Maxwell to Mrs. Eugene M. Landis, 9/5/68, FAA 6174/2; Landis to Maxwell, received by FAA on 9/10/68, handwritten note, FAA 6174/2; Shurcliff to Maxwell, 9/8/68, FAA 6174/2.

66. Bauer to Taylor, 9/11/68, draft letter, NAS; Bauer to Shurcliff, 9/28/68, NAS.

67. John T. Edsall to Seitz, 10/2/68, NAS.

68. Taylor to Dunning, 10/4/68, NAS.

69. "Briefing Notes for John R. Dunning for Use in Connection with NAS Governing Board Meeting on 6 October 1968," NAS.

70. For the anti-SST attitude of some NAS members, see: Richard L. Garwin to Seitz, 9/26/68, and attached Jesse L. Greenstein to Garwin, 8/30/68, NAS.

71. "Summary Minutes: Meeting of the [NAS] Governing Board," 10/6/68, NAS.

72. Bauer to Senator Clifford P. Case, 9/28/68, NAS; Bauer, "Some Thoughts on Human Response to the Sonic Boom," 10/22/68, address to the American Institute of Aeronautics and Astronautics, CL.

73. Shurcliff to Seitz, 11/25/68, and attached draft news release by the Citizens League for release on 12/20/68.

74. Merle A. Tuve (NAS home secretary) to Shurcliff, 12/12/68, CL.

75. Philip Handler to Tuve, 12/17/68, NAS.

76. Phillip M. Boffey, *The Brain Bank of America* (New York, N.Y.: McGraw-Hill Book Co., 1975), pp. 127–128; Shurcliff to Tuve, 12/14/68, NAS.

77. National Academy of Sciences/National Research Council/National Academy of Engineering, *News Report* 19 (2/69):11.

Chapter 17

1. Harold D. Watkins, "SST Funding Bid Left to Nixon," *Aviation Week and Space Technology*, 1/20/69, pp. 30–31; *Wall Street Journal*, 1/13/69, p. 2, NAS Cmte. Newsletter, no. 53, 1/1–16/69, pp. 1–2.

2. Richard J. Kent, Jr., *Safe, Separated, and Soaring* (Washington, D.C., U.S. Government Printing Office, 1980), pp. 172–186; *New York Times*, 7/2/68, p. 21.

3. John E. Robson to Assistant Secretary for International Affairs and Special Programs et al., "Issues Paper—SST Program," 12/11/68, memo, FAA 6174/95; "[Draft] SST Issues Paper," 12/68, FAA 6174/95; Frank W. Lehan to Thomas, "Issues Paper—SST Program," 1/9/69, memo, and attached revisions, FAA 6174/95; Boyd to Thomas and Maxwell, "Review of SST Design," 1/17/69, memo, FAA 6174/95.

4. Arthur F. Burns to John A. Volpe, 1/23/69, FAA 6174/97.

5. President to Volpe, 1/29/69, memo, FAA 6174/97; White House press release, 1/30/69, FAA 6174/97.

6. Volpe to the President, "Supersonic Transport Program," 1/31/69, memo, FAA 6174/97.

7. President to Volpe, 2/5/69, memo, FAA 6174/97.

8. For press reports of Nixon's directive establishing an interagency review committee, see: *New York Times*, 2/8/69, p. 1, *Aviation Daily*, 2/17/69, p. 246, and Harold D. Watkins, "SST Study Committee Formation Slowed," *Aviation Week and Space Technology*, 2/10/69, p. 29, NAS Cmte. Newsletter, no. 55, 2/1–15/69, p. 1; "One for the Road," *Aviation Week and Space Technology*, 1/20/69, p. 15, and *New York Times*, 2/1/69, p. 58C, NAS Cmte. Newsletter, no. 54, 1/17–31/69, p. 1.

9. President to James M. Beggs, 2/19/69, memo, FAA 6174/95; *New York Times*, 2/28/69, p. 77.

10. For the reports by Boeing and the SST Office, see: Calhoun to SS-100 et al., "Outline and Assignments for an SST 'National Benefits' Paper," 2/5/69, memo, FAA 6174/3; Maxwell to George Revercomb, "The SST Program and Related National Benefits," 2/24/69, memo, FAA 6174/97; "National Benefits of the SST Prototype Development," 3/6/69, FAA 6174/95.

11. For the various SST evaluations submitted to the Department of Transportation, see: Charles D. Baker to Thomas and Maxwell, "Supersonic Transport Program," 2/13/69, memo, FAA 6174/97; Charles O. Cary to Thomas, "International Implications of U.S. Decision to Proceed with or Cancel Supersonic Transport Program," 2/5/69, memo, FAA 6174/95; Robert R. Bennett to Lehan, 2/9/69, FAA 6174/95.

12. Maurice H. Stans to Volpe, "Evaluation of SST Program," 2/19/69, memo, FAA 6174/95.

13. *New York Times*, 2/27/69, p. 73M, NAS Cmte. Newsletter, no. 56, 2/17/69, p. 2; John H. Shaffer to Beggs, 2/13/69, FAA 6174/4.

14. For overview of the 2/19/69 meeting, see: "Attendees—SST Briefing," 2/19/69, FAA 6174/97; Calhoun to SS-100 et al., "February 19/1969 Meeting of the Ad Hoc Review Committee," 2/20/69, memo, FAA 6174/97; "Inter-Agency SST Committee" and "Panel Organization Proposals," briefing papers for the 2/19/69 meeting, FAA 6174/97.

15. FAA briefing charts for 2/19/69 Ad Hoc Review Committee meeting, FAA 6174/97; Maxwell to Beggs, "Ad Hoc Review Committee—SST," 2/20/69, memo, FAA 6174/97.

16. "Attendees: SST Meeting—26 February 1969," FAA 6174/97; "Issues Relating to SST," briefing paper for 2/26/69 Ad Hoc Review Committee meeting, FAA 6174/97; FAA briefing charts for 2/26/29 Ad Hoc Review Committee meeting, FAA 6174/97; "SST Economic Update," 2/24/69, FAA 6174/97; Beggs to each Ad Hoc Review Committee member, 2/28/69, FAA 6174/97.

17. Beggs to each Ad Hoc Review Committee member, 2/28/69, and attached "Assignments and Guidance for Working Panels of the Ad Hoc SST Review Committee," FAA 6174/97.

18. Hendrik S. Houthakker to Beggs, 3/3/69, FAA 6174/97.

19. For 3/5/69 meeting, see: "Proposed List of Attendees: SST Ad Hoc Review Commit-

tee Meeting—March 5, 1969," "Agenda for Ad Hoc Review Committee Meeting of 3/5/69," and "Transcript of Proceedings: President's Ad Hoc Review Committee Meeting, 3/5/69," FAA 6174/97.

20. Maxwell to Shaffer, "Participation in Ad Hoc Committee Meeting," 3/6/69, memo, FAA 6174/97.

21. Shaffer to Tillinghast, Lawrence, Halaby, and John H. Crooker, Jr., CAB Chairman, 3/11/69, individual letters, FAA 6174/97.

22. For various airline opinions, see: Thomas to George A. Spater (American Airlines) et al., 2/19/69, telex, FAA 6174/4; United, American, Northwest, TWA, Pan American, and Braniff to Thomas, 2/25/69 (except Braniff: 2/24/69), telexes, FAA 6174/4; Robert F. Six (Continental Airlines) to Thomas 2/25/69, telephone conversation summary, FAA 6174/97; Eastern to Thomas, 2/25/69 (classified as Eastern letter through process of elimination using 10/31/69 *Congressional Record*, pp. H10432–H10446), FAA 6174/95; George A. Spater (American) to Thomas, 2/27/69, G. E. Keck (United) to Thomas, 2/18/69, Harding Lawrence (Braniff) to Thomas, 2/17/69, Halaby (Pan American) to Thomas, 3/1/69, Donald W. Nyrop (Northwest) to Thomas, 3/1/69, Charles C. Tillinghast, Jr. (TWA) to Thomas, 2/28/69, and Six (Continental) to Thomas, 2/26/69, FAA 6174/4.

23. Maxwell to Shaffer, "Airlines' Evaluation and Recommendations—SST," 3/4/69, memo, FAA 6174/97.

24. Thomas to Volpe, "Evaluation and Recommendations Regarding the Supersonic Transport Program," 3/4/69, memo, FAA 6174/97.

25. For the 3/12/69 meeting, see: Beggs to committee members, 3/4/69, FAA 6174/4; James E. Densmore to Beggs et al., "March 12 Meeting of the Ad Hoc SST Review Committee," 3/11/69, memo, FAA 6174/97; "List of Attendees (not definite): SST Ad Hoc Review Committee Meeting—March 12, 1969," FAA 6174/97; agenda for Ad Hoc Review Committee, 3/12/69, FAA 6174/97; "[Draft] Report of SST Ad Hoc Review Committee," 3/19/69, FAA 6174/95; Houthakker to Beggs, "Proposed Construction of SST Prototype," 3/26/69, FAA 6174/95; *New York Times*, 3/16/69, pp. 1, 63; draft reports by the Economic Subcommittee, 3/12/69, the Environmental and Sociological Panel, 3/11/69, the Technological Fallout Panel, and the Balance of Payments Panel, 3/12/69, FAA 6174/97. For the press reports, see: *Wall Street Journal*, 3/12/69, p. 2, *New York Times*, 3/12/69, p. 94C, NAS Cmte. Newsletter, no. 57, 3/1–16/69, pp. 2–3; Harold D. Watkins, "Volpe Seen Backing SST Prototype Work," *Aviation Week and Space Technology*, 3/17/69, p. 30.

26. For the Department of Transportation's view of the Ad Hoc Review Committee, see: Baker to DOT Offices Concerned with SST Review, "Coordination of the Administration's Review of the SST," 2/28/69, memo, FAA 6174/97; Martin Convisser to Beggs et al., 3/14/69, route slip, and attached "Outline of Report to the President: Reevaluation of the SST Program," 3/14/69, FAA 6174/97; Beggs to each member of the Ad Hoc Review Committee, circa 3/19/69, draft report, FAA 6174/95.

27. Maxwell to Houthakker, 3/7/69, and attachments, FAA 6174/97.

28. Maxwell to the record, 3/13/69, memo, FAA 6174/97.

29. *New York Times*, 3/16/69, pp. 1, 63.

30. "SST Ad Hoc Review Committee," circa 3/19/69, draft report, FAA 6174/95.

31. Beggs to each member of the Ad Hoc Review Committee, "SST Ad Hoc Review Committee Report," 3/19/69, memo, FAA 6174/95.

32. Rocco C. Siciliano (Commerce) to Beggs, "Report of the Economic Subcommittee," 3/14/69, memo, FAA 6174/95; Charles W. Harper (NASA) to Beggs, 3/20/69, FAA 6174/95; Harper to Beggs, 3/24/69, and attached "Recommendations by NASA on the SST Program," 3/24/69, FAA 6174/95.

33. Paul A. Volcker (Treasury) to Beggs, 3/18/69, FAA 6174/95; John C. Colman (Treasury) to Beggs, 3/20/69, FAA 6174/95.

34. Houthakker to Beggs, 3/20/69, FAA 6174/95.

35. Lee A. Dubridge (President's Science Adviser and Director of the Office of Science and Technology) to Beggs, "Committee Report," 3/20/69, memo, 6174/97; Dubridge to Beggs, 3/20/69, FAA 6174/95; Russell C. Drew (OST) to Beggs, 3/24/69, and attachment, FAA 6174/95.

36. Russell E. Train (Interior) to Beggs, "SST Ad Hoc Review Committee Report," 3/21/69, FAA 6174/95; Charles C. Johnson, Jr. (Assistant Surgeon General), to Beggs, 3/20/69, FAA 6174/95; Johnson to Beggs, 3/20/69, and attachment, FAA 6174/95.

37. Arnold R. Weber (Labor) to Beggs, 3/21/69, FAA 6174/95; T. C. Muse (DOD) to Beggs, 3/20/69, FAA 6174/95.

38. For 3/25/69 meeting, see: Calhoun to the record, "Plans for Ad Hoc SST Committee," 3/25/69, memo, FAA 6174/97; attendance lists for 3/25/69, SST Ad Hoc Review Committee meeting, FAA 6174/97; Houthakker to Beggs, "Proposed Construction of SST Prototypes," 3/26/69, memo, FAA 6174/95; Johnson to Beggs, 3/26/69, FAA 6174/95; U. Alexis Johnson (State) to Beggs, 3/26/69, FAA 6174/95; Weber to Beggs, 3/26/69, FAA 6174/95.

39. Final report of the SST Ad Hoc Review Committee, 4/1/69, FAA 6174/95.

40. For the OST Ad Hoc Committee, see: Maxwell to Beggs et al., "Request for SST Briefing," 3/11/69, memo, FAA 6174/97; "Agenda Outline—Dubridge Subcommittee Presentations," 3/24/69, FAA 6174/97; Maxwell to Russell Drew, "SST Data," 3/17/69, memo, FAA 6174/97; W. T. Farrish (United) to Maxwell, 3/27/69, telex, FAA 6174/4; Dubridge to the President, "SST," 4/29/69, memo, and attached OST, "Report of the Ad Hoc Supersonic Transport Review Committee," 3/30/69, FAA 6174/97. The OST Ad Hoc Review Committee report is also reprinted in the *Congressional Record-House*, 9/16/71, pp. H554–H557.

41. For demands for a decision to proceed quickly, see: Harold D. Watkins, "Pressure Grows to Delay SST Prototype," *Aviation Week and Space Technology*, 3/31/69, pp. 29–30; *Washington Post*, 3/29/69, p. 3, NAS Cmte. Newsletter, no. 58, 3/17/69, p. 1; *New York Times*, 4/27/69, p. 42; *Aviation Daily*, 5/1/69, pp. 247–248, NAS Cmte. Newsletter, no. 60, 4/17–30/69, pp. 3–4; *Aviation Daily*, 5/1/69, p. 6, NAS Cmte. Newsletter, no. 61, 5/1–16/69, p. 2; Maxwell to Shaffer, "Expiration of Special Default Provisions under Boeing SST Contract," 4/15/69, memo, and Shaffer to Maxwell, handwritten note, FAA 6174/96.

42. For the delay on a decision to go ahead, see: *Washington Post*, 4/15/69, p. A8, NAS Cmte. Newsletter, no. 59, 4/1–16/69, p. 1; *Aviation Daily*, 4/16/69, p. 245, NAS Cmte. Newsletter, no. 60, 4/17–30/69, p. 2.

43. *New York Times*, 4/27/69, p. 42; *Aviation Daily*, 4/16/69, p. 245, NAS Cmte. Newsletter, no. 60, 4/17–30/69, p. 3.

44. For a reduction in SST spending, see: Beggs to Browne and Shaffer, 6/23/69, note, FAA 6174/6; "Breakdowns of SST Funding at $11 Million and $7.5 Million Levels," 6/25/69, FAA 6174/92; Vierling to Browne, "Breakdown of the Current SST $11 Million Fundings," 6/23/69, memo, and attachment, FAA 6174/6; Shaffer to Beggs, "Reductions in SST Program Contractor Funding," 6/27/69, memo, FAA 6174/92; Maxwell to Browne, "Reduction in SST Contract Funding," 7/3/69, memo, 6174/6; C. F. Maas to Chief of Budget Review and Reports Staff, "Revision to September Apportionment Requirement," 8/25/69, memo, FAA 6174/6; R. E. Parsons to Prestemon, 7/24/69, memo, FAA 6174/6.

45. *New York Times*, 6/23/69, p. 78.

46. For the Department of Transportation's lobbying for a go-ahead decision, see: *London Times*, 8/9/69, p. 8, and "Eager to Start," *Aviation Week and Space Technology*, 8/4/69, p. 15, NAS Cmte. Newsletter, no. 67, 8/1–15/69, p. 2; Volpe to the President, "Status Report—Supersonic Transports," 8/15/69, memo, FAA 6174/95; *Aviation Daily*, 8/28/69, p. 312, NAS Cmte. Newsletter, no. 68, 8/16–31/69, p. 4; Thomas to Beggs, "Supersonic Transport Program," 9/15/69, memo, and attached "Draft Memorandum to the President," FAA 6174/95.

47. *Aviation Daily*, 9/17/69, p. 88, NAS Cmte. Newsletter, no. 70, 9/16–30/69, p. 5.

48. *Washington Evening Star*, 8/29/69, NAS Cmte. Newsletter, no. 69, 9/1–15/69, p. 1.

49. *New York Times*, 9/24/69, pp. 1, 92.

50. Harold D. Watkins, "SST Faces Congressional Hurdle," *Aviation Week and Space Technology*, 9/29/69, pp. 16, 18; "The Supersonic 'Go,' " *Aviation Week and Space Technology*, 9/29/69, p. 11.

51. *Washington Daily News*, 9/24/69, p. 28; *Washington Post*, 9/28/69, p. 136.

52. *Washington Evening Star*, 9/24/69, p. A-20; *St. Louis Post Dispatch*, 9/27/69, as reported in *Congressional Record*, 9/29/69, p. E7918.

53. For a congressional reaction to the go-ahead, see: *Washington Post*, 9/26/69, p. A6, and *Wall Street Journal*, 9/24/69, p. 3, NAS Cmte. Newsletter, no. 70, 9/16–30/69, pp. 2–3; *Congressional Record*, 10/1/69, p. H8837, NAS Cmte. Newsletter, no. 71, 10/1–5/69, p. 3.

54. For the fiscal 1970 SST funding request and passage, see: Harold D. Watkins, "SST Faces Congressional Hurdle," *Aviation Week and Space Technology*, 9/29/69, pp. 16–18; *New York Times*, 10/10/69, p. 93; *Aviation Daily*, 10/15/69, p. 260, NAS Cmte. Newsletter, no. 71, 10/1–15/69, p. 2; *Washington Post*, 11/19/69, p. A6, NAS Cmte. Newsletter, no. 73, 11/1–15/69, p. 2; *New York Times*, 12/17/69, p. 109; *Washington Post*, 12/18/69, p. A2, and *Wall Street Journal*, 12/19/69, p. 10, NAS Cmte. Newsletter, no. 76, 12/16–31/69, p. 3; "U.S. Supersonic Transport Status, Problems and Future," 3/11/70, FAA 6174/93.

55. For the release by Reuss of the Ad Hoc Review Committee materials, see: *Aviation Daily*, 10/14/69, p. 252, and *Washington Post*, 10/27/69, p. A6, NAS Cmte. Newsletter, no. 72, 10/16–31/69, pp. 2–3; *New York Times*, 11/1/69, p. 26; *Congressional Record*, 10/31/69, pp. H10432–H10446; "Reuss SST," 11/5/69, Associated Press dispatch, FAA 6174/5.

56. *New York Times*, 11/21/69, p. 46C, NAS Cmte. Newsletter, no. 75, 11/16–30/69, p. 1.

57. *Medical Tribune*, 12/4/69, p. 20, NAS Cmte. Newsletter, no. 76, 12/16–31/69, p. 9.

58. *Congressional Record*, 11/15/69, pp. H10601–H10604.

59. *Aviation Daily*, 10/17/69, p. 278, NAS Cmte. Newsletter, no. 72, 10/16/69, p. 6.

60. John S. Foster, Jr. (DOD) to Volpe, "Supersonic Transport (SST)," 11/2/69, FAA 6174/97.

61. *Aviation Daily*, 11/14/69, p. 83, NAS Cmte. Newsletter, no. 73, 11/1–15/69, p. 3.

62. For the FAA rebuttals, see: FAA to Representative Silvio O. Conte, 11/6/69, draft letter, and attachments, FAA 6174/97.

63. *New York Times*, 12/13/69, p. 34C, NAS Cmte. Newsletter, no. 75, 12/1–5/69, p. 3.

64. Ibid., 12/17/69, p. 109.

65. Baker to Beggs et al., "SST," 10/13/69, memo, and attachment, FAA 6174/5; Francis E. Rundell to Mr. Hilliard Zda, "Why U.S. Carriers Want the SST," 10/21/69, memo, FAA 6174/5.

66. *New York Times*, 11/5/69, p. 28.

Chapter 18

1. For protest letters from members of Congress and private citizens, see the following letters, memos, and dispatch in FAA 6174/5: James Beggs to Representative Clark MacGregor, 9/24/69; Beggs to Representative Frank T. Bow, 9/24/69; Vierling to Senator William Proxmire, 9/26/69; P. V. Siegel (Federal Air Surgeon) to Representative Sidney Yates, 12/9/69; "SST," 11/25/69, AP dispatch, no. 146; G. T. Haugan to SS-100 et al., "Fiscal Year 1970 Senate Hearings; Rebuttal to Senator Proxmire," 12/2/69, memo; Vierling to J. L. Smith, 11/17/69, and attached mailing list; Vierling to James R. King, 11/17/69, and attached mailing list.

2. Walter L. Mazan to Volpe, "SST Program," 11/4/69, memo, FAA 6174/5.

3. For FAA efforts to influence Congress, see the following letters in FAA 6174/5: Secor D. Browne to Senator Robert R. Griffin, 10/2/69, and attached "SST Program"; Vierling to Representative Charles M. Teague, 10/3/69, to Representative Thomas N. Downing, 10/14/69, to Senator George Murphy, 10/14/69, to Senator Hiram Fong, 10/24/69, to Representative Silvio O. Conte, 10/24/69, to Senator Charles H. Percy, 11/4/69; Mariette Risley to Teague, 9/24/69; Teague to Volpe, 9/26/69; Robert F. Bennett (Director of SST Office's Office of Congressional Relations) to Teague, 10/20/69.

4. For FAA public relations efforts, see the following letters in FAA 6174/5: Schaffer to José Gorkin (Parade Publications), 10/4/69; Vierling to Thomas E. Noyes (*Washington Evening Star*), 12/10/69; J. W. McCord to C. L. Blake, 9/23/69; Blake to J. L. Pearce, 12/3/69, and attached abstract; Joseph B. Stover (Air Carrier Service Corporation) to Vierling, 11/18/69.

5. For Boeing's public relations effort, see: W. E. Clothier (Boeing) to Francis E. Rundell, 10/10/69, 11/12/69, 12/10/69, FAA 6174/5.

6. Victor Riesel, "Buzz Aldrin's Earth: Labor Backs Nixon on SST as Soviets-Britons-French Outfly U.S.," *Inside Labor* (syndicated column), 11/11/69, Hall Syndicate, CL.

7. R. Dixon Speas to Vierling, 10/17/69, FAA 6174/5; Shaffer to Speas, 11/10/69, FAA 6174/5.

8. Daniel B. Priest (Air Transport Association) to Vierling, 11/19/69, FAA 6174/5; Vierling to Priest, 11/28/69, FAA 6174/5.

9. Mazan to Volpe, "SST Program," 11/4/69, memo, FAA 6174/5.

10. Natural Resources Council, "Supersonic Transport Resolution," 10/3/69, FAA 6174/5.

11. Malcolm F. Baldwin (Conservation Foundation) to Shurcliff, 10/9/69, CL.

12. Gladys Kessler (Environmental Defense Fund) to Shaffer, 12/9/69, FAA 6174/5.

13. For the founding of Friends of the Earth and its early anti-SST activities, see: "Friends of the Earth (Founded in July 1969 in New York)," CL; Thomas Turner to Shurcliff, 11/12/69, CL; Shurcliff to Gary Soucie, 2/15/70, CL; Soucie to Shurcliff, 1/29/69, CL; William A. Shurcliff, *SST and Sonic Boom Handbook* (New York: Ballantine Books, 1970); Shurcliff to Jonathan P. Ela, 4/4/70, CL; Soucie to *New York Times*, 12/15/69, to Representative Henry S. Reuss, 1/9/69, to Oscar Bakke (FAA), 2/12/70, to Citizens League, 11/21/69, record of telephone call, CL; Friends of the Earth form letter to the President, CL; *New York Times*, 3/5/70, p. 29. For the proposal that suggested producing a pro-SST *Handbook* by the SST Office, see: "Proposal for SST Handbook," circa 3/70, FAA 6174/93.

14. For the Sierra Club's anti-SST activities, see: Roger Marshall to Shurcliff, 10/21/69, and attached forms, CL; Richard C. Sill to Citizens League, 12/8/69, record of telephone call, CL; Sill to Justus Schlichting, 12/21/69, CL; Sill to Students and Project Leaders, Oberlin College Project, 12/22/69, 2/23/70, memos, CL; Shurcliff to Sill, 3/17/70, CL; *Palo Alto Times*, 1/15/70, p. 3; Shurcliff to Phillip Berry, 2/1/70, Berry to Shurcliff, 3/3/70, CL; "If You Put 687 Engines on the Empire State Building It Will Fly: So What?" *Sierra Club Bulletin* (3/70):18–21; Ela to Shurcliff, 3/30/70, CL; Peter Koff to Shurcliff, 4/14/70, message slip-record of telephone call, CL; Shurcliff to the record, 4/22/70, memo, CL.

15. For the formation of the Coalition Against the SST, see: personal interviews; Henry Lenhart, Jr., "Transportation Report/SST Foes Confident of Votes to Clip Program's Wings Again before Spring," *National Journal*, 1/1/71, pp. 43–58.

16. Shaffer to William F. Knowland (Publisher of the *Oakland Tribune*), 1/5/70, FAA 6174/93; Shaffer to William Eaton, 2/2/70, FAA 6174/95.

17. Vierling to Arjay Miller (Dean of Stanford Business School), 1/12/70, FAA 6174/93.

18. Vierling to C. Yarbrough (*Washington Evening Star*), 2/3/70, to James J. Kilpatrick, 4/16/70, to Adam A. Smyser (*Honolulu Star-Bulletin*), 2/16/70, FAA 6174/93.

19. Robert E. Parsons to Harold Kneeland (*Washington Post*), 3/10/70, FAA 6174/93.

20. Vierling to Reginald A. Hubley (Publisher of *Aviation Week and Space Technology*), 1/26/70, to Robert B. Hotz (*Aviation Week and Space Technology*), 3/5/70, FAA 6174/93.

21. For the FAA-Lukens debate, see: Beggs to Shaffer, "SST," 1/26/70, memo, FAA 6174/93; Shaffer to Matthias E. Lukens (President of the Airport Operators Council International and Deputy Executive Director of the New York Port Authority), 1/28/70, 4/15/70, FAA 6174/93; Vierling to Lukens, 2/2/70, FAA 6174/93.

22. For Shurcliff's activities and the SST Office's response, see: Shurcliff to Shaffer, 1/5/70, CL; Vierling to Shurcliff, 1/7/70, 1/23/70, 2/4/70, FAA 6174/93; Shaffer to Shur-

cliff, 1/10/70, FAA 6174/93; Bennett to William E. Timmons (Deputy Assistant to the President for Congressional Relations), "Letter to Dr. William A. Shurcliff, Director of the Citizens League Against the Sonic Boom," 2/13/70, memo, FAA 6174/93.

23. For the SST Office's pro-SST arguments, see the following letters and memos in FAA 6174/93: Bennett to Senator Wallace F. Bennett, 1/9/70; Vierling to Representative William E. Minshall, 4/10/70; Parsons to Representative Glen R. Davis, 3/20/70; Parsons to Senator Jack Miller, 3/26/70; Vierling to Representative Louis Frey, Jr., 4/8/70; Rundell to file, "Telephone Inquiry from the Office of Congressman Talcott (R-California)," 4/13/70, memo; William E. Magruder (Director of SST Development) to Talcott, 4/16/70; Magruder to Representative William S. Mailliard, 4/27/70; Magruder to Senator Robert Parkwood, 5/1/70; Rundell to file, "Telephone Call from Senator Hatfield's Office," 4/7/70, memo.

24. For communication between the SST Office and Capitol Hill, see the following letters, memos, and draft letter in FAA 6174/93; Lockett to Vierling, 2/18/70, memo; Rundell to Acting Director, Supersonic Transport Development, "Meeting with Congressman Reuss' Aide," 3/5/70, memo: Parsons to Assistant Secretary for Environment and Urban Systems, "February 26 Letter from Congressman Reuss re Environmental Aspects of the Supersonic Transport," 3/11/70, memo; Parsons to Eugene L. Lehr, "Comments on Letter from Representative Reuss; Your Memo 3/26," 3/26/70, memo; draft letter re: Friends of Earth, circa 3/70; Rundell, "Record of Telephone Call," 4/30/70, memo; Vierling to Jacqueline Browne, 1/8/70; Vierling to Owen De Ver Sholes, 3/5/70; Vierling to Polly Dyer, 4/16/70; Vierling to Bruce E. Keegan, 4/16/70; Magruder to Darlene Prescott, 4/22/70; Magruder to Robert J. Weggel, 4/20/70; Magruder to Elizabeth Fennell, 4/23/70.

25. Volpe to Russell Train (Chairman of the Council on Environmental Quality), 3/13/70, FAA 6174/93; Beggs to Train, 4/14/70, FAA 6174/93.

26. For Magruder's background and attitude toward the SST, see: Department of Transportation/FAA news release re: biography of William M. Magruder, 4/70, Bauer; and see the following letters in FAA 6174/93: Magruder to J. Prothero Thomas (British Aircraft Corporation), 5/5/70, to E. E. Hood, Jr. (General Electric), 5/5/70, to Mentzer, 5/5/70, to Sam Parsons, 5/5/70, to John C. Brizendine (Douglas Aircraft), 5/6/70.

27. For Earth Day activities, see: "A Word about this Book," *Earth Day—The Beginning* (New York: Bantam, 1970); Northeast SCOPE, "Eco-Directory," circa 11/69, CL; list of environmental action groups in Greater Boston area compiled by *Boston Globe*, circa 2/70, CL; Denis Hayes, "The Beginning," in *Earth Day*; personal interviews; Margaret Mead (President of Scientists' Institute for Public Information) to "Concerned Citizen," 3/70, and attached form, CL; Walter F. Mondale, "Commitment to Survival," in *Earth Day*, pp. 42–47; Adlai E. Stevenson, "Too Little, Too Late," in *Earth Day*, pp. 51–55; Charles A. Hayes, "A Time to Live," in *Earth Day*, pp. 153–155; "Great Divide" to Shurcliff, 4/15/70, CL; letters from various organizations and institutions requesting the *SST and Sonic Boom Handbook*, including those dated 11/14/69, 11/17/69, 10/8/69, 9/9/69, 12/23/69, 2/1/70, CL; letters from high schools and lower schools to the Citizens League requesting SST information for Earth Day activities, including those dated 3/30/70, 3/2/70, 3/31/70, 4/8/70, 3/26/70, 3/11/70, 4/19/70, 2/10/70, 2/9/70, 2/13/70, 2/27/70, CL; letters from colleges and universities to the Citizens League requesting SST information for Earth Day activities, including those dated 3/5/70, 4/6/70, 4/1/70, 3/27/70, 3/24/70, 3/11/70, 3/2/70, 3/19/70, 2/4/70, 2/1/70, 2/20/70, 2/13/70; NEEC, "NEEC General

Ecology Calendar 4/7–4/23/70," and attached "SST Fact Sheet," circa 4/70, CL; Ecology Action Council (UCLA), "The Supersonic Transport (SST): A 'Gross' National Product," 1/15/70, 5/15/70 (revised), CL; Ecology Action Council (UCLA), "Chamber of Quality: Los Angeles Area," 4/1/70, CL; letters from various groups for SST information for Earth Day activities and other purposes, including those dated 4/7/70, 1/7/70, 2/19/70, 3/1/70, 4/16/70, CL.

28. For the May subcommittee hearings, see: *New York Times*, 5/8/70, p. 62, 5/12/70, p. 78; U.S., Congress, Hearings before the Subcommittee on Economy in Government of the Joint Economic Committee, *Economic Analysis and the Efficiency of Government, Part 4—Supersonic Transport Development*, May 7, 11, 12, 1970, pp. III, 896–897, 907 925–927, 967, 997–1007; Lenhart, "Transportation Report," p. 50; personal interviews; *Congressional Record—Senate*, 5/15/70, pp. S7159–S7161.

29. For efforts of the pro-SST forces to counter the impact of the hearings, see the following memo and letters in FAA 6174/93: Magruder to John D. Ehrlichman, "Pan American on the SST," 5/13/70, memo; Magruder to Charles A. Lindbergh, 5/25/70; Magruder to Representative Ken Hechler, 5/20/70; Magruder to Representative Edward P. Boland, 5/20/70; Bennett to Representative Donald D. Clancy, 5/25/70; Magruder to Representative Charlotte T. Reid, 5/20/70; Magruder to Representative Albert H. Quie, 5/21/70; Volpe to Representative Gerald R. Ford, 5/21/70; T. A. Wilson (President of Boeing) to Volpe, 5/20/70; Magruder to William Anders (National Aeronautics and Space Council), 5/20/70; Magruder to Representative William E. Minshall, 5/20/70.

30. For acceleration of protest as a result of the hearings, see: *New York Times*, 5/21/70, p. 34; Philip B. Berry to Sierra Club members, 5/18/70, CL; Rundell to the record, "Telephone Call from Office of Congressman Pickle," 5/12/70, memo, FAA 6174/93; Rundell to SS-1 et al., "Record of Telephone Conversation," 5/13/70, memo, FAA 6174/93.

31. For Magruder's lobbying in the House for the SST funding request and for his compromise with Train, see: Magruder to Ehrlichman, 5/21/70, note, FAA 6174/93; D. McKinny to Mr. Bennett's Office, 5/21/70, note, FAA 6174/93; Magruder to Assistant Secretary Braman, "Environmental Research Planning Conference," 5/25/70, memo, FAA 6174/93.

32. For information on the Massachusetts Committee Against the SST, see: *Christian Science Monitor*, 4/28/70, p. 3; Massachusetts Committee Against the SST, 4/26/70, news release, CL; Peter L. Koff to Shurcliff, 4/28/70, 5/11/70, CL; Massachusetts Committee Against the SST, circa 5/8/70, circular, CL; Massachusetts Committee Against the SST, "Fact Sheet on the SST," 5/70, CL; to Citizens League, 5/6/70, 5/7/70, CL.

33. Town-Village Aircraft Safety and Noise Abatement Committee, *Quotes* 2 (4/70), CL.

34. *Washington Evening Star*, 5/24/70.

35. George Alderson, "Timetable for SST Appropriation (Subject to Delay)," 4/11/70, CL.

36. Coalition Against the SST, "The Facts on the SST," 5/70, CL.

37. Berry to members of the Sierra Club, 5/18/70, CL.

38. For increasing numbers of protest letters, see: Magruder to Jay E. Gould, 5/8/70, and to Representative William S. Broomfield, 5/25/70, FAA 6174/93.

39. For anti-SST lobbying prior to the House SST vote, see: *New York Times*, 5/22/70, p. 62; *Washington Star*, 5/24/70; Alderson to Shurcliff, 5/24/70, CL; personal interviews.

40. "House Funds SST in Close Vote," *Aviation Week and Space Technology*, 6/1/70, pp. 26–27; *New York Times*, 5/28/70, pp. 1, 20, 5/31/70, sec. IV, p. 3.

41. Alderson to SST Collaborators, "House Action on the SST," 5/28/70, memo, CL.

42. Michael McCloskey to Sierra Club members, 6/10/70, postcard, CL.

43. *New York Times*, 6/8/70, p. 73, 6/9/70, p. 82; press release from the Office of Senator Harry F. Byrd, Jr., 6/9/70, CL.

44. Personal interviews.

45. *New York Times*, 8/26/70, p. 26.

46. Alderson to Shurcliff, 7/31/70, CL.

47. Barbara Reid (Legislative Director of Environmental Action) to Shurcliff, n.d., CL.

48. Jack Frankel to Shurcliff, 5/10/70, and attachments, CL; G. E. Schindler (Conservation Chairman of the Sierra Club's Atlantic Chapter) to Shurcliff, 7/25/70, CL; Shurcliff to Schindler, 8/2/70, CL; Shurcliff to Sierra Club, 6/28/70, CL.

49. Alderson to Shurcliff, 6/22/70 and 7/31/70, CL; Tom Turner to Shurcliff, 6/26/70, CL.

50. Alderson to SST Task Force, 6/25/70, memo, CL.

51. For the anti-SST effort by the Federation of American Scientists, see: Marvin I. Kalkstein (Federation of American Scientists) to Shurcliff, 4/25/70, CL; "Federation Opposes the SST," 8/17/70, news release, CL; *Congressional Record—Senate*, 9/29/70, p. S06733; Jeremy J. Stone (Federation of American Scientists) to all U.S. Senators, 9/17/70, and attached "SST Background and Pros and Cons," 9/17/70, CL; Stone to all U.S. Senators, 10/2/70, and attached maps, CL; Stone to all U.S. Senators, 10/1/70, and attached map, CL.

52. *New York Times*, 7/24/70, p. 62.

53. For the SCEP study and its impact, see: Report of the Study of Critical Environmental Problems, *Man's Impact on the Global Environment: Assessment and Recommendations for Action* (Cambridge, Mass.: MIT Press, 1970), pp. 15–18, 67–74, 100–112, 200–212; *Des Moines Register*, 8/2/70; *New York Times*, 8/5/70, p. 34; Coalition Against the SST, "New Facts on the SST," 2/4/71; *Congressional Record—Senate*, 8/5/70, p. S12796.

54. *Honolulu Star-Bulletin*, 8/19/70; *New York Times*, 8/29/70, p. 49.

55. *Des Moines Register*, 8/27/70; *New York Times*, 8/29/70, p. 49.

56. For the statements by prominent economists and their impact, see: *New York Times*, 9/16/70, p. 93; Paul Samuelson, "Statement on the Economics of the SST," CL; Coalition Against the SST, "New Facts on the SST," 2/4/71; *Congressional Record—Senate*, 9/16/70, pp. S15397–S15402; *New York Times*, 9/21/70, p. 42; personal interviews.

57. For protests over the Concorde, see: *Des Moines Register*, 9/2/70, 9/4/70, 9/16/70; Richard Wiggs, *Concorde: The Case against Supersonic Transport* (London: Ballantine/Friends of the Earth, 1971).

58. Eric Schindler (Friends of the Earth—Switzerland) to all FOE Officers, 8/16/70, CL.

59. For communication with Shurcliff and for the idea of a Concorde handbook, see: Edwin Spencer Matthews, Jr. (European Representative of the Friends of the Earth) to Shurcliff, 9/8/70, 9/29/70, 10/16/70, CL; Shurcliff to Matthews, 9/10/70, CL.

60. Town-Village Committee, *Quotes* 2 (8/70–11/70), CL.

61. *New York Times*, 9/4/70, p. 56.

62. For the anti-SST movement in Rhode Island, see: *Providence Evening Bulletin*, 6/1/70, 6/2/70, 6/9/70, 8/3/70, 8/12/70, 9/3/70; Dwight W. Justice (Legislative Chairman of Ecology Action for Rhode Island) to Shurcliff, 8/15/70, CL; *Providence Journal*, 8/9/70, 8/12/70, 8/24/70, 9/4/70; Lenhart, "Transportation Report"; personal interviews.

63. For the anti-SST movement in Iowa, see: *Des Moines Register*, 8/25/70, 8/28/70, 8/29/70, 9/1/70, 9/4/70, 9/7/70, 9/8/70, 9/10/70, 9/11/70, 9/17/70, 9/18/70, 10/5/70.

64. New Mexico Coalition Against the SST, "Help Stop the SST," 8/24/70, CL.

65. Theodore Berland to Shurcliff, 7/9/70, CL.

66. *Honolulu Star-Bulletin*, 9/23/70.

67. Lenhart, "Transportation Report"; personal interviews.

68. Magruder to John C. Pirtle (General Electric), 5/21/70, FAA 6174/93; Magruder to H. W. Withington (Boeing), 5/21/70, and attachment, FAA 6174/93; Magruder to Charles W. Harper (NASA), 6/18/70, FAA 6174/93; Boeing Co., "The United States Supersonic Transport" (9/70).

69. Magruder to Secor Browne (CAB chairman), 5/25/70, and attachment, 5/28/70, and 6/17/80, FAA 6174/93.

70. For Magruder's various advisory councils and research efforts in the environmental analysis areas, see: Magruder to Train, 5/8/70, 6/4/70, FAA 6174/93; Magruder to Assistant Secretary Braman, "National Environmental Policy Act of 1969 (PL 91-190)," 6/24/70, memo, FAA 6174/93; Magruder to William Van Ness, "Environmental R&D Program," 6/30/70, memo, FAA 6174/93; *Des Moines Register*, 8/28/70.

71. For Magruder's attacks on the SST opposition's arguments, see: *Des Moines Register*, 7/14/70, 8/28/70; Richard G. O'Lone, "SST Officials Strive to Blunt Opposition," *Aviation Week and Space Technology*, 8/10/70, pp. 16–17; *New York Times*, 8/29/70, p. 49.

72. For airline SST support, see: Magruder to George A. Spater (American), 6/8/70, FAA 6174/93; Magruder to R. M. Ruddick (United), 6/23/70, FAA 6174/93; Magruder to Major General Clifton F. Von Kann (Air Transport Association), 6/23/70, FAA 6174/93; Magruder to Assistant Secretary for Policy and International Affairs, "Request for Statistical Data," 6/26/70, memo, and attachments, FAA 6174/93.

73. For Magruder's dealings with Congress, see the following letters that he wrote in FAA 6174/93: to Senator Alan Cranston, 5/25/70; to Senator Margaret Chase Smith, 6/4/70; to Representative Ancher Nelson, 6/4/60; to Representative Robert C. McEven, 6/4/70; to Representative G. Robert Watkins, 6/4/70; to Senator Edward J. Gurney, 6/10/70; to Representative Paul N. McCloskey, Jr., 6/12/70; to Representative Clement J. Zablocki, 6/24/70; to Representative Sidney R. Yates, 6/24/70.

74. For SST Office contact with congressional aides, see: Vierling to Anne McFee, 6/18/70, to Paul Duke, 6/23/70, FAA 6174/93; Rundell to the record, 6/10/70, memo, FAA 6174/93; Rundell to SS-1 et al., "Briefing for Senate Aeronautical and Space Sciences Committee Staff," 6/10/70, memo, FAA 6174/93.

75. Rundell to SS-1 et al., 6/11/70, memo, 6174/93.

76. Rundell to SS-100 et al., "Congressional Quarterly SST Fact Sheet," 6/11/70, memo, 6174/93.

77. Magruder to Paul J. C. Friedlander (*New York Times*), 6/8/70, to Joseph G. Harrison (*Christian Science Monitor*), 6/9/70, to Donald B. Miller (*Berkshire Eagle*), 6/24/70, FAA 6174/93.

78. For meetings with various officials to counter adverse publicity, see: Magruder to FAA Administrator, "News System for SST Information," 6/26/70, memo, FAA 6174/93; Magruder to Lukens, 6/12/70, to John Ehrlichman, 6/29/70, FAA 6174/93.

79. Richard G. O'Lone, "SST Officials Strive to Blunt Opposition," *Aviation Week and Space Technology*, 8/10/70, pp. 16–17.

80. Lenhart, "Transportation Report."

81. Richard G. O'Lone, "Volpe Rebuts Critics of SST Program," *Aviation Week and Space Technology*, 6/15/70, p. 28.

82. *Des Moines Register*, 8/28/70.

83. Alderson to Shurcliff, 8/25/70, telegram, CL.

84. "An Urgent Appeal," 9/24/70, Friends of the Earth mailing, CL.

85. For the defecting or undecided senators, see: *New York Times*, 7/29/70, p. 19, 8/26/70, p. 26, 8/28/70, p. 1.

86. Soucie to FOE offices et al., 9/10/70, and attachments, CL.

87. Alderson to Shurcliff, 8/5/70, 9/15/70, CL.

88. For the strategy of the SST proponents and the decision to delay the SST vote until after the November elections, see: *New York Times*, 9/23/70, p. 92, 10/1/70, p. 58, 10/30/70, p. 81; "An Agonizing Appraisal," *Air Line Pilot* 40 (6/71):8–11, 42–43; Lenhart, "Transportation Report."

89. Soucie to Friends of the Earth Directors et al., "SST Status Report," 10/2/70, memo, and attachment, CL; Lenhart, "Transportation Report."

90. *Scottsdale-Daily Progress*, 11/2/70; Owens and Associates Advertising, Inc., "TV Copy, Client—League of Conservation Votes, Subject: SST, Media: KPHO-TV, Date to Run: 11/1–2/70," CL.

91. For the anti-SST campaign in Hawaii, see: *Honolulu Star-Bulletin*, 11/7/70, 9/28/70, 10/22/70, 11/10/70, 12/7/70; Lenhart, "Transportation Report"; personal interviews.

92. For the anti-SST campaign in Iowa, see: *Des Moines Register*, 11/8/70, 11/23/70, 11/26/70, 11/29/70; Lenhart, "Transportation Report"; personal interviews.

93. *Providence Evening Bulletin*, 12/7/70.

94. Statement by Bay Anti-Noise Group (BANG) at 10/28/70 news conference, CL.

95. For the anti-SST campaign in Idaho, see: Gerald A. Jayne (President of the Idaho Environmental Council) and Russell A. Brown (Chairman, SST Study Committee of the Idaho Environmental Council) to selected senators, 11/30/70, CL; Jayne, "Environmental Council Commends Jordan, Raps Church, on SST Vote," 12/14/70, CL; *Idaho Falls Register*, 12/8/70, CL.

96. *Philadelphia Evening Bulletin*, 11/30/70.

97. *Chicago Sun-Times*, 12/1/70, p. 16.

98. For airline fears of the SST, see: *New York Times*, 10/31/70, p. 58; *Providence Evening Bulletin*, 11/14/70; *Philadelphia Evening Bulletin*, 11/30/70; Citizens League Against the Sonic Boom, "Fact Sheet 15c," revised 11/70, CL.

99. *New York Times*, 10/27/70, p. 44; Shurcliff to Moss, 10/28/70, CL.

100. *New York Times*, 10/23/70, p. 82.

101. Tupling, Alderson, Stone (Director of the Federation of American Scientists), Denis Hayes (National Coordinator of Environmental Action), and Carl Pope (Washington Representative of Zero Population Growth) to each senator, 11/27/70, CL.

102. *New York Times*, 9/17/70, p. 93.

103. For the Library of Congress report, the Avco Study, and the three-nation accord, see: *New York Times*, 9/29/70, p. 81, 10/7/70, p. 93; *Honolulu Star-Bulletin*, 9/28/70; Gerald M. Daniels, "SST Environmental Effects: Some Considerations," *Astronautics and Aeronautics* 8 (10/70):22–25; *Des Moines Register*, 10/21/70, 10/27/70.

104. *New York Times*, 10/6/70, p. 46.

105. *Des Moines Register*, 10/27/70.

106. *New York Times*, 12/1/70, p. 92.

107. For the impact of foreign SST programs, see: *Des Moines Register*, 10/13/70, 10/21/70, p. 2; *New York Times*, 11/5/70, p. 94, 11/7/70, p. 88, 11/11/70, p. 88.

108. For the pro-SST efforts of industry and labor, see: *New York Times*, 10/10/70, p. 24, 11/5/70, p. 94, 11/7/70, p. 88, 12/1/70, p. 46; Lenhart, "Transportation Report"; Boeing Company, "USA/SST: Issues and Answers" (1970).

109. *New York Times*, 11/30/70, p. 81.

110. *Des Moines Register*, 11/25/70, p. 2; *New York Times*, 11/15/70, p. 73, 11/28/70, p. 26.

111. For the events prior to the Senate vote, see: *Senate Bill No. S4547*, 91st Cong., 2d sess., calendar no. 1399, submitted by Senator Magnuson on 11/30/70; "An Agonizing Appraisal"; *Congressional Record—Senate*, 12/2/70, pp. S19267–S19268; *New York Times*, 12/2/70, pp. 46, 93, 12/3/70, pp. 1, 93, 12/4/70, pp. 34–35; Lenhart, "Transportation Report."

112. For the Senate vote and its impact, see: *New York Times*, 12/4/70, pp. 1, 34–35, 12/6/70, pp. 47, 65; "Supersonic Shock Wave," *Aviation Week and Space Technology*, 12/14/70, p. 11; "Agonizing Appraisal"; Lenhart, "Transportation Report."

113. For the renewed pro-SST drive and for the formation of the House-Senate Conference Committee, see: *Providence Evening Bulletin*, 12/3/70; *New York Times*, 12/4/70, p. 34, 12/5/70, p. 65, 12/6/70, pp. 1, 44, 12/8/70, p. 20; *Honolulu Star-Bulletin*, 12/4/70; "Senate Defeat Sends SST Funds to Conference," *Aviation Week and Space Technology*, 12/7/70, p. 24; Lenhart, "Transportation Report"; "SST Termination Costs May Exceed $200 Million," *Aviation Week and Space Technology*, 12/14/70, p. 17; "Airline Chiefs Affirm Support of SST Prototype Program," *Aviation Week and Space Technology*, 12/14/70, pp. 19–20.

114. *New York Times*, 12/9/70, pp. 1, 109.

115. Ibid., 12/8/70, p. 20; *Congressional Record—Senate*, 12/7/70, pp. S19495–S19496.

116. "Boeing Sees Cost Rise in SST Slowdown," *Aviation Week and Space Technology*, 12/14/70, pp. 18–19.

117. For House-Senate compromise funding level, see: *New York Times*, 12/9/70, pp. 1, 109, 12/10/70, p. 37, 12/11/70, pp. 1, 34, 37, 12/12/70, pp. 30, 61; "Panel Votes SST $210 Million," *Aviation Week and Space Technology*, 12/14/70, pp. 16–28; Lenhart, "Transportation Report."

118. *New York Times*, 12/16/70, pp. 1, 23.

119. For Proxmire's filibuster and the Senate's rejection of cloture, see: ibid., 12/11/70, p. 37, 12/18/70, pp. 18, 22, 12/20/70, pp. 1, 30; *Providence Evening Bulletin*, 12/14/70; Lenhart, "Transportation Report."

120. For the agreement for a straight SST vote in the 92d Congress, see: *New York Times*, 12/15/70, p. 89, 12/20/70, pp. 1, 30, sec. 4, p. 2, 12/21/70, pp. 1, 26, 12/22/70, pp. 1, 20, 12/25/70, p. 57, 12/30/70, pp. 1, 14, 1/3/70, pp. 1, 33; Lenhart, "Transportation Report."

121. Lenhart, "Transportation Report."

Chapter 19

1. *Michigan State News*, 2/5/71, CL.

2. Alderson to "Cooperating Organizations," 2/6/71, memo, CL; Alderson to "Cooperating Organizations," "Status of SST Campaign," 2/28/71, memo, CL; Coalition Against the SST, "New Facts on the SST," circa mid-2/71, CL.

3. For Shurcliff's activities, see: Shurcliff to Alderson, 1/17/71, 2/6/71, 3/3/71, CL; *Conservation News*, 3/15/71, pp. 9–10, CL; Shurcliff to Environmental Action, 1/23/71, CL; Shurcliff to Alan J. Cushner, Esq., 1/6/71, CL; Shurcliff to Theodore Berland, 2/3/71, CL; Alderson to Shurcliff, 2/23/71, CL.

4. Gary A. Soucie to "Dear Friend of the Earth," 1/16/71, and attached circular, CL.

5. Bulk mailing by New Mexico chapter of Friends of the Earth, circa 1/71, and attached circular and excerpt from *Not Man Apart* entitled, "Senate SST Vote Shows Environmental Muscle," pp. 6–7, CL.

6. Michael McCloskey to Charles H. Shurcliff, 3/3/71, CL.

7. For the anti-SST activities of Common Cause, see: Common Cause, "The Weak Case for the SST and Counter-Arguments to It," 2/9/71, CL; Common Cause, "Report from Washington," vol. 2 (6–7/72), newsletter, CL.

8. Environmental Action, "No Victory on the SST," circa 1/71, leaflet, CL.

9. William A. Shurcliff, "Lobby Groups, Pro and Con" (Chapter 6), *S/S/T and Sonic Boom Handbook*, 1971 edition, pp. 6.1–6.9.

10. For the role of Citizens Against Noise and that group's contact with Shurcliff, see: "Citizens Against Noise: Statement of Purpose, Objectives," circa early 1971, CL; Theodore Berland (Chairman of Citizens Against Noise) to Senator Charles H. Percy, 12/7/70, CL; Berland to Shurcliff, 1/28/71, 3/11/71, 3/28/71, CL; Citizens Against Noise, "Projects for 1971," n.d., CL; Shurcliff to Berland, 2/3/71, CL; various copies of articles on Citizens Against Noise in the Chicago press, CL; Berland to the editor of *Chicago Today*, 1/28/71, CL; list of Citizens Against Noise Steering Committee, n.d., CL.

11. Charles A. Lindbergh to Representative Sidney R. Yates, 2/3/71, CL; *New York Times*, 2/6/71, p. 59.

12. For the SST opponents' reaction to Lindbergh's announcement, see: Shurcliff to Alderson, 2/6/71, CL; Alderson to "Cooperating Organizations," 2/6/71, memo, CL.

13. For state anti-SST noise regulations, see: Shurcliff, "Efforts to Ban SSTs from Individual States of USA," 1/30/71, informal notes, CL; *Washington Post*, 1/5/71, p. A6, NAS Cmte. Newsletter, no. 71-1, 1/1–15/71, p. 10; "Proposed New York Anti-Noise Bill Could Ban Several Aircraft," *Aviation Week and Space Technology*, 1/18/71, p. 23, NAS Cmte. Newsletter, no. 71-2, 1/10–31/71, p. 12.

14. For anti-SST activity in Michigan, see: Michigan Student Environmental Confederation, "Present Status—SST, Washington," circa 2/71, CL; Michigan Student Environmental Confederation, "SST Deferred," press release, 3/18/71, CL; Michigan Student Environmental Confederation, "Students Endorse Michigan Anti-SST Legislation," press release, 2/8/71, CL; Governor William G. Milliken to Representative John Conyers, Jr., et al., circa 2/71, telegram, CL.

15. *SCOPE* [Student Council on Pollution and Environment] Newsletter, no. 5, (2/71):10.

16. Town-Village Committee, *Quotes*, 3 (1/71), "SST Funding," statement, 2/1/71, CL.

17. *New York Times*, 3/3/71, p. 35.

18. For SST developments in Hawaii, see: *Honolulu Star-Bulletin*, 1/5/71, 2/25/71; Shurcliff, "Efforts to Ban SSTs."

19. For SST developments in California, see: Shurcliff, "Efforts to Ban SSTs"; Jim Jacobson (Chairman of the San Diego Citizens Committee to Ban the SST) to "Any Organization or Group—Social, Civic, Minority, Labor Environmental, Religious, Cultural, Political, etc.," 2/10/71, memo, CL.

20. For the SST protest in New England, see: Shurcliff, "Efforts to Ban SSTs"; Peter Koff to "New England Anti-SST'ers," 2/11/71, memo, CL.

21. "Scientists Argue Environmental Effects," *Aviation Week and Space Technology*, 12/17/70, p. 21; Boeing Company, "Air Quality as Affected by Air Transportation Including the Supersonic Transport—Revision D," 1/8/71, contract No. FA-SS-67-3, document No. D6A11867-1.

22. *New York Times*, 1/14/71, p. 73.

23. Department of Transportation, "The Supersonic Transport: White Paper," 2/4/71 (revised version of "The SST and the National Interest," a presentation by William M. Magruder to the Executive Committee of the National Aeronautic [SIC] Association, 8/12/70).

24. *New York Times*, 2/4/71, p. 69; *Washington Post*, 1/14/71.

25. Leo L. Beranek (Chairman of the SST Community Noise Advisory Committee) to Magruder, "Supersonic Transport Noise," 2/5/71, memo, R. Bauer files; Boeing Company, "Proposed Agenda: SST Community Noise Advisory Committee Meeting," 2/4/71, R. Bauer files.

26. John C. Schettino (Executive Secretary of the SST Community Noise Advisory Committee) to Ray Bauer, 2/8/71, R. Bauer files; *New York Times*, 2/23/71, pp. 1, 73.

27. Schettino, "SST Noise—Will It Really Be a Problem?" 2/71, R. Bauer files.

28. For the SST opposition's reaction to the Beranek Committee's findings, see: Alderson to "Cooperating Organizations," "Status of SST Campaign," 2/28/71, memo, CL; Coalition Against the SST, "SST 'Quiet Engine' Is a Hoax," 2/25/71, press release, CL; "GE Drops SST Engine Afterburner," *Aviation Week and Space Technology*, 2/15/71, p. 26; Shurcliff to Alderson, 3/3/71, CL.

29. *Honolulu Star-Bulletin*, 3/3/71.

30. For the greater White House role in the pro-SST drive, see: *Washington Post*, 1/14/71; "Washington Roundup," *Aviation Week and Space Technology*, 2/8/71, p. 15; *Washington Evening Star*, 2/24/71, p. A-1, NAS Cmte. Newsletter, no. 71-4, 2/16–28/71, p. 2; Harold D. Watkins, "Nixon Support Spurs SST Funding Drive," *Aviation Week and Space Technology*, 2/15/71, p. 26, 31; *New York Times*, 1/30/71, p. 11.

31. For the pro-SST drive by industry, see: *Washington Post*, 1/14/71; Shurcliff, "Lobby Groups, Pro and Con," *SST and Sonic Boom Handbook*, 1971 edition, pp. 6.1–6.9; Harold D. Watkins, "Nixon Support Spurs SST Funding Drive," *Aviation Week and Space Technology*, 2/15/71, p. 26, 31; Committee for an American SST, "Action USA/SST," 2/22/71, brochures; "Campaign for SST Funding Accelerating," *Aviation Week and Space Technology*, 3/1/71, pp. 45–46.

32. *New York Times*, 2/23/71, p. 73.

33. Ibid., 3/7/71, sec. IV, p. 31, 3/23/71, p. 42.

34. Ad Hoc Committee of the Technical Committee of the American Institute Aeronautics and Astronautics, "The Supersonic Transport: A Factual Basis for Decisions," 3/1/71.

35. For the impact of foreign SST efforts, see: *New York Times*, 2/11/71, p. 90, 3/1/71, p. 15; "Concorde Is Coming," *Aviation Week and Space Technology*, 2/8/71, p. 11

36. For Alderson's warning of intensified pro-SST lobbying, see: Alderson to "Cooperating Organizations," 2/6/71, 2/28/71, memos, CL; Alderson to Shurcliff, 2/23/71, CL.

37. *New York Times*, 2/12/71, p. 28.

38. Alderson to Shurcliff, 2/23/71, CL.

39. For the testimony in the House subcommittee hearings, see: *New York Times*, 3/2/71, p. 69, 3/3/71, p. 87, 3/4/71, p. 70, 3/6/71, p. 31; "An Agonizing Appraisal," *Air Line Pilot* (6/71):8–11, 42–43; U.S., Congress, House, Hearings before a Subcommittee of the Committee on Appropriations, *Civil Supersonic: Aircraft Development (SST)*, 92d Cong., 1st sess., March 1–3, 1971. For James McDonald's findings and testimony, see also: Phillip M. Boffey, *The Brain Bank of America* (New York: McGraw-Hill, 1975), pp. 133–137; Lydia Dotto and Harold Schiff, *The Ozone War* (Garden City, N.Y.: Doubleday, 1978), pp. 39–43.

40. *New York Times*, 3/6/71, p. 31.

41. Ibid., 3/7/71, p. 54.

42. For the Senate committee hearings, see: U.S. Congress, Senate, Hearings Before the Committee on Appropriations, *Civil Supersonic Transport*, 92d Cong., 1st sess., March 10–11, 1971; *New York Times*, 3/11/71, pp. 73–74, 3/14/71, p. 21.

43. *New York Times*, 3/18/71, pp. 1, 28.

44. For the House SST vote and its immediate impact, see: ibid., 3/19/71, pp. 1, 24, 3/22/71, p. 36, 3/23/71, p. 13; Donald C. Winston, "Alternate SST Funding Considered,"

Aviation Week and Space Technology, 3/22/77, p. 16; Michigan Student Environmental Confederation, "SST Defeated," 3/18/71, press release, CL; "Naysayers Never Die," *Aviation Week and Space Technology*, 3/22/71, p. 1.

45. For the actions immediately before the Senate vote and for the Senate vote itself, see: "An Agonizing Appraisal," *Air Line Pilot* (6/71):8–11, 42–43; *New York Times*, 3/20/71, pp. 1, 58; 3/24/71, pp. 86–87, 3/25/71, pp. 1–24; *Honolulu Star-Bulletin*, 3/22/71, p. A-12, 3/25/71.

46. For the aftermath of the Senate vote, see: *New York Times*, 3/25/71, pp. 24–25, 3/26/71, pp. 1, 23; "SST Termination Process Begins," *Aviation Week and Space Technology*, 3/29/71, pp. 14–16; Robert L. Murdy (Department of Transportation) to Gordon E. Thiel (General Electric), 4/13/71, and attached "Action Items Resulting from Government/GE Termination Planning Meeting of 8 April 1971," FAA 6174/96; Murdy to Matthew E. Brislawn (Boeing), 4/13/71, and attached "SST Contracts," FAA 6174/95.

47. For the attempt to revive the SST program with the Ford amendment, see: "SST Termination Process Begins," *Aviation Week and Space Technology*, 3/29/71, pp. 14–61; *New York Times*, 5/11/71, pp. 22, 38, 5/13/71, pp. 1, 26, 27, 5/14/71, pp. 1, 67, 5/15/71, p. 60, 5/18/71, p. 78, 5/20/71, pp. 1, 18–19, 5/21/71, pp. 1, 78.

Chapter 20

1. *New York Times Index 1969–1972* (New York: New York Times, 1970–1973).

2. "SST Termination Process Begins," *Aviation Week and Space Technology*, 3/29/71, pp. 14–16; "Letter from Seattle," *Aviation Week and Space Technology*, 4/12/71, p. 60; "Toward a Technological Appalachia," *Aviation Week and Space Technology*, 3/29/71, p. 9.

3. "SST Postmortem," *Aviation Week and Space Technology*, 4/5/71, p. 9; "The Ecological Problem," *Aviation Week and Space Technology*, 4/12/71, p. 11.

4. Congressional Record—Extensions of Remarks, 4/6/71, pp. E2709–E2710.

5. William A. Magruder, "SST: Lessons in All That Pain," *Astronautics and Aeronautics* 9 (7/71):16–18; *New York Times*, 10/26/71, p. 29.

6. Ibid., 12/15/71, p. 105.

7. Ibid., 7/13/72, p. 71.

8. Wilderness Society, *Wilderness Report* 8 (4/71):2.

9. Michael McCloskey to Shurcliff, 3/25/71, telegram, CL.

10. Ibid., 5/11/71, CL.

11. George Alderson to Shurcliff, 3/29/71, CL.

12. Shurcliff to Alderson, 4/1/71, CL.

13. Shurcliff to Phillip S. Berry and Michael McCloskey, 4/17/71, CL.

14. Shurcliff to Friends of the Earth, Ltd., London, 4/18/71, CL.

15. Graham Searle to Richard Wiggs, 4/29/71, CL; Searle to Shurcliff, 4/29/71, CL.

16. For data on Shurcliff's activities during the immediate post-SST defeat period, see: Theodore Berland (Chairman of Citizens Against Noise) to Shurcliff, 4/2/71, CL; Shurcliff to Berland, 4/6/71, 4/24/71, CL; Shurcliff to Berry and McCloskey, 4/17/71, CL;

Shurcliff to Friends of the Earth, Sierra Club, CAN, et al., 5/5/71, CL; various Citizens Against Noise letters in response to Citizens League inquiries, CL.

17. For Shurcliff's anti-SST activities from mid-1971 onward, see: Alderson to Shurcliff, 12/31/71, handwritten letter, CL; Shurcliff to Berland, 6/27/71, 10/8/71, CL; Shurcliff to Mrs. Carol U. Berstein, 11/14/71, CL; Friends of the Earth (South Africa) to Richard Wiggs and Shurcliff, 1/27/73, CL; Citizens League Against the Sonic Boom, "Reasons Why Airlines Have Decided Not to Buy Concorde SSTs," 2/4/73, and revision of 6/27/73, CL; Shurcliff to Clifford Deeds (Director of the Town-Village Aircraft Safety and Noise Abatement Committee), 4/4/73, 11/8/73; Deeds to Shurcliff, 12/5/73, CL; Deeds to Representative John W. Wydler, 8/16/73, CL; Shurcliff to Alderson, 10/27/73, and 1/19/73, CL; Alderson to Shurcliff, 5/28/73, postcard, CL; Stewart M. Brandborg (Executive Director of the Wilderness Society) to Anthony Wedgwood Benn (British Secretary of State for Industry), 5/30/74, CL; Alderson to Shurcliff, 1/17/73, 1/24/73, CL. For Shurcliff's role in the 1973 anti-SST effort in New York State and for the entire New York SST situation, see: "Memorandum Re: New York Aircraft Noise Bill," circa 1/28/73, CL; David R. Brower (President of Friends of the Earth) and Alfred S. Forsyth (Sierra Club's Atlantic Chapter) to New York State Legislators, circa 2/73, CL; Malcolm Moore (Coordinator of the Coalition Against Aircraft Noise) to Shurcliff, circa late 1/73, handwritten note, CL; Moore to Shurcliff, 2/1/73, 4/16/73, and 5/17/73, CL; Shurcliff to Moore, 2/10/73, CL; Deeds, "A.5851/S. 3802—Aircraft Noise Bill," circa 2/30/73, announcement, CL; A.5851 and S.3802, 2/20/73, CL; "Initial Cooperating Groups of the Coalition [Against Aircraft Noise]," as of 3/6/73, CL; Moore to "Cooperating Groups," 3/6/73, 3/27/73, 4/4/73, 4/18/73, memos, CL; *Guardian*, 3/15/73; *New York Post*, 3/6/73, p. 10; Coalition Against Aircraft Noise, list of cooperating groups, 3/22/73, CL; Coalition Against Aircraft Noise, revised list of cooperating groups, 4/4/73, CL; "Additional Cooperating Groups," 4/30/73, CL; Moore to several legislators, n.d., CL; Coalition Against Aircraft Noise, 4/12/73, news release, CL; New York State Americans for Democratic Action, circa 4/16/73, news release, CL; Coalition Against Aircraft Noise, "Statement of Purpose," circa end of 1972, CL; Coalition Against Aircraft Noise to All [New York State] Senators, "Aircraft Noise Bill, Senate 3802-b; B. C. Smith and others," 4/26/73, memo, CL; Peggy Berscheid (Friends of the Earth Membership Secretary) to Shurcliff, 5/16/73, CL. For Shurcliff's support of Citizens Against Noise, see: Berland to Shurcliff, 7/5/72, 10/11/72, 12/13/72, 4/20/73, 10/25/73, CL; Berland to *Travel & Leisure, American Express*, 6/13/73, CL; Berland to Mrs. Joseph M. Zoghby, 6/22/72, CL; Shurcliff to Berland, 4/26/72, 6/27/72, 10/6/72, CL; Shurcliff to several people, 6/29/72, CL; Berland to Dr. W. H. Ferry, 7/5/72, CL; Berland to Ms. Carol U. Bernstein, 7/11/72, CL; Berland to Alderson, 8/15/72, CL; Shurcliff to CAN, 12/6/72, CL; Berland to "Members and Friends," n.d., CL; Berland to Letters-to-the-Editor, *Chicago Daily News*, 10/25/73, CL; CAN "Alert," "Stop the Concorde—Now!" circa 3–4/75, CL.

18. For Alderson's anti-SST activities after the 1971 SST defeat in Congress, see: Alderson to Shurcliff, 1/7/72, 2/29/72, 3/7/72, 6/27/72, 12/23/72, CL; Alderson to John H. Shaffer, 12/31/71, CL; Shurcliff to Alderson, 3/3/72, 3/14/72, 2/31/72, 7/1/72, 12/30/72, 2/3/73, 8/27/73, CL; "Model Bill for State Regulation of Airport and Community Noise," circa 5/72, CL; Alderson to SST Cooperators, "First Firm Order for Concorde," 5/26/72, memo, CL; Friends of the Earth, "Concorde Noise Data," 5/26/72, information sheet, CL; Michael McCloskey to Peter Koff, 6/2/72, CL; Alderson to "SST Cooperators," 11/2/72, memo, CL; Alderson to Shurcliff, circa 11/12/72, handwritten note, CL; Alderson to National Cooperators, SST, "New York Effort against Concorde," 12/14/72,

memo, CL; Alderson to Shurcliff and Malcolm Moore, memo, CL; Alderson to FAA Office of the General Counsel, 2/27/73, CL; Coalition Against Aircraft Noise, list of "Cooperating Groups," 5/1/73, CL; Moore to Cooperating Groups, 5/15/73, memo, CL; Moore to Cooperating Groups, 6/5/73, memo, and attached draft letter to the Port Authority of New York and New Jersey, 6/1/73, CL; Coalition Against Aircraft Noise, "Summary Report," 6/5/73, CL; Alderson to Richard Wiggs, 9/11/73, CL; Alderson to Russell E. Train (Administrator of the Environmental Protection Agency), 9/18/73, CL.

19. *Wall Street Journal*, 5/25/72, p. 34.

20. "The Black Arithmetic of the Concorde," *Business Week*, 2/10/73, p. 29; David Harris, "Concorde Sales Details Secret 'In Public Interest,'" *Daily Telegraph*, 12/22/72, p. 6; Anti-Concorde Project information sheet, reprinted in R. Berry O'Brien, "Concorde Deposits, Reclaimed," *Daily Telegraph*, 7/21/73, p. 1, CL; Andrew Wilson, "BOAC Says Concorde 'Too Noisy,'" *Observer*, 6/24/73, p. 1.

21. The discussion of the history of American Concorde policy from about 1970 through the Concorde decision of Secretary of Transportation William T. Coleman, Jr., in February 1976, is based on the following sources: Douglas Ross, "The Concorde Compromise: The Politics of Decision-Making," *Bulletin of the Atomic Scientists* (3/78):46–51; Peter Gillman, "Supersonic Bust: The Story of the Concorde," *Atlantic Monthly* (1/77):72–81; U.S. Department of Transportation, Federal Aviation Administration, "Concorde Supersonic Aircraft," (Draft) Environmental Impact Statement (3/75); U.S. Department of Transportation, *The Secretary's Decision: Concorde Supersonic Transport* (Washington, D.C.: Department of Transportation, 2/4/76); U.S. Department of Transportation, "Concorde Granted Limited Schedule in U.S. for Demonstration Period," 2/4/76, news release, DOT-09-76; U.S. General Accounting Office, *The Concorde—Results of a Supersonic Aircraft's Entry into the United States*, CED-77-131 (9/15/77).

22. For the discussion on stratospheric effects, in addition to the sources in note 21 above, see: *New York Times*, 11/6/71, p. 29, 11/16/72, p. 10, 11/19/72, p. 50, 1/22/75, p. 41; Harold S. Johnston, *Science* 173 (1971):517; Town-Village Committee, *Quotes* 5 (1/73):4, vol. 5 (3/73):2, vol. 7 (4/75):3; Paul Langer, "Scientists Sure Supersonic Plane Problems Can Be Solved," *Boston Evening Globe*, 3/5/74, p. 24; A. J. Grobecker, S. C. Coroniti, and R. H. Cannon, Jr., *Report of Findings: The Effects of Stratospheric Pollution by Aircraft* (Washington, D.C.: Department of Transportation, 1974); Allen L. Hammond, "Ozone Destruction: Problem's Scope Grows, Its Urgency Recedes," and "Public Credibility on Ozone," *Science*, 3/28/75, pp. 1181–1183; Thomas M. Donahue, letter, *Science*, 3/28/75, pp. 1143–1145; Alan J. Grobecker, letter, *Science*, 3/28/75, pp. 1143–1145; Luther J. Carter, "Deception Charged in Presentation of SST Study," *Science*, 11/23/75, p. 861; William H. Allen, letter, *Science*, 3/19/76, pp. 1124–1125; National Academy of Sciences, Climatic Impact Committee, *Environmental Impact of Stratospheric Flight* (Washington, D.C.: National Academy of Sciences, 1975); Lydia Dotto and Harold Schiff, *The Ozone War* (Garden City, N.Y.: Doubleday, 1978), pp. 43–119; U.S., Congress, Office of Technology Assessment, *Impact of Advanced Air Transport Technology, Part I: Advanced High-Speed Aircraft* (Washington, D.C.: U.S. Government Printing Office, 1980), pp. 89–90.

23. For the immediate response to Secretary Coleman's decision by anti-SST groups, by members of Congress, by the Port Authority, and by British Airways and Air France, see: "The Concorde Furor," *Newsweek*, 2/16/76, pp. 16–21; *Congressional Record—House*, 2/5/76, p. H748; *Congressional Record—Senate*, 2/4/76, pp. S1172–S1176, 2/5/

76, p. S1330, 2/6/76, pp. S1463–S1464; *Boston Evening Globe*, 3/26/76; *Washington Post*, 5/23/76, pp. B1, B4; *New York Times*, 5/24/76, pp. 1, 10.

24. For the results of the FAA's noise monitoring and public opinion surveys at Dulles and for the debate over these activities, see: FAA, "FAA Releases First Concorde Monitoring Report," 6/11/76, news release, no. 76-55; FAA, "FAA Releases Concorde Monitoring Report for Month of June," 7/9/76, news release, no. 76-56; Thomas Love, "Public Opinion of Concorde Has Improved," *Washington Star*, 5/23/77, reprinted in *Congressional Record—State*, 6/7/77, p. S9016; U.S. General Accounting Office, *The Concorde—Results of a Supersonic Aircraft's Entry into the United States*, 9/15/77, CED-77-131. Moreover, complaints about the Concorde and its noise levels generally continued at the same high level during the rest of 1976 and early 1977. See: *Aviation Daily*, 7/14/76, p. 253, 8/25/76, p. 311, 9/16/76, p. 86, 2/16/77, p. 254; "Concorde Noise," *Aviation Week and Space Technology*, 11/22/76, p. 23.

25. *Aviation Daily*, 8/17/76, p. 257, 9/16/76, p. 82; "Pollution Unit Sets SST Emission Rules," *Aviation Week and Space Technology*, 8/23/76, p. 26.

26. For the economic and financial data and estimates on the Concorde, see: *Aviation Daily*, 9/20/76, p. 97, 10/4/76, p. 183, 10/12/76, p. 222, 10/14/76, p. 239, 2/11/77, p. 230; Rosaline K. Ellingsworth, "Concorde Economics Keyed to Utilization," *Aviation Week and Space Technology*, 8/23/76, pp. 28–34.

27. *Aviation Daily*, 9/27/76, p. 141.

28. Robert Walters, "Selling the Concorde," *Parade*, 8/29/76, p. 17; *New York Times*, 5/10/77, pp. 1, 54.

29. For the various court actions concerning the Port Authority's ban, see: *New York Times*, 5/12/77, pp. A1, B6, 8/18/77, pp. 1, 38, 9/30/77, pp. 1, 13, 10/18/77, pp. 1, 28; "SST Plea Denied," *Aviation Week and Space Technology*, 8/23/77, p. 25.

30. For Concorde landings and noise levels at JFK, see: *New York Times*, 10/20/77, pp. 1, 41, 10/21/77, p. A1, 10/22/77, p. 24, 11/23/77, p. 39; *Aviation Daily*, 11/4/77, p. 26; "Concordes Begin Service to New York," *Aviation Week and Space Technology*, 11/28/77, pp. 14–15. For the actions and views of local anti-Concorde groups at the time, see: William Claiborne, "Concorde's 'ROAR,' " *Washington Post*, 4/18/77; *New York Times*, 5/16/77, p. 57; *Aviation Daily*, 11/9/77, p. 55, 11/17/77, p. 98, 11/18/77, p. 109, 11/21/77, p. 116.

31. For SST economic performance since 1976, see: *New York Times*, 10/21/77, p. A21, 7/28/78, pp. D1, D5, 2/23/79, p. D3, 8/11/79, pp. 1, 26; "Concordes Begin Service to N.Y.," *Aviation Week and Space Technology*, 11/28/77, pp. 14–15; "Concorde's Troubles," *Newsweek* (European edition), 8/21/78, pp. 36–37.

32. Richard Edmonds, "Low on Fuel, SSTs Break Hold on JFK," *New York Daily News*, 12/22/77, p. C5.

33. For Concorde-generated sonic booms during the 1976–1979 period, see: *Boston Evening Globe*, 9/9/76, p. 13; Deborah Shapley, "East Coast Mystery Booms: A Scientific Suspense Tale," *Science*, 3/31/78, pp. 1416–1417; *New York Times*, 2/12/79, p. B7.

34. "Marketing Observer," *Business Week*, 4/10/78, p. 94.

35. *Aviation Daily*, 5/8/78, p. 46; *Wall Street Journal*, 1/16/80, pp. 1, 25, 9/16/80, p. 42, 1/8/81, p. 22, 1/14/81, p. 30. *New York Times*, 8/11/79, pp. 1, 26, 7/25/80, sec. IV, p. 1, 7/1/81, sec. IV, p. 5, 1/9/81, sec. IV, p. 3, 1/10/81, p. 31.

36. For the history of the TU-144 in the 1970s, see: "Russians Term TU-144 Flights Start of Supersonic Service," *Aviation Week and Space Technology*, 1/5/76, p. 25; "The USSR's SST Aeroflop," *New Scientist*, 12/9/76, p. 572; *New York Times*, 8/18/75, 11/2/77, p. A3, 8/11/79, pp. 1, 26; "Concordski into Service," *New Scientist*, 11/3/77, p. 273; *Aviation Daily*, 11/23/77, p. 130; "A Shaky Lift-Off for 'Concordski,' " *Business Week*, 12/12/77, pp. 58–59.

37. For various American proposals during the 1970s to build or at least consider developing a second-generation SST, see: *Aviation Daily*, 9/11/76, pp. 83, 86, 10/22/76, p. 287; Boeing Commercial Airplane Company, "Challenges Facing an Advanced U.S. SST Program," 7/7/78; L. T. Goodmanson and A. Sigalla, "The Next SST—What Will It Be?" paper for the AIAA/SAE 13th Propulsion Conference, Orlando, Florida, 7/11–13/77; F. O. MacFee, "The Next 35 Years," *Headlines* (an in-house publication by General Electric, Aircraft Engine Group, Lynn and Everett, Massachusetts), 9/28/77; "Hypersonic Transport Studied," *Aviation Week and Space Technology*, 11/14/77, p. 54; *New York Times*, 12/12/78, pp. C1, C4; "New Look at Supersonics," *Aviation Week and Space Technology*, 4/30/79, p. 21; Office of Technology Assessment, *Impact of Advanced Air Transport Technology*.

38. *New York Times*, 12/12/78, pp. C1, C4.

39. See listings in note 1 to chapter 1.

40. For a specific discussion of the managerial lessons of the SST conflict, see: Mel Horwitch, "Managing a Colossus," *Wharton Magazine* (Summer 1979):24–31; Mel Horwitch and C. K. Prahalad, "Managing Multi-Organization Enterprises: The Emerging Strategic Frontier," *Sloan Management Review* (Winter 1981), pp. 3–16. For a summary of current thinking on managing large-scale technological endeavors like the SST program, see: Mel Horwitch, "Designing and Managing Large-Scale, Public-Private Enterprises: A State-of-the-Art Review," *Technology in Society* 1(Fall 1979), pp. 179–192.

41. For various studies and reports on post-1970 attitudes toward technology, see: "Science Still in Public Favor," *Science*, 10/26/73, p. 369; National Science Board, *Science at the Bicentennial: A Report from the Research Community* (Washington, D.C.: Government Printing Office, 1976), pp. 71–92; Phillip Abelson, "Communicating with the Public," *Science*, 11/5/76, p. 565; Richard W. Anderson and W. Lipsey, "Energy Conservation and Attitudes toward Technology," *Public Opinion Quarterly* 42 (1978):17–30; *New York Times*, 4/9/79, pp. A1, D9. See also the listings in note 1 to chapter 1. For the purest reflection of the new antitechnology attitide, see: J. McDermott, "Technology: The Opiate of the Intellectuals," *New York Review of Books*, 7/31/69, pp. 25–35. See also: Rosalind Petchetsky, "Statement on the Relationship Between Technology and Social Change: Toward a Critique of 'Technologist' Ideology," Conference of the American Political Science Association, Chicago, 9/8/71. For the rise of technology assessment and other attempts to institutionalize the environmentalists' concerns, see: Walter A. Hahn and Rosemary A. Chalk, *The Technology Assessment Act of 1972*, 73-41SP (Washington, D.C.: Congressional Research Service, 2/6/72); Anne Hessing Cahn and Joel Primack, "Technological Foresight for Congress," *Technology Review* (3–4/73):39–48; Edward Wenk, Jr., "Technology Assessment in Public Policy: A New Instrument for Social Management of Technology," *Proceedings of the IEEE* 63 (3/75):371–379; John C. Cobb, letter, *Science*, 11/12/76, p. 674; Leon Lipson, letter, *Science*, 11/26/76, p. 890. For the continued concern and strength of the environmentalists, see: Russell E. Train, "The Environment Today," *Science*, 7/28/78, pp. 320–324; "Breeder, Arms Sales Are

Targets of New Lobby Group," *Science*, 1/14/77, p. 160. But widespread environmental concern also appeared to decline during the 1970s. See: Riley E. Dunlap and Don A. Dillman, "Decline in Public Support for Environmental Protection: Evidence from a 1970–1974 Panel Study," *Rural Sociology* 41 (Fall 1976):382–390; Riley E. Dunlap, Kent D. Van Lierre, and Don A. Dillman, "Evidence of Decline in Public Concern with Environmental Quality: A Reply," *Rural Sociology* 44 (Spring 1979):204–212. For the declining appeal of the confrontational approach by environmentalists and for the new emphasis on implementation and reaching agreement with potential adversaries, see: *New York Times*, 6/13/77, p. 22; David Sleaper, "Earth Day Birthday: After the Alarms," *Boston Sunday Globe*, 4/22/77, p. A7; "From Protest to Policymaking," *Business Week*, 3/26/79, pp. 58–59; transcript of "Meet the Press" interview of Denis Hayes, 5/8/77, pp. 4–5. For the evolution and life cycles of various protest movements see the listings in note 4 to chapter 1. For the person who, after being coordinator of the coalition, returned in the mid-1970s to become involved in the anti-nuclear-breeder-reactor campaign, see: Joyce Teitz Wood to Shurcliff, 4/19/75, CL: Shurcliff to Joyce Teitz Wood, 4/22/75, CL.

42. Najeeb Halaby, *Crosswinds: An Airman's Memoir* (Garden City, N.Y.: Doubleday, 1978); personal interviews.

43. Edward J. Burger, *Science at the White House: A Political Liability* (Baltimore, Maryland: The Johns Hopkins University Press, 1980), pp. 111–113; *New York Times*, 9/12/77, p. 20.

44. Ibid., 12/4/80, sec. III, p. 10; Tracy Kidder, "Tinkering with Sunshine," *Atlantic Monthly* (10/77):70–83; Shurcliff to Leonard Brown (Vietnam Peace Parade Committee), 7/8/73, CL; personal interviews.

45. Personal interviews.

46. *Seattle Post-Intelligencer*, 6/20/74, p. A3.

Index